For Dee

CORRUPT RESEARCH

The Case for Reconceptualizing Empirical Management and Social Science

Raymond Hubbard

Drake University

Los Angeles | London | New Delhi
Singapore | Washington DC

Los Angeles | London | New Delhi
Singapore | Washington DC

FOR INFORMATION:

SAGE Publications, Inc.
2455 Teller Road
Thousand Oaks, California 91320
E-mail: order@sagepub.com

SAGE Publications Ltd.
1 Oliver's Yard
55 City Road
London EC1Y 1SP
United Kingdom

SAGE Publications India Pvt. Ltd.
B 1/I 1 Mohan Cooperative Industrial Area
Mathura Road, New Delhi 110 044
India

SAGE Publications Asia-Pacific Pte. Ltd.
3 Church Street
#10-04 Samsung Hub
Singapore 049483

Acquisitions Editor: Vicki Knight
Editorial Assistant: Yvonne McDuffee
Production Editor: Laura Barrett
Copy Editor: Sarah J. Duffy
Typesetter: C&M Digitals (P) Ltd.
Proofreader: Eleni Georgiou
Indexer: Judy Hunt
Cover Designer: Michael Dubowe
Marketing Manager: Nicole Elliott

Printed in the United States of America

Library of Congress Cataloging-in-Publication Data

Hubbard, Raymond

Corrupt research : the case for reconceptualizing empirical management and social science / Raymond Hubbard, Drake University.

pages cm
Includes bibliographical references and index.

ISBN 978-1-5063-0535-6 (pbk. : alk. paper)

1. Errors, Scientific. 2. Science—Methodology.
3. Missing observations (Statistics) 4. Fraud. I. Title.

Q172.5.E77H83 2016
001.4—dc23
2015023135

This book is printed on acid-free paper.

15 16 17 18 19 10 9 8 7 6 5 4 3 2 1

Table of Contents

Preface

I spent most of the 1970s in graduate school, studying first geography and then economics. During this time I was taught, quite informally it must be emphasized, about how we conceive of, or "do," science. That is, like just about everyone in the social and management sciences, I was initiated into what I call in this book the *significant difference* paradigm. This paradigm, whose monopolistic grip on graduate training is alive and well today, espouses the virtues of hypothetico-deductivism as the means by which science progresses. Succinctly, one comes up with a research problem, then proposes through appeals to various theories a number of empirically testable hypotheses which help solve it. And to be de rigueur, these hypotheses must be amenable to tests of statistical significance. All of this was suitably impressive to an apprentice like me, and I became adept at the precise numbering and articulation of hypotheses as well as the nuts and bolts of performing said statistical tests in deciding on their merits. As taught, knowledge accumulation looked to be a rather clinical, detached, cut-and-dried matter. I didn't realize at the time that this seemed rather formulaic, but even if I had, wasn't science all about *method* anyway?

Prompted by ongoing concerns such as the true scientific status of the management and social sciences fields, their apparent inability to provide solutions to many practical problems, and the limitations of their primary analytical methods, I grew increasingly disenchanted with the epistemological legitimacy of the significant difference model. So much so that I became utterly convinced that research conducted in this paradigm—along with the publication biases it actively promotes favoring "original" research in single works with statistically significant results, and against null outcomes and replications—is little more than pseudoscientific pedantry generating corrupt empirical literatures.

This book defends the above claim. It does so by exposing the flaws of the significant difference paradigm across three broad areas—philosophical, methodological, and statistical. At the same time, an alternative conception of science, the *significant sameness* paradigm, is introduced and contrasted in a head-to-head manner with the significant difference counterpart on these same three dimensions. It is shown to be a superior account of how real science makes headway. The thesis put forward here, then, is that we must change the way in which we conceive of the scientific enterprise itself and replace the notion of significant difference with that of significant sameness in graduate education in the social and management sciences. Absent such a change of direction, these disciplines will continue to flounder as so-called sciences and remain unable to produce knowledge relevant to their constituencies.

Being almost totally at odds with academic conventional wisdom, I fully expect the ideas expressed in this book to suffer a harsh reception. As a matter of fact, I encountered a taste of this a few years ago when I submitted some of these ideas in condensed form for publication consideration to a prestigious journal. The editor of this journal replied via e-mail in June 2008: "I am hard pressed to decide who on our editorial board would support publishing it. Consequently my inclination is to decline to send it out for review. . . . My motive is just to try to avoid spending time in a round of reviewing whose outcome can be anticipated." This, of course, is the crux of the problem—monopolism. So I am prepared for a negative backlash from the academic community. A worse response from this quarter, however, as discussed in Chapter 8 and illustrated by the editor just quoted, would be to simply ignore criticism of the status quo and carry on with business as usual. My hoped-for reaction is some support and further discussion of the contents of this book, for it is no screed: All issues raised are backed by detailed empirical evidence, numerous examples, and argumentation.

I see the book as a text/reference/resource aimed at faculty and graduate students in the management and social sciences—one that, upon exposure, these groups should find difficult to ignore. It presupposes that the reader has had solid introductory courses in both statistics and research methods. An extensive reference list also allows those interested to explore further any of the topics addressed herein.

Being multidisciplinary in reach, I have tried to incorporate in this book examples from all areas of the management and social sciences (as well as from biomedicine) illustrating the numerous problems involved in the execution and interpretation of research in the significant difference paradigm. However, this coverage is uneven. I have taught marketing for over 30 years, so this area, understandably, receives more attention than others in the business disciplines. Likewise, psychology comes in for extensive coverage among the social sciences. This is because, while guilty of the same research faults as others abiding by the tenets of the significant difference model, scholars in this field, both historically and at the present time, seem more committed to the need to rectify them. I've no doubt that readers from all of the disciplines included in these pages can think of many more pertinent examples, both pro and con, and I would be delighted to hear from them on this score.

I wish to express my thanks to those whom I have worked with and learned from over the years on various aspects of the material contained in the book. These include Stuart Allen, Scott Armstrong, Susie Bayarri, Rod Brodie, Eldon Little, Ken Meyer, Rahul Parsa, and Dan Vetter. A nod to SQB for her comment on confidence intervals. Special gratitude is owed to Murray Lindsay. I am also grateful to the people at SAGE who helped make this book a reality, especially Vicki Knight (acquisitions editor), Yvonne McDuffee (editorial assistant), Laura Barrett (production editor), and Sarah Duffy (copy editor). Finally, I want to thank my wife, Dee, to whom the book is dedicated, for her love, assistance, and perseverance.

SAGE
50
YEARS

About the Author

Raymond Hubbard is Professor Emeritus of Marketing at Drake University, in Des Moines, Iowa. He holds a BSc (economics) Honors degree from the University of London; an MSc in geography from the University of the West Indies, in Kingston, Jamaica; and an MA in economics and a PhD in geography from the University of Nebraska, Lincoln. He taught previously at the State University of New York, Fredonia, and held visiting positions at the University of Washington, Seattle, and at the University of Auckland, in New Zealand. His research interests include applied methodology and the sociology and history of knowledge development in the management and social sciences. He has published numerous articles on these topics in journals in these fields. He is a lifelong supporter of Sunderland AFC and a Cornhusker fan since the early 1970s. He can be reached at drabbuhyar@aol.com.

INTRODUCTION

> If we now look at established procedures in the physical sciences we find that the scientist begins to believe that (s)he is winning when (s)he gets reproducible results from several experiments done under various conditions, perhaps with different instruments at different sites, etc. Looking for reproducible results is a search for significant *sameness*, in contrast to the emphasis on the significant *difference* from a single experiment. (Nelder, 1986, p. 113)

> I propose that the future vitality and success of our profession depends on making sure our research-based knowledge is relevant and useful. This will require the Academy of Management . . . to be far more engaged with the real world than has traditionally been the case. (Cummings, 2007, p. 355)

Complaints about the usefulness of academic management and social science research in dealing with real-world issues are common. In addition to Cummings's (2007) observations in the introductory quotation, Van de Ven and Johnson (2006) cite a number of recent studies challenging the relevance of scholarly management research for solving practical problems. This view is supported by Ghoshal (2005, p. 75), T. G. Gill (2010, p. 1), and Pfeffer (2007, p. 1334). Indeed, Bennis and O'Toole (2005, p. 99) claim that academic publishing is seen as a "vast wasteland" as far as business people are concerned. Given Micklethwait and Wooldridge's (1996, p. 12) premise that much management theory is bedeviled by obfuscation, jargon, and faddishness, such impressions are understandable.[1]

In marketing, November (2004) advises practitioners to ignore academic research. Reibstein, Day, and Wind (2009, p. 1), meanwhile, in a *Journal of Marketing* guest editorial called "Is Marketing Academia Losing Its Way?", are worried about an alarming and growing gap between the interests and priorities of marketing academicians and the needs of marketing executives. Or take the area of management accounting, where Otley (2003, p. 319) comments that "we have done very little sound work in this field, and we have certainly failed to influence practice in a significant way."

Even economics, the discipline which the other business and social science fields often seek to equal, "positively extols esoteric irrelevance" (Ormerod, 1997, p. 20), "create[s] more confusion than clarification" (Hossein-zadeh, 2014, p. 44), and is "useless" as a means of understanding a capitalist economy (Keen, 2001, p. 10). Evidence of this discontent was seen when Hayek (1989, p. 3) took the opportunity in his Nobel Memorial Lecture, provocatively titled "The Pretence of Knowledge," to berate the economics discipline for making "a mess of things." It continues to be witnessed in Akerlof and Shiller's (2009, p. xi) charge, two more Nobel laureates, that "ignorance of how the economy really works has led to the current state of the world economy, with the breakdown of credit markets and threat of collapse of the real economy in train." Still another Nobel laureate, Stiglitz (2010, p. xx), blames the economics profession for helping to precipitate this crisis. As does Madrick (2014). Indeed, Shiller (cited in Fox, 2009, p. 232) calls the efficient market hypothesis, the backbone of academic and policy thinking no less, the most remarkable error in the history of economic theory.[2]

The contributions to useful knowledge made by the social sciences as a whole were roundly lampooned by Andreski (1972) in his *Social Sciences as Sorcery*. Three years later Elms (1975, p. 967) talked of a "crisis of confidence" in social psychology brought on by, among other things, a demand for relevant research. The following decade Shweder and Fiske (1986, p. 1) weighed in that

> there has been in the social sciences, at least in recent years, a vague sense of unease about the overall rate of progress of the disciplines. A . . . literature has emerged . . . either challenging the scientific status of social research or expressing concern about the accomplishments of the social sciences. Some have even talked of a "crisis" in social inquiry.

Yet again in the next decade, Loftus (1996, p. 161) disclosed:

> I have developed a certain angst over the intervening 30-something years [since entering the psychology discipline in 1964]—a constant, nagging feeling that our field spends a lot of time spinning its wheels without really making much progress. This problem shows up in obvious ways—for instance, in the regularity with which findings seem not to replicate.

And based on what they see as an unprecedented level of anxiety concerning the reliability of research findings in psychology, Pashler and Wagenmakers (2012, p. 528) ask once more whether there is a crisis of confidence in the field. They answer in the affirmative. Finally, while sympathetic, Flyvbjerg (2001, pp. 1–2) acknowledges the assaults on the credibility of the social sciences as sciences. In this respect, Bauer (1994, p. 128) goes so far as to say that "in the social sciences, little is known or predictable that is deeper than triviality or different from commonsense knowledge." Bauer's position is seconded by Taagepera (2008, p. 236): "The ruling emperor of social sciences has no clothes. His quantitative garb is largely make-believe."

Especially pertinent is that some investigators are troubled by the news that despite the enormous amount of data-based research taking place, there is nevertheless a paucity of empirical generalizations—which is to say empirical regularities, "stubborn facts," or phenomena—in the management and social sciences. In Barwise's (1995, p. G30) view, a very weak definition of an empirical generalization is "*any empirical observation which has been found to generalize.*" Bass (1993, p. 2) comments that an empirical generalization "is a pattern or regularity that repeats over different circumstances and that can be described simply by mathematical, graphic, or symbolic methods." Subsequently, Bass and Wind (1995, p. G1) explained that an empirical generalization is a "pattern that repeats but need not be universal over all circumstances," while Ehrenberg (1995, p. G20) considers them to be "merely data-based regularities." Lastly, Shelby Hunt (1991, p. 113) explains that "an empirical regularity is a statement summarizing observed uniformities or relationships between two or more concepts or variables." The common thread running through these definitions is the idea that results are essentially *repeatable* over a wide range of conditions (e.g., organizations, geographic areas, time periods, measurement instruments, methods of data collection, researchers).[3]

Barwise (1995, pp. G30–G31) takes these definitions further when elaborating on five characteristics that "good" empirical generalizations or facts should possess. The first is *scope;* they are not universal, but nonetheless hold under a variety of different conditions. The second is *precision;* they describe a phenomenon that has been witnessed several or many times, and the more specific that description the better. The third is *parsimony;* they are uncluttered by erasing a number of variables that might have mattered. The fourth is *usefulness;* they are of considerable benefit to practitioners. And the fifth is a *link with theory;* they stimulate theory construction because their relative persistence deserves an explanation.

Unfortunately, the dearth of facts in the business and social sciences prompted Leone and Schultz (1980, p. 11) to assert that marketing's knowledge base is "more marsh than bedrock," and Armstrong and Schultz (1993) to conclude that marketing does not possess a body of "principles." In economics, Hicks (1979, pp. 1–2) admits: "There are very few economic facts which we know with precision. . . . There are few economic 'laws' which can be regarded as at all firmly based." Similarly, Keuzenkamp (2000, p. 1) concedes that if *Econometrica* were to publish a single issue containing well-established facts, "it might be very thin indeed." And while defending the field, Randall Collins (1989, p. 125) is sensitive to the accusation that after 100 years of research, sociology is dismissed as a science in many quarters because it has no findings (facts) or valid generalizations.

Unsparing criticisms, these. But accurate nonetheless. For, as told above, it is inescapable that there is growing concern over the lack of advancement in the management and social sciences. Moreover, the oft-invoked excuse that the disappointing progress is attributable to the comparative youth of the disciplines involved is wearing thin. This lack of forward momentum in the social and business sciences is by now a decades-long all-too-familiar refrain, one which will echo well into the future unless fundamental changes are wrought in these disciplines. To this end I am of the belief that progress in the behavioral and business disciplines will not come about short of a total reconceptualization of how we think about science. This book lays out what this reconceptualization must involve—a relinquishing of the emphasis on the idea of *significant difference* and a commitment to the notion of *significant sameness.*

The thesis of this book, expanding on Hubbard and Lindsay (2013a, 2013b), is that a crucial reason for the scarcity of useful knowledge is because members of the social and management sciences subscribe overwhelmingly to a single methodological paradigm, one that revolves around the idea of significant *difference.* With few exceptions (such as those doing qualitative research) the significant difference paradigm monopolizes graduate business and social science education. Consequently, those in these research communities have been taught that this approach describes *the* scientific method.

This is unfortunate because the significant difference paradigm militates against the procurement of facts and the theories which could be built around them. By and large, this model offers a poor description of how science works. For instance, it rests on a simplistic conception of knowledge, in which theories are articulated over an extremely short period of time. This paradigm sees the research process as one of testing these hastily assembled theories following the hypothetico-deductive model of explanation. The goal is to produce statistically significant ($p \leq .05$) outcomes, in the main on isolated data sets, in what must be presented as "original" or "novel" contributions if the manuscript is to be published. Explicit attention to external validity (generalizability) considerations for the most part is cursory, ignored, and/or taken to be handled satisfactorily within the statistical model of generalization—from sample *to* population—courtesy of random sampling. This latter belief, coupled with the widespread opinion that good theories should produce universal laws, feeds the propensity to overgeneralize the results of these unique studies to other contexts and time periods. The legacy of this monopolistic paradigm, when seen together with the well-known editorial-reviewer biases against publishing "negative" (i.e., $p > .05$) results and replication research, is an empirical literature consisting almost entirely of unverified, fragile results whose role in the development of cumulative knowledge is of the shakiest kind.

It is important to add that this monopoly enjoyed by the significant difference paradigm in the management and social sciences—and the counterproductive research attitudes and behaviors it sanctions—is no straw man (Hubbard & Lindsay, 2013b, pp. 1394–1395). Rather, it is a faithful description of the research culture in these areas, and it codifies bad science (Hubbard & Lindsay, 2013b).

Monopolies seldom are desirable, particularly when it comes to scientific inquiry. There are alternative pathways to knowledge acquisition, not just that touted by the significant difference tradition. A major one of these, labeled significant *sameness* by the eminent statistician John Nelder in an introductory quotation, is developed in greater depth throughout this book.[4] In doing so it is shown how the significant sameness paradigm views the conception of knowledge as messy: Data rarely speak for themselves; universal generalizations are impossible because of the intrinsically contingent nature of relationships; and model uncertainty is a fact of life and cannot be addressed by even the most sophisticated statistical manipulations in a single data set, but which can be resolved by gathering additional data via well-designed studies. This paradigm is concerned with developing theory in research programs whose first aim is the discovery of empirical regularities, followed by increasingly deeper and higher level generalizations, involving many sets of data over an extensive period of time. The role of statistical significance testing is marginalized. Instead the goal is to determine whether significant sameness is found between initial and subsequent studies as defined by *overlapping confidence intervals* around the parameter(s) of interest. The significant sameness paradigm emphasizes the importance of well-designed replications because replication is the only way to establish empirical regularities. These replications systematically probe, often using purposive sampling, the scope and limits of quantitative findings *across* relevant (sub)populations. It should not be thought, however, that significant sameness means "brute empiricism." On the contrary, in keeping with a *critical realist* philosophy, theory construction is invigorated by seeking explanations for these regularities (and exceptions), and this process depends heavily on the use of abductive inference as well as deductive and inductive reasoning.

The logical positivists' quest for certain knowledge is chimerical. Rather, consistent with the significant sameness viewpoint, the most that can be done is for the scientific community to continue eradicating mistakes in our knowledge. This is achieved by ruling out competing explanations for a phenomenon via the accumulation of evidence derived from critically testing our theories.

This book shows that the significant sameness model, viewed as a whole, represents a new and superior way for designing research and analyzing results in the management and social sciences than is offered by the significant difference approach. Yet an important caveat is necessary at this juncture. Calling the significant sameness paradigm—the need to uncover first empirical regularities and then the theories to account for them—a new approach to the establishment of knowledge is true enough when applied to modern-day social and management science.[5] But it is commonplace in the physical sciences, where it has been responsible for much of the knowledge development in these areas.

In addition it must be stressed that reservations concerning the viability of the significant difference paradigm documented in this book go far beyond the endemic problems of researchers misinterpreting the results and capabilities of statistical significance tests. This topic does, however, receive in-depth coverage where appropriate.

It must further be emphasized that too many academicians in the social and management sciences seem not to be concerned with attempting to provide useful knowledge for those making social and business policy decisions. For the most part, scholarly priorities are attuned to securing career advancement within the "publish or perish" academic world. This means an unbending fealty to the significant difference paradigm, and with it the lack of relevant, applicable knowledge produced by the management and social sciences now and in the future.

The book is organized around the head-to-head contrasting of the two conceptions of science over a number of broad dimensions—philosophical, methodological, and statistical. These are summarized in Table 1-1.

Table 1-1 Contrasting the Significant Difference and Significant Sameness Paradigms

Categories	*Significant Difference*	*Significant Sameness*
Philosophical		
Conception of knowledge	*Unproblematic.* Centers on rejecting the null hypothesis at the $p \leq .05$ level to establish facts.	*Problematic.* Data rarely speak for themselves. Proof in science is impossible. The focus, instead, needs to be on the scholarly community gradually weeding out errors by eliminating rival explanations for a phenomenon on the basis of the accumulation of evidence obtained from critically testing theories.
Model of science	Almost exclusive attention on *testing,* rather than developing, theory via the hypothetico-deductive method. Logical positivist/empiricist orientation.	*Developing* theory using inductive enumeration to identify and generalize empirical regularities over many data sets. These regularities, in turn, are accounted for using abductive inference. Consistent with a critical realist philosophy.
	A single study can produce rational knowledge.	Many studies within research programs are necessary to develop and establish theory. And this takes a great deal of time.

(Continued)

Table 1-1 (Continued)

Categories	*Significant Difference*	*Significant Sameness*
	Good theory produces (or should produce) *universal* generalizations.	Science deals with *restricted*, not universal, generalizations that possess extensive empirical backing and known boundary conditions.
Role of "negative" results	Rarely published; considered to reflect poorly on the researcher rather than on nature.	Crucial in establishing boundaries on findings. Negative results also are a heuristic for developing better (deeper) theory.
Methodological		
Importance of replication	Incidental and rarely done. Considered to be an inferior kind of research. "Novel" or "original" research is all-important.	Defines this paradigm. Protects literature from specious results. More important, replication is the only vehicle available for discovering empirical generalizations and placing bounds on their application. In addition, the entire validity generalization process is based on replication research.
Definition of replication success	Statistical significance in the same direction as the earlier study.	Significant sameness as revealed by overlapping confidence intervals around point estimates.
Conception of generalization/ external validity	*Statistical* generalization following the representative model (i.e., from sample to population). Emphasis on random sampling.	*Empirical* generalization across data sets (subpopulations), often using purposive sampling.
Statistical		
Model uncertainty	Often ignored, further contributing to the primacy placed on statistics in conferring knowledge status.	Explicitly acknowledged. Statistics is subordinate to research design. Model uncertainty can be overcome only by examining many sets of data.

Categories	*Significant Difference*	*Significant Sameness*
Nature of predictions	*Qualitative*. This is all that managerial and social science theories are capable of offering.	*Quantitative*. Notion of predictive precision and severe tests is crucial in dealing with model uncertainty.
Role of *p*-values	Lies at the heart of the entire research process. Statistical significance is considered to be the essential criterion for establishing knowledge claims.	Used as a heuristic, not as an objective measure of knowledge.
Role of effect sizes	Beginning to be reported in some business and social science disciplines.	Routinely reported and interpreted.
Role of confidence intervals	Rarely used in business and social science disciplines.	Routinely reported and interpreted.
Individual Researcher Philosophy		
	Centered on personal academic career advancement in a publish-or-perish environment. Publishing papers and accumulating citations are of the utmost importance. Contributions to knowledge are a secondary concern.	Centered on knowledge development in research programs. Career prospects stem from knowledge discoveries, not publication and citation counts.

Source: Adapted from Hubbard, Raymond and R. Murray Lindsay (2013a), "From Significant Difference to Significant Sameness: Proposing a Paradigm Shift in Business Research," *Journal of Business Research*, 66 (September), p.1379 with permission from Elsevier.

In a nutshell, this book demonstrates that the significant difference paradigm is philosophically suspect, methodologically impaired, and statistically broken. As such, even on its own terms it is a model of corrupt research to be discarded. Aggravating matters, the significant difference paradigm is embedded in an academic social structure whose publication biases complete the institutionalizing of this corruption. While no route to knowledge generation is perfect, the significant sameness approach avoids the above problems by offering an alternative, and better, perspective on the conduct of management and social science.

Contrasts between the two paradigms begin on philosophical grounds. Accordingly, Chapter 2 outlines the intellectual cornerstones of the significant difference model.

Notes

1. See also J. Gill and Whittle (1993, p. 281) and Pfeffer and Sutton (2006b, p. 13) in this regard.
2. Ariely (2008, 2009) and Kahneman (2011) are worth reading in this context.
3. See also Bass (1995) and Uncles and Wright (2004, p. 5).
4. Ehrenberg's (e.g., 1993a, 1993b; 1995) contributions on this topic also are foundational.
5. As will be shown in Section 8.4, however, what is not generally known is that the idea of significant sameness was instrumental in the 19th century evolution of both the statistics and social science disciplines. Thereafter, it lost favor.

PHILOSOPHICAL ORIENTATION—SIGNIFICANT DIFFERENCE

> As you know, the notion of statistical significance underpins marketing research. (Burns & Bush, 2010, p. 504)

> My thesis, however, is that the hypothetico-deductive method . . . actually retards the progress of science. (Locke, 2007, p. 868)

2.1 Introduction

This chapter describes some key philosophical maxims—conception of knowledge, model of science, and the role of "negative" results—characteristic of the significant difference research paradigm. Thus, Section 2.2 illustrates that the conception of knowledge development shared by those in this paradigm is too simplistic, namely, rejecting the null hypothesis at the .05 level or better. This viewpoint continues unabated as the epitome of methodological rigor among management and social science researchers, something captured when examining the growth of empirical research and statistical significance testing in these areas from the end of World War II through 2007. It also accounts for the unfortunate habit of overgeneralizing the results of single studies with $p \leq .05$ results to other contexts and time periods.

The model of science adopted by members of the significant difference school, *hypothetico-deductivism,* is presented and critiqued in Section 2.3. In the main it is a framework shown to be philosophically untenable. It also is an inaccurate description of how real scientists actually behave.

The role that "negative" or null results (i.e., findings which do not conform with predictions, but more usually are thought to be statistically insignificant, or $p > .05$, results) play in the significant difference paradigm is recounted in Section 2.4 and constitutes the final third of this chapter. Such results mostly are viewed with suspicion and/or hostility, which has led to a bias against their publication. The consequences of this publication bias are explored. They include dire threats to the objectivity, integrity, and self-correcting nature of science itself through the production of empirical literatures contaminated with errors and outright falsehoods. Concluding comments are made in Section 2.5.

2.2 Conception of Knowledge

2.2.1 Unproblematic: Rejection of the Null Hypothesis, H_0

The conception of knowledge in this paradigm is taken to be largely unproblematic; a knowledge claim is made when a researcher can reject the null hypothesis (H_0) at $p \leq .05$. This methodological imprimatur was handed down by arguably the greatest statistician of all time, Sir Ronald A. Fisher. The origin of the *p*-value is not due to Fisher, but rather to Karl Pearson (1900), who introduced it in his χ^2 test. Yet there is no doubt that it was Fisher who was responsible for popularizing statistical significance testing and *p*-values. He did this via multiple editions of his books *Statistical Methods for Research Workers* (1925) and *The Design of Experiments* (1935)—books that Savage (1954, p. 275) called the two most influential in the development of statistics in the 20th century.

Fisher used discrepancies in the data to reject what he dubbed the null hypothesis, that is, he calculated the probability of the data (x) on a true null hypothesis, or $Pr(x \mid H_0)$. Formally, $p = Pr(T(X) \geq T(x) \mid H_0)$. *P* is the probability of getting a test statistic T(X) greater than or equal to the observed result, T(x), in addition to more extreme ones, conditional on a true null hypothesis, H_0, of no effect or relationship. So the *p*-value is a measure of the (im)plausibility of the actual observations (as well as more extreme and unobserved ones) obtained in an experiment or other study, assuming a true null hypothesis. The rationale is that if the data are seen as being rare or highly discrepant under the null hypothesis, this constitutes *inductive evidence* against H_0.[1]

As for the sanctity of the .05 level, Fisher (1966, p. 13) simply commented: "It is usual and convenient for experimenters to take 5 per cent. as a standard level of significance." With respect to his famous disjunction (Fisher, 1959, p. 39), a $p \leq .05$ indicates that "*either* an exceptionally rare chance has occurred, *or* the theory of random distribution [null hypothesis] is not true." And rejections at lower levels, such as $p < .01$, $p < .001$, and so on, are said to furnish even stronger evidence against H_0.

For Fisher (1926, p. 504) the *p*-value from a statistical test plays an important epistemic role by helping to certify scientific knowledge: "A scientific fact should be regarded as experimentally established only if a properly designed experiment *rarely fails* to give this [$p \leq .05$] level of significance." He cemented this conviction when proclaiming: "Every experiment may be said to exist only in order to give the facts a chance of disproving the null hypothesis" (Fisher, 1966, p. 16). Moreover, Fisher (1973, p. 46) held that the *p*-value is an "objective" measure of evidence against the null hypothesis:

> The feeling induced by a test of significance has an objective basis in that the probability statement on which it is based is a fact communicable to and verifiable by other rational minds. The level of significance in such cases fulfils the conditions of a measure of the rational grounds for the disbelief [in the null hypothesis] it engenders.

Fisher thought that statistics could play a central role in promoting inductive inference, that is, drawing inferences from the particular to the general, from samples to populations. For him, "inductive inference is the only process known to us by which essentially new knowledge comes into the world" (Fisher, 1966, p. 7), and he saw *p*-values from significance tests as being evidential (Hubbard & Bayarri, 2003, p. 172).

Finally, Fisher (1970, p. 2) championed the use of statistical analysis as an important vehicle for elevating disciplinary status: "Statistical methods are essential to social studies, and it is principally by the aid of such methods that these studies may be raised to the rank of sciences." So the use of statistical methods, and significance tests in particular, is presented as a key route to scientific progress.

In light of the above endorsements, researchers nervous about their scholarly credentials enthusiastically adopted Fisher's promise of a seemingly "objective" and "scientific" criterion for justifying knowledge claims. In fairness, Fisher (e.g., 1955, p. 74, 1966, p. 13) warned that the conclusions drawn from a test of significance are *provisional*, especially in the early stages of a research program. He saw the test of significance as one more piece of evidence, to be used with other relevant pieces, for assessing the merits of a hypothesis (Cochran, 1974, p. 1461). In addition, his daughter and biographer, Joan Fisher Box (1978, pp. 135–136), confides that her father despaired over the thoughtless application of his methods, though I am unaware of his making any public criticisms of such carelessness. Having said this, it is easy to see from the above how Fisher's words of caution were completely undone by the force of his own rhetoric. So much so that, in reflecting more than 50 years ago on the then influence of Fisher's (1925) *Statistical Methods for Research Workers* on applied research, Yates (1951, p. 33) already was lamenting that investigators often regarded the results of a test of significance to be the ultimate objective of an experiment. Today, this tendency is well nigh universal. Nelder (1985, p. 238), whose views are shared by Guttman (1985), Nester (1996), and Vickers (2010, p. 54), among other statisticians, makes this point bluntly: "The grotesque emphasis on significance tests in statistics courses of all kinds . . . is taught to people, who if they come away with no other notion, will remember that statistics is about tests for significant differences." Below I supply evidence confirming Nelder's misgivings. But first I wish to address the increase over time in empirical research published in the social and management sciences, something which would seem to augur well for the discovery of linchpins to scientific advance, namely, empirical generalizations.

2.2.2 The Growth of Empirical Research: 1945–2007

In this section I track over time the amount of empirical research published in the social and management sciences, beginning with the former.[2] Table 2-1 charts the proportion of empirical work in four social sciences: geography, political science, psychology, and sociology.[3] The data were obtained by inspecting all articles and research notes in a number of leading journals from each of these areas to determine whether they were empirical. Specifically, I content-analyzed a randomly selected single issue of each journal for every year from 1945 through 2007. Leading journals were targeted because they might be expected to mirror best research practices within their disciplines. The parentheses contain the journals included in each social science: geography (*Annals of the Association of American Geographers, Economic Geography, Professional Geographer*), political science (*American Journal of Political Science, American Political Science Review, Public Administration Review*), psychology (*Journal of Applied Psychology, Journal of Comparative Psychology, Journal of Consulting and Clinical Psychology, Journal of Educational Psychology, Journal of Experimental Psychology: General, Psychological Bulletin, Psychological Review*), and sociology (*American Journal of Sociology, American Sociological Review, Social Forces*).[4]

Table 2-1 is based on the examination of 10,874 papers, of which 7,928 (72.9%) are empirical. It reveals interesting patterns in the growth of empirical research. For example, with some minor exceptions, all four social sciences show an inexorable rise in the percentage of articles devoted to data-based research. Comparing 1945–1949 with 2000–2007, these figures grew as follows: geography (34.5%–69.6%), political science (2.4%–64.9%), psychology (70.5%–91.0%), and sociology (34.1%–90.8%).

Table 2-2 shows the increase over time in the percentage of empirical work published in the five business fields of accounting, economics, finance, management, and marketing.[5] They rest on the investigation of 15,505 papers, some 9,710 (62.6%) being empirical. As with the social sciences, prestigious journals were selected from the management areas. Unlike the social sciences, however, management disciplines are of more recent vintage; only the economics journals and two others (*The Accounting Review* and *Journal of Marketing*) spanned the entire time period 1945–2007. The journals representing each field, together with their initial dates of publication, are given in parentheses: accounting (*The Accounting Review, Journal of Accounting and Economics*, 1979; *Journal of Accounting Research*, 1963), economics (*American Economic Review, Economic Journal, Journal of Political Economy, Quarterly Journal of Economics, Review of Economics and Statistics*), finance (*Journal of

Table 2-1 The Growth of Empirical Research in the Social Sciences: 1945-2007

	Geography			*Political Science*		
Years	*Total*[a]	*Empirical*[b]	*%*[c]	*Total*	*Empirical*	*%*
1945-2007	1,262	689	54.6	1,722	857	49.8
1945-1949	58	20	34.5	82	2	2.4
1950-1959	161	48	29.8	170	28	16.5
1960-1969	212	105	49.5	218	80	36.7
1970-1979	240	132	55.0	329	171	52.0
1980-1989	189	127	67.2	319	201	63.0
1990-1999	211	124	58.8	339	203	59.9
2000-2007	191	133	69.6	265	172	64.9
	Psychology			*Sociology*		
Years	*Total*	*Empirical*	*%*	*Total*	*Empirical*	*%*
1945-2007	6,004	5,058	84.2	1,886	1,324	70.2
1945-1949	261	184	70.5	164	56	34.1
1950-1959	809	688	85.0	330	172	52.1
1960-1969	1,129	1,005	89.0	300	210	70.0
1970-1979	1,216	1,045	85.9	295	224	75.9
1980-1989	988	774	78.3	295	228	77.3
1990-1999	888	713	80.3	274	227	82.8
2000-2007	713	649	91.0	228	207	90.8

[a]Total refers to the total number of articles and research notes.

[b]Empirical refers to the number of empirical articles and research notes.

[c]% refers to the percentage of empirical articles and research notes.

Finance, 1946; *Journal of Financial Economics*, 1974; *Journal of Financial and Quantitative Analysis*, 1966; *Journal of Money, Credit and Banking*, 1969), management (*Academy of Management Journal*, 1958; *Administrative Science Quarterly*, 1956; *Human Relations*, 1947; *Journal of Management*, 1975; *Journal of Management Studies*, 1964; *Organizational*

Table 2-2 The Growth of Empirical Research in the Management Sciences: 1945-2007

	Accounting			*Economics*			*Finance*		
Years	*Total*[a]	*Empirical*[b]	*%*[c]	*Total*	*Empirical*	*%*	*Total*	*Empirical*	*%*
1945-2007	1,481	717	48.4	4,594	2,315	50.4	1,768	1,116	63.1
1945-1949	60	5	8.3	202	43	21.3	24	7	29.2
1950-1959	146	17	11.6	531	160	30.1	60	30	50.0
1960-1969	248	42	16.9	671	262	39.0	125	64	51.2
1970-1979	298	116	38.9	927	425	45.8	431	203	47.1
1980-1989	311	218	70.1	928	514	55.4	450	283	62.9
1990-1999	229	173	75.5	793	526	66.3	381	289	75.9
2000-2007	189	146	77.2	542	385	71.0	297	240	80.8

	Management			Marketing		
Years	*Total*	*Empirical*	*%*	*Total*	*Empirical*	*%*
1945-2007	4,502	3,351	74.4	3,122	2,234	71.5
1945-1949	38	13	34.2	125	48	38.4
1950-1959	189	99	52.4	221	81	36.7
1960-1969	523	265	50.7	424	205	48.3
1970-1979	981	741	75.5	659	502	76.2
1980-1989	1,082	856	79.1	659	502	76.2
1990-1999	945	759	80.3	524	433	82.6
2000-2007	744	618	83.1	510	463	90.8

[a]Total refers to the total number of articles and research notes.

[b]Empirical refers to the number of empirical articles and research notes.

[c]% refers to the percentage of empirical articles and research notes.

Behavior and Human Decision Processes, 1966; *Strategic Management Journal,* 1980), and marketing (*Journal of Consumer Research,* 1974; *Journal of Marketing, Journal of Marketing Research,* 1964).

A picture similar to that encountered in the social sciences is apparent in the management areas. Here, too, an almost monotonic increase in the percentage of empirical work is observed in all five domains. Between 1945–1949 and 2000–2007, the frequency of published research that is empirically based rose in the following manner: accounting (8.3%–77.2%), economics (21.3%–71.0%), finance (29.2%–80.8%), management (34.2%–83.1%), and marketing (38.4%–90.8%).

In both the social and management sciences, empirical research is systematically displacing nonempirical/conceptual papers. Pointedly, in some disciplines this research has attained almost monopolistic status.[6] Yet despite this explosion in data-based research, few empirical regularities are discernible (Bamber, Christensen, & Gaver, 2000; Hubbard & Lindsay, 2002; Mick, 2001; Rossi, 1997).

Take marketing as an example of this state of affairs. Andrew Ehrenberg and his colleagues (e.g., Ehrenberg & Bound, 1993; Ehrenberg & England, 1990) have produced fine examples of empirical generalizations in the areas of buyer behavior and price elasticities. Bass (1993, 1995) also has had success with regard to new product diffusion. Yet it is highly doubtful that these stubborn facts would have emerged were it not for the research orientation—they were actively seeking regularities—and tenacity shown by the authors themselves. Sadly, these are the exceptions and not the rule. The lack of repeatable empirical facts, Barwise (1995, p. G30) reminds us, means that much marketing practice and teaching is based on only anecdotal evidence. In this sense, academic marketing research may have generated a collective output that is little or no better than Ries and Trout's (1993) popular press offering *The 22 Immutable Laws of Marketing,* which is entirely anecdotal in content. It is difficult to build a science relevant to the practice of marketing, or any other discipline—see Dawes's (1994) *House of Cards: Psychology and Psychotherapy Built on Myth* as a conspicuous example—that consists primarily of hearsay.

Lest the above account come across as excessively grim, we would do well to consult Armstrong and Schultz's (1993) attempt to find useful, empirical marketing principles. Specifically, they examined some 566 normative statements about products, pricing, place, or promotions (the four Ps, or marketing mix, and quintessentially core elements of the marketing curriculum) selected from nine marketing textbooks. Armstrong and Shultz reported that none of these statements had empirical support. They also found, based on the agreements among four raters, that only

20 of the 566 statements qualified as meaningful principles. When 20 marketing professors were asked how many of the 20 meaningful principles were correct, useful, surprising, and had empirical support, none met all four criteria. Disturbingly, Armstrong and Schultz discovered that 9 of the 20 principles were judged to be nearly as correct when their wording was reversed.

2.2.3 The Hegemony of Statistical Significance Testing: 1945–2007

The Social Sciences

The social sciences are awash with tests of statistical significance. It is almost as if, absent such tests, a research paper is somehow "unscientific." In this section I present evidence on the spread of statistical significance testing in the social sciences. I begin with psychology, a field that contained the earliest consumers of such methods.

Psychology. Given the crucial role that experimentation plays in the discipline, it is not surprising that psychologists became Fisherian converts long before those in the other social sciences. Joan Fisher Box (1978, p. 130) writes that her father's *Statistical Methods for Research Workers* (Fisher, 1925) did not receive a single favorable review. This may have been true in Great Britain. It was not true in the United States, where Harold Hotelling (1927, p. 412) pronounced it to be of "revolutionary importance."

Indeed, at Columbia University, Hotelling was one of a triumvirate of players in Rucci and Tweney's (1980) phylogeny of the spread of Fisher's methods in psychology in America. The other two were George Snedecor at Iowa State College, whom Fisher Box (1978, p. 313) describes as the "midwife in delivering the new statistics to the United States," and Palmer Johnson at the University of Minnesota. Hotelling studied with Fisher at Rothamsted Experimental Station in 1929, and Johnson learned from Fisher when the latter was a professor of eugenics at University College, London (his appointment there was made in 1933). Snedecor became a disciple and arranged for Fisher to lecture at Iowa State College in the summers of 1931 and 1936. Interestingly, because Fisher is not easy to read (Savage, 1976, p. 443), Snedecor's (1934, 1937) own expositions of the great man's work were well received among psychologists (Lovie, 1979, p. 169).

Each of these three important promoters of Fisherian methods in the United States passed on the message to their students in the 1930s and beyond. For example, while at Stanford, Hotelling instructed the psychologist Quinn McNemar, who, in turn, taught others in the discipline, such as Lloyd Humphreys (Northwestern), David Grant (Wisconsin), Allen Edwards (Washington), and so on. Likewise, Snedecor's influence is evident in Edward Lindquist's (1940) popular book preaching the Fisherian gospel in education and psychology.[7]

After World War II, the Neyman–Pearson conception of significance testing claimed ascendancy over Fisher's approach among mathematical statisticians. But this was not the case in social science statistics textbooks, where an anonymous hybridization of Fisher's and Neyman–Pearson's methods typically are showcased (Gigerenzer, 1993; Gigerenzer et al., 1989; Goodman, 1993; Hubbard & Bayarri, 2003, 2005). This important issue is visited in some detail in Chapter 7.

I examined the publication incidence of empirical research using tests of statistical significance in the seven leading American Psychological Association journals listed in the previous section. This required the inspection of 5,058 empirical papers and research notes, of which 4,477 (88.5%) used significance tests.

By way of background, Gigerenzer and Murray (1987, ch. 1) allege that between about 1940 and 1955 an "inference revolution" took place in psychology. In this time inferential statistics, and especially *p*-values, were broadly adopted and eventually institutionalized as the single method of inductive inference. It was during this approximate period that rejection of the null hypothesis at $p \leq .05$ gradually infiltrated psychology textbooks on statistical analysis (Halpin & Stam, 2006; Huberty, 1993; Huberty & Pike, 1999).

The content of psychology journals strongly supports the notion of a 1940–1955 inference revolution. For example, two colleagues and I (Hubbard, Parsa, & Luthy, 1997), with data gathered from the *Journal of Applied Psychology,* showed that whereas 25.0% of empirical work published in this journal between 1940 and 1945 employed *p*-values, this number more than doubled to 59.7% during 1950–1954. Further confirmation of an inference revolution in psychology is provided by Patricia Ryan and me in our analysis of 12 American Psychological Association journals (Hubbard & Ryan, 2000). We found that while only 4.0% of empirical work in these journals reported *p*-values for 1935–1939, the corresponding numbers for 1940–1944 and 1950–1954 were 22.7% and 71.7%, respectively.

Beginning as they do in 1945, my data do not permit an additional empirical check of Gigerenzer and Murray's (1987) inference revolution hypothesis. What they do clearly reinforce for 1945–1949, however, is the rapidity with which *p*-values were deified by psychologists; some 62.5% of empirical studies featured this index in their accounts (see Figure 2-1 and Table 2-3). This figure jumped to 81.3% during 1950–1959 and 86.9% in the following decade. A relentless—though by now subdued because the upper limit is approaching—increase in this number is observed thereafter: 1970–1979 (90.4%), 1980–1989 (91.9%), 1990–1999 (91.9%), and 2000–2007 (95.4%). These figures are largely corroborated by Parker's (1990) investigation of the growth of statistical significance testing in *Perception & Psychophysics* over the period 1966–1990.

Gigerenzer and Murray (1987) write that prior to the inference revolution, psychologists relied mostly on assorted, nonstandardized ways for making inductive inferences. These included the presentation of copious descriptive data for individual subjects or small groups. Occasionally, mechanistic criteria for gauging significance, such as three times the probable error and critical ratios, were utilized (see Hubbard, Parsa, & Luthy, 1997, for details). L. D. Smith, Best, Cylke, and Stubbs (2000), in an article

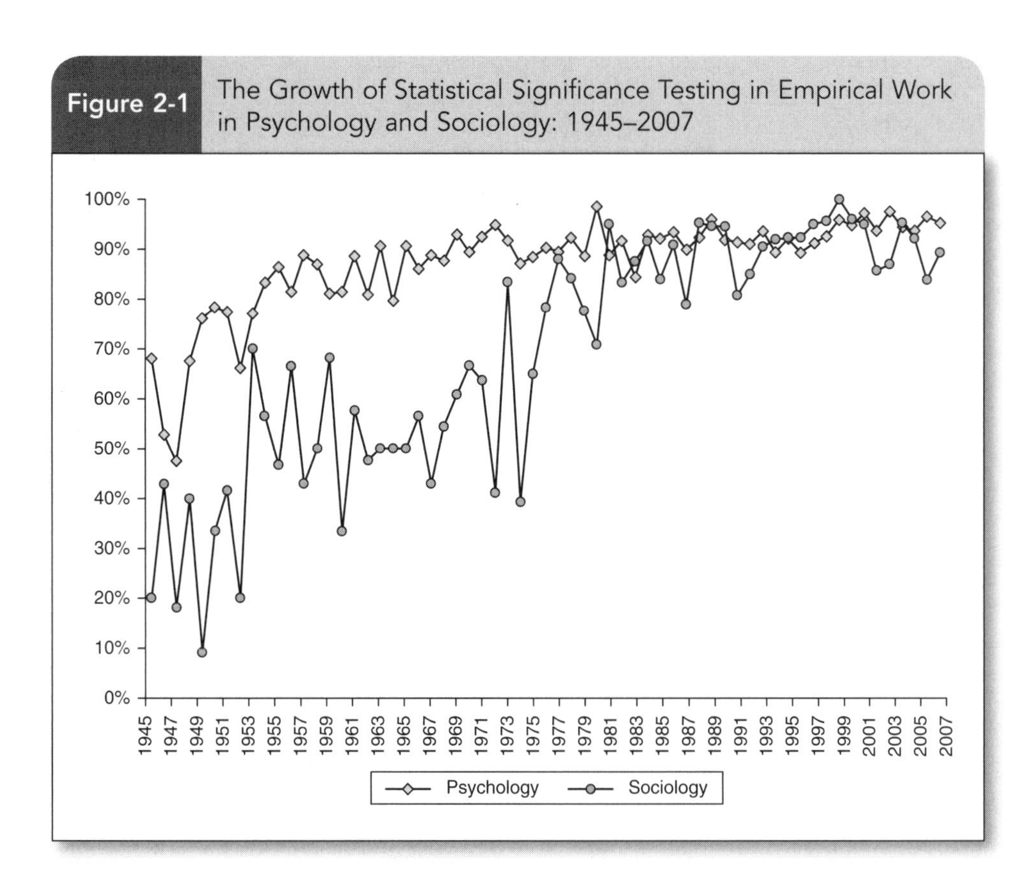

Figure 2-1 The Growth of Statistical Significance Testing in Empirical Work in Psychology and Sociology: 1945–2007

Table 2-3 The Growth of Statistical Significance Testing in the Social Sciences: 1945-2007

	Geography		*Political Science*	
Years	*Number*[a]	*%*[b]	*Number*	*%*
1945-2007	270	39.2	620	72.3
1945-1949	0	0.0	0	0.0
1950-1959	2	4.2	4	14.3
1960-1969	17	16.2	24	30.0
1970-1979	52	39.4	99	57.9
1980-1989	74	58.3	146	72.6
1990-1999	68	54.8	186	91.6
2000-2007	57	42.9	161	93.6
	Psychology		*Sociology*	
Years	*Number*	*%*	*Number*	*%*
1945-2007	4,477	88.5	956	72.2
1945-1949	115	62.5	15	26.8
1950-1959	559	81.3	87	50.6
1960-1969	873	86.9	107	51.0
1970-1979	945	90.4	154	68.8
1980-1989	711	91.9	198	86.8
1990-1999	655	91.9	208	91.6
2000-2007	619	95.4	187	90.3

[a]Number refers to the number of empirical articles and research notes using tests of statistical significance.

[b]% refers to the percentage of empirical articles and research notes using tests of statistical significance.

tellingly called "Psychology Without *p* Values," note that graphical techniques were a common means of conveying information in psychology's early days. Of particular interest, no consensus existed about which were the appropriate ways for drawing inferences from data, and, importantly, this was not seen to be a problem among members of the profession

(Gigerenzer & Murray, 1987). Finally, it is sobering to observe that some of the most respected theoreticians and experimentalists in psychology—H. Ebbinghaus, Wolfgang Köhler, Jean Piaget, B. F. Skinner, S. S. Stevens, Edward L. Thorndike, and Wilhem Wundt—had no use for inferential statistics, preferring instead to exercise their own judgment (Gigerenzer & Murray, 1987, p. 26; Smith et al., 2000, p. 260). It would be beneficial to resurrect such thinking.

Sociology, Political Science, and Geography. Data on the frequency of statistical significance testing were collected for the period 1945–2007 for sociology, political science, and geography using the same journals listed earlier. For sociology, this involved a total of 1,886 articles, 1,324 (70.2%) of which are empirical, with 956 (72.2%) of the latter using tests of significance. Corresponding numbers for political science are 1,722 total articles, with 857 (49.8%) empirical, and 620 (72.3%) of the empirical papers using significance tests. Finally, for geography, 1,262 articles in total, 689 (54.6%) empirical, and 270 (39.2%) invoking significance tests.

Over time, sociologists' zeal for *p*-values is exceeded only by the psychologists' (see Figure 2-1 and Table 2-3). Whereas none of the data-based works in political science and geography during 1945–1949 involved statistical tests, 26.8% already did so in sociology. This figure almost doubled in sociology during the 1950s (50.6%), while on average it remained low in political science (14.3%) and geography (4.2%). Nonetheless, a steady increase in statistical significance testing is seen in Figure 2-2 and Table 2-3 for political science, such that for the period 2000–2007 this discipline's count (93.6%) just eclipses that for sociology (90.3%).[8] For this same time period, geography (42.9%) yields a comparatively modest number, perhaps because the area of spatial statistics offers fewer testing options than its aspatial counterpart.

The Management Sciences

Figures 2-3 and 2-4 and Table 2-4 depict the almost uniform rise of empirical articles and research notes employing statistical significance tests in the business sciences for 1945–2007. I begin the discussion with economics, by far the oldest business discipline.

Economics. Keuzenkamp and Magnus (1995, p. 16) quip that if economists have natural constants, then the best known is .05. But it was not always this way. Morgan (1990, p. 235), for example, remarks that econometric work in the 1920s and 1930s was centered chiefly on the measurement of phenomena and that the role of inference was not viewed as

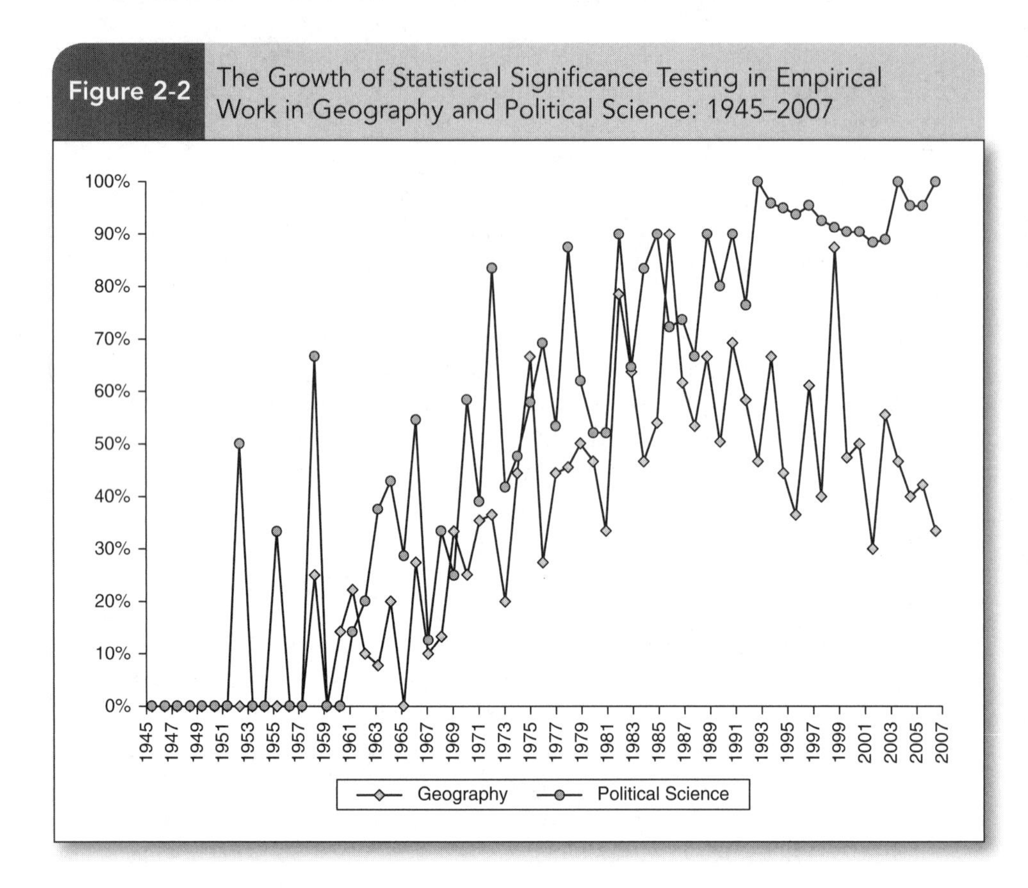

Figure 2-2 The Growth of Statistical Significance Testing in Empirical Work in Geography and Political Science: 1945–2007

important. This philosophy preceded, and subsequently was echoed in, the motto of the Cowles Commission for Research in Economics, created in the United States in 1932, that "Science is Measurement." Things changed, however, with the publication of Trygve Haavelmo's (1944) 115-page paper "The Probability Approach in Econometrics." In this paper Haavelmo gave a lengthy explanation of the Neyman–Pearson ideas on hypothesis testing, and urged that economic theories be cast as formal statistical hypotheses. Haavelmo's article greatly influenced the work of the Cowles Commission. It led to a new motto for the Commission in 1952: "Theory and Measurement." Now, the gathering of data would be dictated by neoclassical economic theory.

Haavelmo's article also marked a reorientation in empirical economics from theory *development* to theory *testing* (Keuzenkamp, 2000, pp. viii, 160–162; Morgan, 1990, pp. 257, 263–264; Ziliak & McCloskey, 2008, p. 113). This shift in emphasis was aided by another Cowles Commission member (and director from 1948 to 1954), Tjalling Koopmans, who

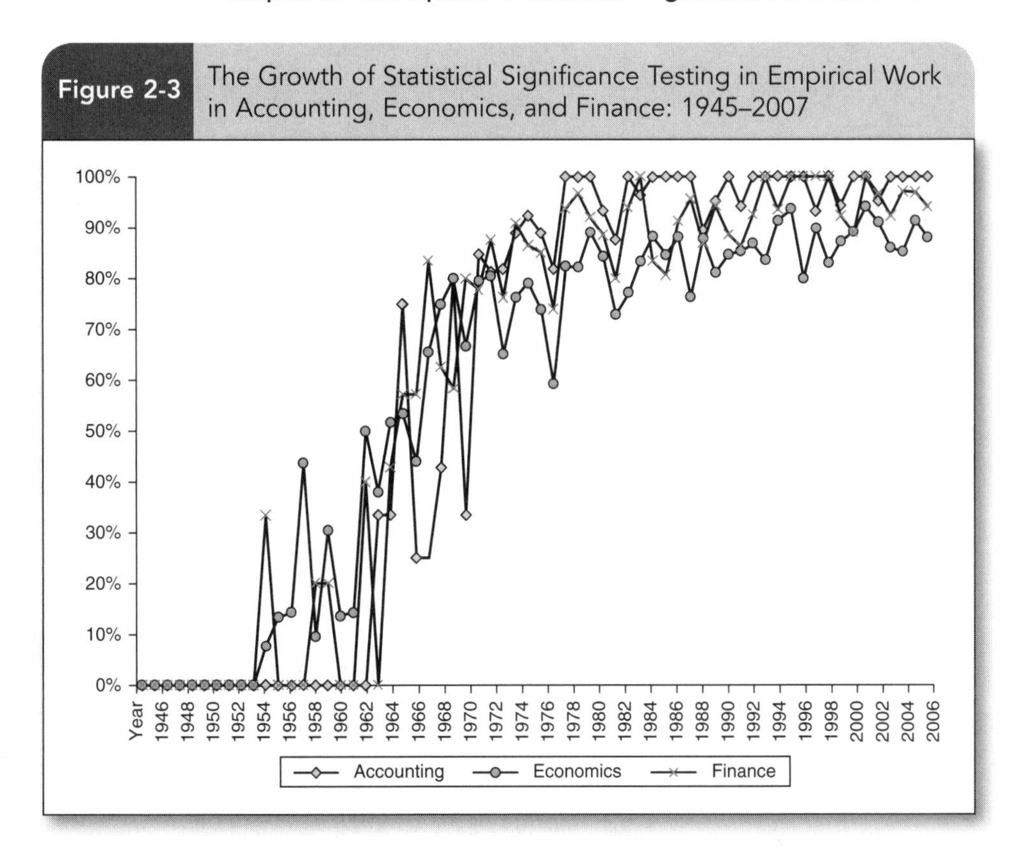

Figure 2-3 The Growth of Statistical Significance Testing in Empirical Work in Accounting, Economics, and Finance: 1945–2007

argued that the focus on measurement error had resulted in the neglect of sampling error. In what I regard as a critical mistake, over the years economists, through their obsession with significance testing, have de facto downplayed measurement error and elevated sampling error to dominance. Today this is personified in Hendry's (1980, p. 403) exhortation that "the three golden rules of econometrics are test, test, test."

Employing the same procedures used with regard to the social science literatures, I traced the growth of significance testing in economics from 1945 through 2007. The data are based on a content analysis of the five prestigious economics journals listed previously. While econometrics textbooks almost exclusively present the Neyman–Pearson model of hypothesis testing, these journals instead are saturated with Fisherian *p*-values, as shown in Figure 2-3 and Table 2-4. All told, some 4,594 articles were reviewed, with 2,315 (50.4%) being empirical. Of the data-based articles, 1,697 (73.3%) employed Fisherian significance testing.

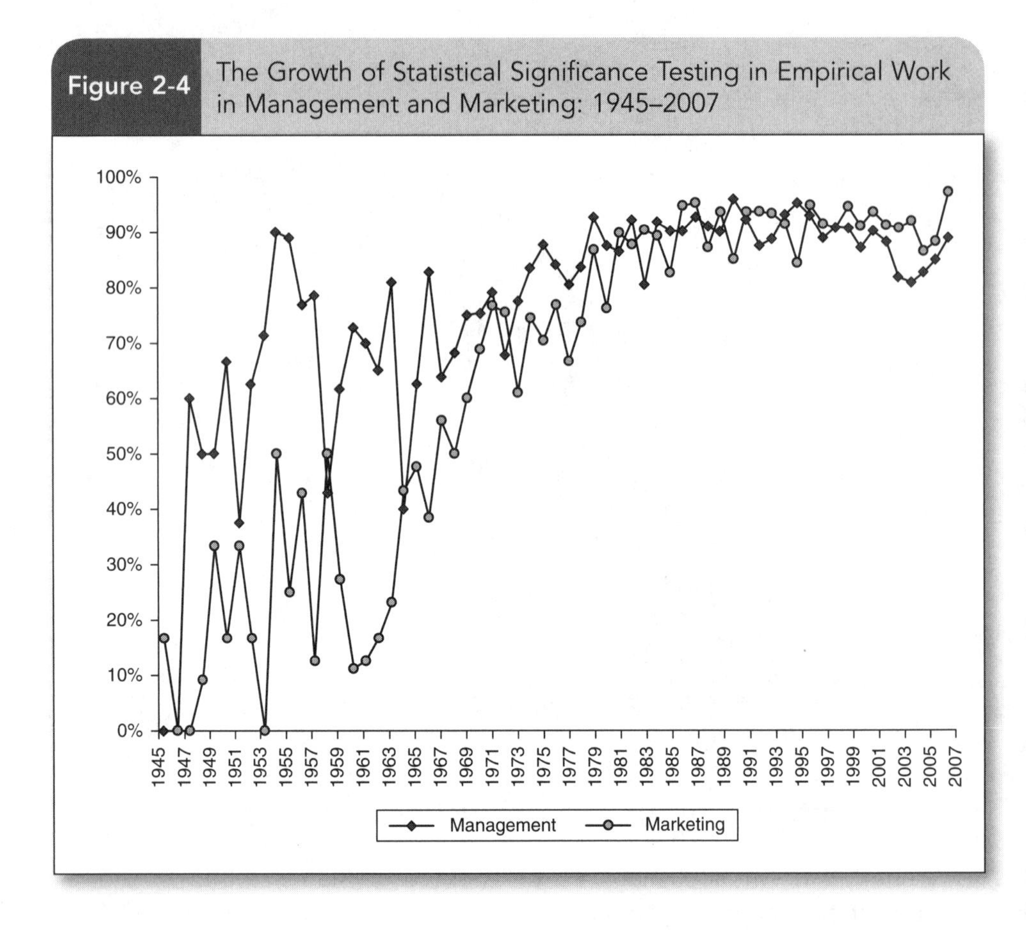

Figure 2-4 The Growth of Statistical Significance Testing in Empirical Work in Management and Marketing: 1945–2007

In my sample, no statistical significance tests were performed during the period 1940–1945. Thereafter, a steady increase in their visibility is apparent. For instance, the decade 1950–1959 saw 13.8% of empirical articles featuring tests of significance, probably in response to the Cowles Commission's call for theory testing. This figure rose abruptly to 52.3% during the 1960s. Even during the 1970s (74.6%), when the contribution of econometrics was being questioned (Keuzenkamp, 2000; Morgan, 1990), growth in the use of such testing nevertheless occurred. And it continued to do so in the 1980s (83.1%), the 1990s (85.7%), and for 2000–2007 (89.1%). Economists take Hendry's (1980) mantra to heart.

As with psychology, in the earlier days in economics there was variation in the tests carried out by researchers, and no ground rules for presenting results were enforced (Morgan, 1990, p. 157). This altered, of course, when conventional wisdom made the reporting of *p*-values for all intents and purposes mandatory.

Table 2-4 The Growth of Statistical Significance Testing in the Management Sciences: 1945-2007

	Accounting		*Economics*		*Finance*		*Management*		*Marketing*	
Years	*Number*[a]	*%*[b]	*Number*	*%*	*Number*	*%*	*Number*	*%*	*Number*	*%*
1945-2007	643	89.7	1,697	73.3	967	86.6	2,854	85.2	1,792	78.2
1945-1949	0	0.0	0	0.0	0	0.0	7	53.8	6	12.5
1950-1959	0	0.0	22	13.8	3	10.0	67	67.7	23	28.4
1960-1969	18	42.9	137	52.3	35	54.7	185	69.8	88	42.9
1970-1979	100	86.2	317	74.6	173	85.2	606	81.8	351	73.3
1980-1989	211	96.8	427	83.1	253	89.4	764	89.3	443	88.2
1990-1999	170	98.3	451	85.7	275	95.2	696	91.7	395	91.2
2000-2007	144	98.6	343	89.1	228	95.0	529	85.6	423	91.4

[a]Number refers to the number of empirical articles and research notes using tests of statistical significance.

[b]% refers to the percentage of empirical articles and research notes using tests of statistical significance.

Accounting, Finance, Management, and Marketing. With the exception of economics, many of the other business journals originated during or after the hardware and software needed to make the computation of statistical tests effortless were routinely available. So indoctrination into the significant difference paradigm had already taken place.

"Historically," however, there are some counterintuitive empirical results in the spread of statistical significance testing in the business fields. Granted, the absolute numbers are small, nevertheless it is the "softer" disciplines of management (53.8%) and marketing (12.5%) that used significance tests during 1945–1949, whereas accounting, economics (as noted earlier), and finance saw zero usage. Management's high percentage for this time is due to the fact that only one journal, *Human Relations,* is in the database. And *Human Relations* has a pronounced psychology bent.

By the 1960s statistical significance testing in empirical research was firmly ensconced in accounting (42.9%), finance (54.7%), management (69.8%), and marketing (42.9%). Yet the 1970s witnessed a dramatic upsurge in such work: accounting (86.2%), finance (85.2%), management (81.8%), and marketing (73.3%). For 2000–2007, the numbers grew again: accounting (98.6%), finance (95.0%), management (85.6%), and marketing (91.4%). There is little room left for the further expansion of statistical significance testing in management science empirical studies; the practice has long since had a stranglehold in these fields.

2.2.4 Summarizing the Dominance of the Statistical Significance Test

Fisher (1959, p. 76) reminisced that "the common tests of significance . . . have come to play a rather central part in statistical analysis." As has been shown, even this strong assertion is, in fact, an understatement; these tests are considered to be almost the *only* legitimate way of making inductive inferences from numerical data. The test of significance is no mere statistical technique, but the glue that holds together the entire research process. It largely dictates how we formulate hypotheses; design questionnaires; organize experiments; and analyze, report, and summarize results (Hubbard & Armstrong, 2006, p. 114). To illustrate this, the test of statistical significance has come to be seen as a mark of scientific rigor (Lindsay, 1995, p. 35), as the sine qua non of scientific respectability (Cowles, 2001, p. 179), as the centerpiece of inductive inference (Hubbard & Ryan, 2000, p. 678), as an objective and universally defined standard of scientific demonstration (Gigerenzer et al., 1989, p. 108), as an automaticity of inference (Goodman, 2001, p. 295), and simply "as an end, in and of itself" (Cicchetti, 1998, p. 293).

In the significant difference paradigm statistical inference and scientific inference are interchangeable; asterisks outweigh substance.[9] To judge the validity of this account it is necessary only to raise the question: "How would one analyze data if the significance test was outlawed?"[10] I suspect that many researchers in the management and social sciences would be hard pressed to respond.

2.2.5 Overgeneralizing the Results of Single Studies

Another facet of the unproblematic conception of knowledge justification embedded in the significant difference paradigm needs to be made clear. This is the tendency for researchers to place far more confidence in the conclusions of single works—predominantly empirical in composition and sporting $p \leq .05$ outcomes, although not confined to this genre—than is warranted. Such studies are perceived as having an aura of finality about them, as if they have largely settled the matter of the research topic at hand (see, e.g., De Long & Lang, 1992, p. 1258). Perhaps because of this there is an accompanying proneness to overgeneralize the scope of the findings of single works (Gauch, 2003, pp. 254–255; Hubbard & Lindsay, 1995, p. 52; Starbuck, 2006, pp. 34–35), something Wells (2001, p. 494) denounces as "The Perils of $N = 1$." More broadly, Tversky and Kahneman (1974, p. 1124) call this phenomenon of generalizing on the basis of insufficient evidence the "representativeness heuristic" (see also Cassidy, 2009, p. 195).

Incredibly, this penchant for overgeneralizing the applicability of initial results from one-off studies persists even in those all-too-rare situations where later works somehow manage to circumvent the barriers against publishing replications and reexaminations (see Chapter 5) and rebut, often decisively, earlier findings. Five examples from the marketing, and one from the psychology, literatures presented below epitomize this inclination.

Chronologically, the first example is the study on subliminal advertising carried out at a New Jersey drive-in movie theater in 1956 whose results have been referred to many times in marketing and consumer behavior textbooks (cf. Wilkie, 1986). This study, publicized in the press, claimed to show how the subliminal messages "Hungry? Eat Popcorn" and "Drink Coca-Cola" flashed repeatedly on the screen for 1/3,000 of a second (far below the level of the limen, i.e., our ability to consciously perceive stimuli) boosted the sales of popcorn and Coca-Cola by 58% and 19%, respectively. These dramatic findings from an investigation notable for its absence of scientific controls have found no subsequent support in research addressing the connection between subliminal advertising and buyer behavior (see, e.g., Moore, 1982; Rosen & Singh, 1992; Theus, 1994).

Regardless, Kerin, Hartley, and Rudelius (2013, p. 118) report that customers spend $50 million a year for audiotapes containing subliminal messages to help them stop smoking, lose weight, and improve their self-esteem. Further, about two thirds of U.S. consumers say that subliminal messages are hidden in commercials, and roughly half believe that this technology can cause them to buy things against their will.

A second example is Julian Simon's (1979) reanalysis of the experimental data employed in Zielske's (1959) authoritative contribution on the remembering and forgetting of advertisements. Zielske's determination as to which method of scheduling advertisements is more effective—spaced or pulsed—is equivocal, and his published results pictured only idealized representations of the raw data. Simon's reanalysis of the latter points to a definitive conclusion: A spaced advertising schedule is more dollar-effective than one that is pulsed. Frustratingly, despite Simon's reworking of the raw data, it is still Zielske's idealized findings which are more likely to be discussed in the classroom. For example, in a convenience sampling of 15 consumer behavior textbooks published either for the first time or in subsequent editions between 1982 and 1988 (thereby allowing enough time for Simon's results to be included), only two cite both his work and Zielske's, and one of these fails to mention their different conclusions (Hubbard, Brodie, & Armstrong, 1992, p. 5). Thirteen of the 15 textbooks do not cite Simon's article, while 10 of them feature Zielske's experiment.[11] Reflecting this disparity, Google Scholar shows that Zielske ($n = 196$) has attracted more than six times as many citations as Simon ($n = 30$).

Levitt's (1960) hugely influential publication on "marketing myopia" is the third example. This is when executives are said to become overly enamored with their products, thereby losing sight of the underlying needs of their customers and the poor decisions which inevitably stem from this. Specifically, Levitt was concerned with what he considered to be the short-sighted thinking of those in charge of the U.S. railroad and film industries. As Morris (1990, p. 279) explains, Levitt's argument, "which has become familiar to almost every teacher and probably most students of marketing," was that railroad managers saw their business as running trains and did not anticipate a market shift toward freeway and air transport. Likewise, Levitt opined, movie moguls were focused on the cinema and were unprepared for the competition posed by television.

After painstaking research 30 years later, Morris (1990) was able to demonstrate that it was government regulations, not deficient marketing practices nor unwise management, which prevented the railroads from moving into alternative modes of transportation. Similarly, Morris showed that regulations stymied the movie studios from encroaching on the television market. In the final analysis, Levitt's (1960) anecdotal offering that the film and railroad industries "failed to be marketing oriented is unsubstantiated by the historical evidence" (Morris, 1990 p. 282). Facts notwithstanding, it is still Levitt's conjecture which holds sway in the textbooks and classrooms. As a bizarrely lopsided indicator of this, Levitt's article has managed an astonishing total of 2,201 citations (Google Scholar), while Morris's, with a mere 3, is long since forgotten.

The fourth example is from Gorn (1982), who published an article stating that product preferences can be classically conditioned through a single pairing with background music. Kellaris and Cox (1989) believed that Gorn's results may have been due to demand artifacts. This concerned them because Gorn's article was drawing notice, being cited in the *Social Sciences Citation Index* 34 times between 1982 and 1988. However, the work was also being referred to in consumer behavior textbooks, at least one of which used it as a basis for declaring that classical conditioning of product preferences is "well established and widely used." In three well-designed experiments, Kellaris and Cox failed to replicate Gorn's results. Despite this, according to Google Scholar, Gorn (n = 603) continues to outperform Kellaris and Cox (n = 158) by a factor of almost 4 to 1 when it comes to garnering citations.[12]

The fifth example is provided by Bottomley and Holden (2001). In circumstances they labeled "unique," these authors conducted a study to determine the empirical generalizability of Aaker and Keller's (1990) model stating that consumers evaluate brand extensions on the basis of the quality of the original brand, the degree of fit observed in the parent and extension categories, and the interaction between the two. Like Gorn's (1982), Aaker and Keller's paper made an impact in intellectual circles, being cited some 92 times from 1990 to 2001, and their findings similarly were appearing in marketing textbooks (Bottomley & Holden, 2001, p. 494).

What makes Bottomley and Holden's (2001) paper unique is that they were able to gain access to Aaker and Keller's (1990) original data set as well as seven others from studies attempting to replicate the latter's work.[13] This is noteworthy because, to my knowledge, it represents the most comprehensive example of independent replication research in academic marketing. It also is anomalous because, as shown in Section 5.2, replication research seldom is seen in marketing.

In their reanalyses of the eight data sets, Bottomley and Holden (2001) were able to resolve many of the conflicting results between them. They also cautioned that a key lesson from their work is the danger of drawing "firm conclusions about theory on the basis of only one study" (p. 494).

An example from psychology completes the discussion of the risks involved in putting too much faith in the results of one-off studies. Kelley and Blashfield (2009) trace the history of Broverman, Broverman, Clarkson, Rosenkrantz, and Vogel's (1970) article on sex bias in the mental health profession. After analyzing questionnaire returns, Broverman and her colleagues wrote that a double standard was discernible among mental health clinicians in how they viewed men and women. In particular, they found that assessments of mental health were biased in favor of men.

Kelley and Blashfield (2009, p. 123) note that Broverman et al.'s (1970) study has had an enormous impact on the thinking of a generation of psychologists and mental health experts, along the way becoming one of the most highly cited papers in psychology. Using *Science Citation Index* and *Social Science Citation Index* databases, Kelley and Blashfield discovered that Broverman et al. had attracted an amazing 934 citations since publication of their article. To lend perspective on this result, Kelley and Blashfield also checked on the citations gathered by Stephen and Christine Abramowitz, the two most prolific authors on sex bias during the 1970s. Their best-known paper (Abramowitz, Abramowitz, Jackson, & Gomes, 1973) acquired 58 citations. Impressively, Kelley and Blashfield reveal, Broverman et al. have collected more citations than the top-cited articles from some of the field's 20th century luminaries, including Hans Eysenck ($n = 866$), B. F. Skinner ($n = 590$), and Robert Sternberg ($n = 226$).

And yet a critique by Stricker (1977) as well as replications by Phillips and Gilroy (1985) and Widiger and Settle (1987), each published in well-known psychology journals, have brought to light "fatal methodological flaws" in Broverman et al.'s (1970) work (Kelley & Blashfield, 2009, p. 126). Baffling to Kelley and Blashfield (2009, p. 128) are how Broverman et al.'s findings continue to be accepted in the field:

> It is almost as if these critical articles [i.e., Stricker, 1977; Phillips & Gilroy, 1985; Widiger & Settle, 1987] were never published. Researchers citing Broverman et al. seem to be unaware that the central conclusions from that article are erroneous.

Pursuing this lead, Kelley and Blashfield add that Broverman et al. were cited some 53 times between 2000 and 2008 alone, while plaudits earned for works over their entire lifespans by scholars attempting to correct the written record on this topic suffer in comparison: Stricker ($n = 82$), Phillips and Gilroy ($n = 26$), and Widiger and Settle ($n = 33$). Baffling indeed.

Given their criticisms over the apparent neglect among psychologists about issues relating to safeguarding the literature, it will be instructive to see how well Kelley and Blashfield's (2009) paper itself is received. While still early days, Google Scholar informs us that as of February 2013 they have been cited only 3 times.

Findings based on single works can be highly misleading. They serve to underscore Kendall's (1961, p. 5) warning that "the pathway of knowledge is littered with the wreckage of premature generalization." But the seriousness of the lessons for knowledge development in the examples presented above go well beyond this reproof. By continuing to dwarf the influence of the very studies exposing their fallacious results, these works offer compelling evidence that science in the significant difference paradigm is not self-correcting. Like urban legends, erroneous results maintain a life of their own. This chilling message about whether we can trust what we read in the literature is reinforced explicitly in Sections 2.4, 4.4.1, 4.4.2, 5.4, 5.5.2, 7.4.1, and 8.3.2, and implicitly elsewhere throughout this book. We need many more replications, and far greater recognition (e.g., citations) and rewards (e.g., promotion, tenure) for those performing such essential tasks.

2.3 Model of Science—Hypothetico-Deductivism

This paradigm is linked inextricably with the hypothetico-deductive (H–D) model of scientific explanation. According to the philosopher Hausman (1992, p. 304), the essence of this method is captured in four steps:

1. *Formulate* a model or theory, *T*.
2. *Deduce* a prediction, hypothesis, or some other empirical proposition, *P*, from *T* in conjunction with a number of other auxiliary propositions. The latter would include descriptions of initial conditions, other relevant information (theories), and *ceteris paribus* ("other things being equal") qualifiers.
3. *Test P*, because *T* can be evaluated only indirectly in the H–D model.
4. *Judge* whether *T* is confirmed or disconfirmed on the basis of whether *P* turned out to be "true" or "false."

The H–D conception of the scientific method has dominated a large part of 20th century philosophical thinking. It is anointed in a number of preeminent works. These include Braithwaite's (1953) *Scientific Explanation,*

Popper's (1959) *The Logic of Scientific Discovery,* Nagel's (1961) *The Structure of Science,* and Hempel's (1965) *Aspects of Scientific Explanation.* Recall how Fisher's declaration that the use of statistical analysis could raise the status of a discipline found a receptive audience among applied researchers. This same promise of scientific respectability also was offered by proponents of the H–D model. Consider Braithwaite's (1953, p. 9) assurance: "It is this hypothetico-deductive method applied to empirical material which is the essential feature of a science; and if psychology or economics can produce empirically testable hypotheses, *ipso facto* they are sciences." Small wonder that researchers in these areas rallied to the cause.

As chronicled in Section 2.2.3, the pages of the leading management and social science journals are flooded with the empirical (statistical significance) testing of hypotheses, to the virtual exclusion of other means of data analysis. Economists typically are in the vanguard of compliance with the H–D method, and more apt to use formal/mathematical analyses to deduce propositions from some axiomatic system. Researchers in finance and accounting strive to emulate their colleagues in economics. In the "softer" areas, such as management, marketing, anthropology, geography, political science, social psychology, and sociology, the "deduction" of hypotheses tends to be more ad hoc—what Meehl (1990, p. 199) terms a "loose derivation chain." Mostly this follows from a review of pertinent literatures that would seem to make the hypotheses offered plausible.

Over time, the primacy reserved for theory and theory *testing* has resulted in the following commonly held opinions and/or behaviors:

- Hypothesis generation (discovery) and testing (justification) are viewed as quite separate and distinct activities, with philosophers of science focusing on the latter (Suppe, 1977). This same situation holds in the social science and business disciplines where the majority of scholars, as noted above, continue to be almost totally absorbed with justification rather than with discovery. Depending on one's philosophical orientation, the objective is to either confirm (logical positivism/empiricism) or falsify (Popper, 1959, 1963) theories. In contradistinction, the process of theory development has received much less attention (Haig, 2005; Hubbard & Lindsay, 2002; Hunter, 2001; Wells, 2001).
- It is the application of the "scientific method" which distinguishes science from non-science. And the H-D model, with its proclivity for theory testing, is seen widely as the very embodiment of *the* scientific method (Hubbard & Lindsay, 2013b, p. 1394). Therefore, if the correct methodological recipe is adhered to, this makes the output "science." In keeping with this outlook, science progresses by following an orderly, mechanistic, sanitized protocol. Apostles of the significant difference paradigm reject Tukey's

(1980) claims that science does not begin with tidy questions, nor end with tidy answers. For them, each research question (hypothesis) is framed in meticulous detail—H_1, H_2, H_3, H_{3a}, H_{3b}, H_{3c}, H_4, and so on. And the findings are reported, via *p*-values, with a seemingly impressive degree of precision, as if addressing decisively the topic(s) at hand in *that* moment. All research endeavors involve exact questions and answers.

- A single study (e.g., Broverman et al., 1970) can yield immediate and lasting knowledge. Such beliefs are revealed in the idea of the "crucial experiment," implying that theories can be conceived, evaluated, and finalized over a very short period of time (Bauer, 1994; Box, 1994; Haig, 2014; Lakatos, 1970).[14] This is seen in Hausman's (1992) point 1 above, where a theory is simply "formulated." It is why Locke (2007, p. 867) bemoans that the H–D approach demands premature theorizing, a consequence of which is that "theories tend to be grounded in myths and superstitions" (Van de Ven, 2007, p. 17) rather than facts. One manifestation of this view is the lack of replications that are conducted across the management and social sciences (see Chapter 5). Another is that traditional statistics courses bypass the development of theories and concentrate only on testing models in the "one-shot" context (Box, 1994; Chatfield, 1995; Ehrenberg, 1968). Still another, T. Clark, Floyd, and Wright (2006, p. 655), T. G. Gill (2010, p. 295), Hambrick (2007, p. 1346), Helfat (2007, p. 185), and McGrath (2007, p. 1373) admonish, is the requirement by editors of high-quality journals that papers must make a *theoretical* contribution, even when this comes at the expense of those that report interesting and well-documented facts. On this account, facts are subordinate to theory.
- A good theory produces (or should produce) universal generalizations (Aram & Salipante, 2003; S. R. Clegg & Ross-Smith, 2003; Flyvbjerg, 2001; Krebs, 2001; Starbuck, 2006; Teigen, 2002).[15] This idea of laws being universal finds modern expression in Hempel's (1965) deductive-nomological, or covering law, model of scientific explanation. Here, causation is interpreted in Humean terms as one of the constant conjunction of events yielding universal or lawlike empirical regularities. Bishop (2007, pp. 318–320, 401; see also Easton, 2002; Manicas & Secord, 1983; Tadajewski, 2008; Yu, 2006) argues that this discredited positivist philosophy and its quest for apodictic knowledge endures as the received view in the social sciences. Views like this explain why the conception of generalization, articulated in Chapter 6, follows the "representative model" of inferential statistics (Cook & Campbell, 1979). They also help to account for the propensity to overgeneralize results (Bamber et al., 2000; Wells, 2001) noted earlier, and to display a bias against publishing negative findings (Hubbard & Armstrong, 1992, 1997; Lindsay, 1994; Sterling, Rosenbaum, & Weinkam, 1995).
- Theory testing in this paradigm is carried out predominantly on single data sets. As such, considerations of *internal validity* are paramount (D. T. Campbell & Stanley, 1966)—can I reasonably conclude from this

data set that changes in the dependent variable are, in fact, attributable to the manipulation of the independent variable(s)? *External validity*, or generalizability, issues usually are ignored or downplayed (Laurent, 2000; McQuarrie, 2004; Rogers & Soopramanien, 2009; Rozin, 2009; Steckler & McLeroy, 2008; Wells, 1993, 2001; Winer, 1999) or else consigned to the "conclusions" section, where an appeal is made for future research to address this issue.

- It is imperative, from the standpoint of publishing the manuscript, that these one-shot studies using single data sets find statistically significant results that buttress the "originality" or "novelty" claims made in the work. Once this creative idea is accepted for publication, the researcher switches to another novel topic. In other words, what should be seen as the first step in possibly establishing the tenability of a result all too often is accepted as the last word on the research problem. In this manner, the worshiping of original research with statistically significant outcomes guarantees the prevention of a cumulative body of knowledge against which to judge the credibility of future results. Nelder (1986, p. 112), in his presidential address to the Royal Statistical Society, disparaged this practice, calling it the "cult of the isolated study." This cult is ubiquitous in the management and social sciences. Its legacy is highly damaging, namely, a literature composed chiefly of fragmented, one-off results whose contributions to knowledge are of the most speculative kind. As an example, why should anyone care about the fact that intrinsic motivation was found ($p \leq .05$) to improve salesforce satisfaction if this has been found only once, for 56 salespeople in two companies in the United Kingdom in 1995 (Hubbard & Lindsay, 2002, p. 386).
- Good science preaches the virtues of *descriptive* correlational analyses. Fair enough. The downside is that insufficient attention is directed at the search for causal *explanations* (Bunge, 1997, 2004), the latter defined as "clarifying the mechanisms through which and the conditions under which the causal relationship holds" (Shadish, Cook, & Campbell, 2002, p. 9). While both description and explanation are vital for knowledge accumulation, it is the discovery of these mechanisms that is scientists' only protection against the "problem of confounders," something which is always a threat in observational (nonrandomized) studies (Hubbard & Lindsay, 2013b, p. 1395).

Given the above beliefs, it is understandable why tests of statistical significance have enjoyed such a privileged status in conventional social and business research methodology.

Unfortunately, the H–D model of science on which the behavioral and business disciplines rest squarely is deeply flawed. On the one hand it is susceptible to the Duhem–Quine argument, which states that one never tests scientific hypotheses by themselves (Hausman's, 1992, point 2 at the beginning of Section 2.3). When deducing a hypothesis from some theory, numerous auxiliary propositions and initial and boundary conditions are conjoined with the deduced hypothesis. Accordingly, if a hypothetical prediction "fails" (e.g., the null is not rejected), this could be due to the falsity of one or more of the auxiliary propositions, and so on. In this manner we can continue to defend the failed hypothesis by blaming endlessly the support cast, thus leading to a state of "infinite regress" over the truth/falsity of the deduced hypothesis. On the other hand, the H–D model also is vulnerable to the fallacy of affirming the consequent. If a hypothetical prediction "succeeds" (e.g., the null is rejected), there can be other mutually incompatible hypotheses that entail identical predictions. The H–D approach is unable to discriminate between these competing alternative hypotheses, even if some of them are theoretically absurd (Ketokivi & Mantere, 2010, p. 318). The upshot, econometricians Darnell and Evans (1990, p. 37) observe, is that no attempts to test theories—whether by confirmation or falsification—can ever be conclusive.

Because of this, philosophers of science (e.g., Hausman, 1992; Nola & Sankey, 2007) have largely forsaken the epistemological warrant proffered by the H–D model; Psillos (1999, p. 174) labels it crude, Glymour (1980a, p. 322, 1980b, p. 36) calls it hopeless, Gorski (2004, p. 28) says it doesn't work, and Gower (1997, p. 15) states that it cannot ratify the acceptance of any scientific belief. Of course, no path to knowledge procurement is infallible. What is jarring, however, is that while the H–D method faces severe philosophical difficulties, one finds little evidence of such concerns in the behavioral and business disciplines. At the same time the steadfast hewing to the dictates of hypothetico-deductivism places the conduct of research in these same areas at odds with practicing investigators in all the natural sciences (Barwise, 1995, p. G32). In other words, members of the significant difference paradigm accede to a conception of knowledge development that neither philosophers nor practitioners of science endorse. No matter; to them the H–D model remains sacrosanct.

2.4 The Role of "Negative" ($p > .05$) Results

2.4.1 Publication Bias and the Credibility of Empirical Findings From Individual Studies

Many researchers in this paradigm are convinced that an editorial-reviewer bias exists in the management and social (and biomedical) sciences against publishing so-called negative or null results (see, e.g., Bakan, 1966; Bakker, van Dijk, & Wicherts, 2012; Doucouliagos & Stanley,

2013; Fanelli, 2009; Feige, 1975; C. J. Ferguson & Brannick, 2012; Gerber & Malhotra, 2008; Greenwald, 1975; Hubbard, 1995b; Hubbard & Armstrong 1992, 1997; Ioannidis, 2005b; Ioannidis & Trikalinos, 2007; Kruskal, 1978; Lindsay, 1994; Masicampo & Lalande, 2012; Mayer, 1993; Nosek, Spies, & Motyl, 2012; Salsburg, 1985; Simmons, Nelson, & Simonsohn, 2011). By the latter are meant results incompatible with predictions, but more broadly regarded as statistically insignificant, or $p > .05$, outcomes.[16] For instance, almost 50 years ago, when commenting on a former editor (Arthur W. Melton) of the *Journal of Experimental Psychology,* Bakan (1966, p. 427) had this to say: "His clearly expressed opinion that non-significant results should not take up the space of the journals is shared by most editors of psychological journals." Such views have persisted, indeed intensified, over the years, allowing Scott Maxwell (2004, p. 148) to broadcast without fear of contradiction that "typical editorial practices virtually [mandate] statistical significance as a prerequisite for publication," meaning that $p \leq .05$ findings must be secured "at all costs" (Fanelli, 2010, p. 1; Ioannidis, 2012, p. 647).

Fisher (1966) had an early hand in fanning this publication bias in two ways. First, he wrote that "the null hypothesis is never proved or established, but is possibly disproved, in the course of experimentation" (p. 16). Given the asymmetry in this proposition, it is easy to see why editors, reviewers, and researchers could interpret "insignificant" results as being, at best, inconclusive and thus less worthy of publication. In comparison, the rejection of H_0 comes across as definitive. Second, when selecting, arbitrarily, the .05 level to demarcate statistical significance, Fisher instructed experimenters to be "prepared to ignore all results which fail to reach this [.05] standard" (p. 13). So researchers are left with the distinct impression that $p > .05$ results are irrelevant.[17]

At issue is that this publication bias against null findings may seriously distort the content of an empirical literature, even to the degree where its probity cannot be taken for granted. To see this, consider the three options available to an investigator who obtains initially $p > .05$ results. First, null results are not as likely to be submitted for publication. Second, if submitted, null results are more likely to be rejected for publication. Third, confronted with null results the researcher opts to continue with the topic by searching for $p \leq .05$ outcomes. These three scenarios, all of which compromise trust in the scientific enterprise, are addressed in turn.

Negative Results Are Less Likely to Be Submitted for Publication

For example, Greenwald (1975), using intentions data from 36 *Journal of Personality and Social Psychology* authors, estimated the probability of null results being submitted for publication to be .06, compared with .59 for those with "significant" results. Coursol and Wagner's (1986) inspection of 609 responses from a survey of counseling psychologists found that while 82% of the articles reporting positive outcomes were submitted for publication, this figure was only 43% for those with neutral or negative findings. The fact that scholars are less willing to offer manuscripts with $p > .05$ findings for publication misrepresents the volume of empirical work carried out in any given research area (C. J. Ferguson & Heene, 2012, p. 556; Gerber & Malhotra, 2008, p. 5).

Negative Results Are Less Likely to Be Published

Part of the explanation for this is that manuscripts with null findings often are considered to reflect poorly on the researcher's skills rather than on nature, and so once again are thought to be undeserving of publication (see, e.g., Hubbard & Lindsay, 2013b, p. 1394). There is some justification for this. For example, research in both the social (Bakker et al., 2012; J. Cohen, 1988; S. E. Maxwell, 2004; Ottenbacher, 1996) and management (Cashen & Geiger, 2004; T. D. Ferguson & Ketchen, 1999) sciences reveals that many studies have inadequate statistical power to reject a false H_0, a topic of sufficient concern as to necessitate further discussion in the appendix to this chapter. Alternatively, a null finding may simply indicate the absence of any substantive effect in the population (Nickerson, 2000, p. 261).

There is, however, weighty evidence to show that null results are less likely to be published than their non-null peers. Steven Kerr, James Tolliver, and Doretta Petree (1977), for example, surveyed 429 editors and advisory board members of 19 leading management and social science journals to elicit common reasons for manuscript acceptance or rejection. They reported that even when a manuscript was judged to be otherwise competent and of current interest to the field, $p > .05$ results markedly lowered the likelihood of acceptance. Parallel results were found in a survey of 268 manuscript reviewers for Canadian psychology journals (Rowney & Zenisek, 1980). In addition, Atkinson, Furlong, and Wampold (1982) asked 101 consulting editors of two psychology journals to evaluate three versions of a manuscript that differed only with respect to the level of statistical significance attained. The statistically insignificant and almost significant versions were more than three times as likely to be rejected for publication than was the statistically significant one. Epstein (2004) corroborated Atkinson et al.'s findings in a similar study in the field of social work.

Many authors share the same beliefs as those held by editors and reviewers. For instance, 61% of authors who had published empirical articles in various education and psychology journals in 1988 were of the opinion that only research yielding statistically significant findings would be published (Kupfersmid & Fiala, 1991).

Six previous empirical studies from the social sciences, five employing databases from psychology and one from sociology, suggest the existence of a publication bias against null outcomes. For example, Sterling (1959) examined 362 empirical papers published in the 1955 issues of the *Journal of Clinical Psychology, Journal of Experimental Psychology, Journal of Social Psychology,* and the 1956 issue of the *Journal of Comparative and Physiological Psychology.* He found that only 2.7% of those using significance tests failed to reject the null hypothesis. Smart's (1964) analysis of these same four journals in 1962 showed that 8.7% reported null results. Bozarth and Roberts (1972) determined that only 6% of 1,046 articles using statistical tests in the *Journal of Consulting and Clinical Psychology, Journal of Counseling Psychology,* and *Personnel and Guidance Journal* between January 1967 and August 1970 were unable to reject the null hypothesis. Greenwald's (1975) estimate, based on a content analysis of a single annual (1972) issue of the *Journal of Personality and Social Psychology*, was 12.1%. Sterling et al. (1995), using 1986–1987 data, repeated the earlier study by Sterling (1959) and discovered that only 4.4% of articles failed to reject H_0. The proportion of negative results was higher in sociology. Following a review of the *American Journal of Sociology, American Sociological Review,* and *Social Forces* from July 1969 to June 1970, for instance, Wilson, Smoke, and Martin (1973) found that 19.5% (15/77) of empirical articles had findings which were not statistically significant at the $p \leq .05$ level.

In the business fields, Lindsay (1994) surveyed all empirical budgeting and control articles published in three major accounting journals—*Accounting, Organizations and Society; Journal of Accounting Research*; and *The Accounting Review*—during the period 1970–1987. This procedure identified a total of 38 usable empirical studies, of which 6 (16%) were classified as yielding negative results. And based on a random sample of articles from three leading marketing journals—*Journal of Consumer Research, Journal of Marketing,* and *Journal of Marketing Research*—it was estimated that some 7.8% of the manuscripts published between 1974 and 1989 were unable to reject the null hypothesis (Hubbard & Armstrong, 1992). Moreover, this number has been declining over time; for 1974–1979 it was 11.4%, while for 1980–1989 it was 5.7%.

A recent article incorporating all major disciplines supports the existence of a publication bias in favor of positive and against negative results. Fanelli's (2010, p. 2) study, consisting of a random sample of 1,316 articles from 49 U.S. states (Delaware excepted) and the District of Columbia, showed that only 17.6% of papers told of negative findings. For 5 states this value ranged between 2% and 10%, while investigators from another 8 states recorded zero negative outcomes. Of interest, Fanelli also provides information that more successful researchers—those with a greater publication output—report fewer null results than their less prolific colleagues.

Evidence from the medical literature also suggests a publication bias against null findings. As an example, Simes (1986) reported that whereas the pooled results for published trials for a particular treatment of ovarian cancer showed statistically significant benefits, the pooled results of registered trials (which included both published and unpublished studies) evaluating the same treatment did not. In like manner, Dickersin, Chan, Chalmers, Sacks, and Smith (1987) contacted 318 authors of published clinical trials to see if they had been involved with any unpublished ones. Responses from 156 of them yielded 271 unpublished and 1,041 published trials; while only 14% of the unpublished reports favored the test therapy, this figure was 55% for the published results.

The truth of the matter, as Fanelli (2012) shows, is that negative results are disappearing from the physical and biological sciences, and especially from the management and social sciences, in most countries. This is a great concern for her, owing to the fact that "negative data . . . are crucial to scientific progress, because this latter is only made possible by a collective self-correcting process" (p. 892).

Negative Results Obligate Researchers to Search for $p \leq .05$ Outcomes

On finding null results the investigator decides to persevere with the topic at hand. Greenwald (1975) asked his sample if they were likely to conduct an exact or modified replication of their work following an original full-scale test of their main hypothesis. If the original result was statistically significant, the probability was .36; if an insignificant result was obtained, the probability was .62.

This perseverance, remember, is driven almost solely by the need to offer statistically significant findings to the editor if the manuscript is to have a chance of appearing on the printed page. Under circumstances like these, it is reasonable to speculate that some individuals will be drawn to questionable research behaviors aimed at providing editors-reviewers with what they want to see.

One such questionable behavior in this context is data mining. An activity which consists of tirelessly reworking the data in a search for statistically significant outcomes, data mining is said to encourage the proliferation of Type I errors (erroneous rejections of the null hypothesis or false-positive results) well in excess of their nominal .05/.01 levels (see Chatfield, 1995; Denton, 1985, 1988; Feige, 1975; Hubbard & Vetter, 1996; Lindsay, 1994, 1997; Lovell, 1983). Or as Leamer (1983, pp. 36–37) put it in an article drolly titled "Let's Take the Con Out of Econometrics," after fitting endless models to data the researcher carefully selects from the bramble of computer output the one she or he enshrines as a rose.

Greenwald (1975, p. 15) believes the underestimation of Type I errors to be "frightening, even calling into question the scientific basis for much published literature." Wilson et al. (1973) had earlier come to this same conclusion with respect to findings in sociology. Fears about the damaging impact of such practices on the integrity of results continue undiminished (cf. C. J. Ferguson & Heene, 2012, p. 558; Pashler & Wagenmakers, 2012, p. 528), culminating in literatures "infested with error" (Holcombe & Pashler, 2012, p. 355).

A variation on this data mining theme is what Norbert Kerr (1998, p. 196) refers to as HARKing (Hypothesizing After the Results are Known), or presenting post hoc (based on one's statistically significant results) hypotheses as if they were a priori in nature. Such a practice, of course, cuts the legs out from under the logic sustaining the H-D model of explanation. Never mind; according to Kerr, HARKing, which raises fortuitous results to center stage, is prevalent in academe. In backing this claim, he tells that a survey of 156 behavioral scientists in social psychology, clinical/community psychology, and sociology revealed that about 45% of them had personally observed, and 55% suspected, instances of HARKing among colleagues. Kerr's results have found emphatic validation. Based on a 20% response rate from a survey of 1,940 management faculty at 104 PhD-granting business schools in the United States, Bedeian, Taylor, and Miller (2010, p. 716) found that 92% reported knowledge of researchers within the previous year developing hypotheses after the results are known.

Moreover, being driven mostly by the need to offer $p < .05$ results supporting one's ideas, HARKing contributes directly to the problem of confirmatory bias, while simultaneously eliminating an opportunity to falsify a theory (Leung, 2011, p. 475). This indicates again that science following a significant difference philosophy tilts toward the production of suspect empirical literatures.

In concert with the above strategies, persevering with a research topic might include the duplicitous reporting by investigators of only those results confirming ($p < .05$) their expectations and/or the suppression of "inconvenient" data. Fanelli (2009, p. 6) estimates that some 9.5% of researchers are complicit in this regard. Her valuation seems low in comparison with others. Thus, for example, Bedeian et al. (2010, p. 716) discovered that 77.6% of their survey respondents indicated awareness of management researchers selecting only those data supporting hypotheses, while 59.6% claimed to know of scholars discarding observations felt to be inaccurate. John, Loewenstein, and Prelec's (2012, p. 527) online survey of 5,963 psychologists at American universities—response rate 36.1% (2,155)—put these same estimates at 67% and 62%, respectively.

Empirical support for other common deceptive research habits include withholding methodological specifics or results, using another's ideas without permission or giving credit, not revealing data that contradict the author's own earlier work, and publishing the same data or results in two or more outlets (see Bedeian et al., 2010, p. 716, for details). Further sins of omission and commission involve failing to report all dependent measures and all study conditions, ceasing data collection after obtaining the desired result(s), rounding down *p*-values, and claiming to have predicted an unexpected outcome (see John et al., 2012, p. 527, for details).[18]

The ultra-competitive "publish or perish" environment typical of the significant difference school may induce some researchers to go so far as to commit fraud in their efforts to get into the presses, say, by fabricating or falsifying data. Evidence, and not just high-profile cases such as that of the Dutch social psychologist Diederik Stapel (Carpenter, 2012, p. 558), supports this grave concern. For instance, responses from 250 of 663 (38%) accounting faculty who published extensively in that field's 30 top journals found 4% of them admitting to falsifying results, while further believing that 21% of their literature is so tainted (Bailey, Hasselback, & Karcher, 2001, p. 35). Comparable findings emerged from answers to a survey of 1,000 economists (234 returns) administered at the January 1998 meetings of the American Economic Association in Chicago: 4.4% said they have falsified research data, and they thought that about 6% of articles published in the best economics journals are based on such input (List, Bailey, Euzent, & Martin, 2001, p. 166). As another example, in a meta-analysis of 18 surveys, when scholars were asked, anonymously, whether they had ever fabricated or falsified data, some 2% said yes (Fanelli, 2009, p. 6). When survey questions were framed in relation to the behavior of fellow professionals, 14% of respondents said they had personal knowledge of a colleague who had fabricated or falsified data (Fanelli, 2009, pp. 6–7). Replies to John et al.'s (2012, p. 527) poll of psychologists indicates that about 9% of them confessed to falsifying data. Meanwhile, Bedeian et al. (2010, p. 719) write that 26.8% of U.S. management faculty answering their questionnaire reported knowledge of researchers fabricating outcomes.

John et al. (2012, p. 524) label questionable research practices the "steroids" of scientific competition, artificially bettering the careers of those who adopt them while penalizing scholars who play by the rules. An initial reaction might be that this is an odd and inappropriate analogy. It is, however, a peculiarly apt one conveying as it does the ruthlessness attending a "win at all costs" outlook pervasive in both the sporting and academic arenas.

In view of the sensitivity of the subject matter, few would be surprised if the estimates on the incidence of fabricating/falsifying data and other dubious research practices documented above turn out to be low. On this topic, John et al. (2012, p. 524) suggest that participating in questionable research behaviors "may constitute the de facto scientific norm." Honig, Lampel, Siegel, and Drnevich (2013, p. 2) agree when commenting that rising trends of ethical misconduct are not attributable to individual lapses but tell instead of "systemic problems that are deeply embedded in the institutional fabric of the modern research process." This is not hyperbole; compelling evidence bolsters John et al.'s and Honig et al.'s conjectures. For example, Masicampo and Lalande's (2012, pp. 2272–2273) analysis of the distribution of 3,627 *p*-values found in the August 2008 issues of the *Journal of Experimental Psychology: General, Journal of Personality and Social Psychology,* and *Psychological Science,* along with the preceding 11 issues of each, shows them to be much more common immediately below the arbitrary .05 level than would be expected on the basis of their occurrence in other ranges. This anomaly, which they attribute to undue emphasis on the attainment of $p \leq .05$ outcomes, was present in all three journals. Similar results indicative of the warping influence of the .05 level on the distribution of *p*-values are conspicuous in Gerber and Malhotra's (2008) examination of 46 data-based articles published in the 2003–2005 issues of the *American Sociological Review* and *Sociological Quarterly,* and the 2003–2006 issues of the *American Journal of Sociology.*[19] They concluded that "the hypothesis of no publication bias can be rejected at approximately the 1 in 10 million level" (p. 3).

Findings like these accentuate the insidious effects that editorial-reviewer biases against $p > .05$ results can have on the integrity of a discipline's contributions to knowledge. Perversely, then, the current academic incentive structure rewards the publication of nonreplicable findings (Hartshorne & Schachner, 2012, p. 1), a conclusion so disheartening that some (e.g., Ioannidis, 2012; Sovacool, 2008; Stroebe, Postmes, & Spears, 2012, p. 681) contest as mythical the idea that science is self-correcting. This same verdict was reached earlier in Section 2.2.5.

2.4.2 Publication Bias and the Credibility of Meta-analyses

Of immediate concern, owing to the publication bias against negative results, meta-analyses (i.e., attempts to quantitatively summarize the empirical literature on a particular issue) conducted under the aegis of the significant difference paradigm may not be as helpful as first imagined. They can even lead us astray. I agree with Sohn (1996, p. 229) that caution should be exercised in crediting meta-analyses as vehicles for knowledge discovery. This is because, as shown quite dramatically above, their databases (the available literatures constituting their input) may be untrustworthy. In fact, sometimes the findings of a meta-analysis may not be replicable (see, e.g., Allison & Faith, 1996; Bullock & Svyantek, 1985; Felson, 1992; Kilpatrick, 1992), thereby eviscerating its purpose.

Proliferation of False Positives

Behaviors motivated by a "hunt" for statistically significant differences (Salsburg, 1985, p. 220), or what Imrey (1994, p. 68) calls *p*-varication, contribute to Rosenthal's (1979, p. 638) well-known *file drawer problem* where "journals are filled with the 5% of the studies that show Type I errors, while the file drawers back at the lab are filled with the 95% of the studies that show nonsignificant (i.e., $p > .05$) results." Needless to say, a field whose literature is corrupted with false-positives is one whose credence is in danger (Simmons et al., 2011, p. 1359).

Regrettably, information based on examinations of various meta-analyses carried out in the social science and medical areas reveals them to be victims of said corruption. Illustrative of this condition, Bakker et al.'s (2012, p. 543) review of 13 meta-analyses encompassing 281 primary studies in a spectrum of fields in psychology uncovered biases and/or an excess of $p \leq .05$ outcomes in 7 of them. Christopher Ferguson and Michael Brannick's (2012, pp. 122–124) conclusions, following an inspection of 48 meta-analyses selected from leading American Psychological Association (*American Psychologist, Developmental Psychology, Journal of Abnormal Psychology, Journal of Consulting and Clinical Psychology, Journal of Personality and Social Psychology, Psychological Bulletin*) and Association of Psychological Science (*Perspectives on Psychological Science, Psychological Science*) journals over the period 2004–2009, also are bothersome. They determined that publication bias was apparent in 25% of these meta-analyses. In a similar vein, Ioannidis and Trikalinos (2007, p. 245) saw exaggerated frequencies of statistically significant results in 6 of 8 biomedical meta-analyses (with 55–155 studies in each one). What confirms the suspiciousness of these overly optimistic findings is the implausibly high occurrence of $p \leq .05$ results given the low levels of statistical power characteristic of meta-analyses undertaken in these disciplines (see, e.g., Bakker et al., 2012, p. 543; Cafri, Kromrey, & Brannick, 2010). Outcomes of this sort bear out Ioannidis's (2011, p. 16) grievance that an epidemic of false claims "is rampant in the social sciences and . . . particularly egregious in biomedicine."

Inflation of Effect Sizes

A related drawback with respect to the publication bias against negative findings and the usefulness of meta-analyses must be raised. Because unpublished (file-drawer) works—thought to be at least 50% of the psychology literature (Bakker et al., 2012, p. 544)—often are underrepresented in meta-analyses, published effect size estimates are inflated. For example, 10 of the 12 education and psychology meta-analyses examined by Mary Smith (1980) showed average effect sizes in published journal accounts to be 33% higher than those reported in theses and dissertations. McLeod and Weisz's (2004) comparison of youth psychotherapy outcomes reported in journals versus unpublished dissertations likewise breeds skepticism. They revealed effect sizes in the former to be twice the size of those in the latter. Continuing, Shadish, Doherty, and Montgomery (1989) surveyed a random sample of 519 members of organizations involved with family and marital psychotherapy outcomes to see if they possessed file drawer studies on the issue. After analyzing 375 responses, they intimated that there may be almost as many of these unpublished works as there are published studies and dissertations. They concluded that population effect sizes of published works are about 10% to 40% larger than those based on unpublished research. Information on the efficacy of psychological, educational, and behavioral treatment casts additional doubt on the veracity of meta-analytic effect size estimates. Lipsey and Wilson's (1993, pp. 1194–1195) dissection of a subset of 92 (from 302) meta-analyses permitting a comparison of effect sizes in published versus unpublished venues in these areas shows published accounts to be some 13% higher than file-drawer appraisals. A simulation study by David Lane and William Dunlap (1978) also shows disquieting results.

They found that editorial biases in favor of .05 and .01 outcomes led to published average effect sizes being one-half to one standard deviation greater than their true values.

Summing up, publication bias against negative outcomes arising from tests of statistical significance directly contributes to the creation of an empirical literature whose believability is suspect. Howard et al. (2009, p. 148) go beyond this in their withering assessment that "because of the widespread use of NHST [null hypothesis significance testing] in psychological research, it is therefore possible that all extant research literatures are systematically misleading and some demonstrably wrong." Similarly, Doucouliagos and Stanley's (2013, p. 332) meta-analysis of findings from a broad range of areas in economics led them to comment that "all summaries of empirical economics . . . must be regarded with some skepticism" and that "reports of economic facts are greatly exaggerated." Presumably, views like these underlie Christopher Ferguson and Moritz Heene's (2012, p. 558) position that seldom have they run across a meta-analysis which has resolved a controversial debate in an area.[20]

2.5 Conclusions

Affiliates of the significant difference model have an unquestioning impression of how knowledge is gained. It is enough to be able to reject the null hypothesis at the $p \leq .05$ level in single studies and then generalize these unique findings to other circumstances and time periods. Further, negative ($p > .05$) results are viewed with disdain. Given these beliefs, it is easy to comprehend how the use of statistical significance testing has come to occupy a vise-like grip on empirical research performed in the management and social sciences. Devoid of such testing, data-based research in these areas is considered to be insufficiently rigorous. The insistence on $p \leq .05$ outcomes leads researchers eager to generate them into improprieties which impugn the validity and reliability of published findings. That just about all empirical articles in the behavioral and management sciences are able to come up with statistically significant results in the face of typically underpowered designs (see the appendix to this chapter) is highly improbable. In some instances, such findings are just too good to be true (Francis, 2012a, p. 585, 2012b, p. 151; Schimmack, 2012, p. 561). This makes it almost certain that fallacious results are entering the literature at troubling rates (Asendorpf et al., 2013, p. 113; Pashler & Harris, 2012, p. 535). And in areas like biomedicine this manipulation of *p*-values in the quest for publishable results can harm patients (Stang, Poole, & Kuss, 2010, p. 225) and cost lives (Ziliak & McCloskey, 2008).

Followers of the significant difference paradigm endorse wholeheartedly a model of explanation—hypothetico-deductivism (HARKing excepted)—that for the most part has been abandoned by philosophers of science. It is a model that rarely describes the work of practicing scientists as is sometimes falsely portrayed in textbooks.

In short, the significant difference paradigm legislates bad science (Hubbard & Lindsay, 2013b). The philosophical foundations of the significant sameness model, when juxtaposed with those used to rate the merits of the significant difference paradigm—conception of knowledge, model of science, and the role of "negative" results—open a far sturdier route to knowledge development in the management and social sciences. These foundations are laid out in Chapter 3.

Notes

1. The idea that rare occurrences constitute evidence against a hypothesis has a pedigree dating back to the first "significance test" by John Arbuthnot in 1710 concerning the birth rates of males and females in London, and is continued in the work of Mitchell, LaPlace, and Edgeworth, among others. See Baird (1988), Cowles (2001), Gigerenzer et al. (1989, pp. 79–84), Hacking (1965, pp. 74–80), Huberty and Pike (1999, pp. 1–4), and Sauley and Bedeian (1989, pp. 336–338) for synopses of this early history of statistical testing.

2. The reader will appreciate that the data presented in this chapter have been amassed over a period of several years.

3. The arbitrariness of later classifying economics as a management rather than a social science is recognized.

4. Note that all journals except *Professional Geographer* (1949) and *American Journal of Political Science* (1957), with inaugural issues in parentheses, cover the entire 1945–2007 time period.

5. For every year, two randomly chosen issues of management and marketing journals were included in the database.

6. A growing imbalance between empirical and nonempirical/conceptual articles is unfortunate because the latter often have a greater impact on the scholarly community (MacInnis, 2011, p. 151), at least as measured by citations (for details, see Hubbard, Norman, & Parsa, 2010; Yadav, 2010). In addition, some practitioners may find conceptual pieces readable, interesting, and helpful.

7. See also Shirley Martin's (2011) informative telling of the spread of Fisherian methods in experimental dissertations in education under the guidance of Lindquist at the University of Iowa and Johnson at Minnesota during World War II.

8. This last figure for sociology is almost identical with Leahey's (2005, p. 3) estimate that 91% of empirical articles published in the *American Journal of Sociology* and *American Sociological Review* for the period 1995–2000 used tests of statistical significance.

9. In Guttman's (1985, p. 3) words, many researchers don't comprehend the distinction between statistical inference (an aspect of what he calls the scaffolding of science) and science (the building or structure) itself. Thus, they believe that they are doing science when they are dealing merely with part of the scaffolding. Which is to say they are not building anything. In this connection see also Bolles's (1962) telling of the differences between statistical versus scientific hypotheses, as well as Meehl (1967, p. 107, 1990, p. 202).

10. A book edited by Harlow, Mulaik, and Steiger (1997) titled *What If There Were No Significance Tests?* confronts just such an issue.

11. Upsettingly, Hubbard, Brodie, & Armstrong (1992, p. 5) add, Simon's reappraisal of Zielske's data took years to get into print. He was so disenchanted with the "saga," as he named it, of trying to publish his results that he thought it necessary to tabulate in an appendix to his eventual publication the events propelling such an unusual course of action. The contents of the appendix do not paint a picture of scholarly cooperation on Zielske's part. See Chapter 5 for more evidence of this lack of collaboration among researchers.

12. All Google Scholar citations reported in this section were accessed in February 2013.

13. It is unique in another way not addressed by Bottomley and Holden (2001). That they were actually able to get these eight data sets is quite remarkable given that, as detailed in Section 5.5.2, researchers in the social and management sciences are not especially cooperative in such matters.

14. Keuzenkamp (2000, p. 160) proposes that the idea of the "crucial experiment" probably is rare in physics—it may even be a scientific fiction (Hubbard & Lindsay, 2002, p. 393)—never mind in the social sciences.

15. According to Giere (1999, p. 89), the notion that laws of nature must be true statements of universal form had its origins in theology:

> Here there can be no serious doubt that, for Descartes and Newton, the connection between laws of nature and God the creator and lawgiver was explicit. Nor can there be any doubt that it was Newton's conception of science that dominated reflection on the nature of science throughout the eighteenth century, and most of the nineteenth as well.

Needless to say, God's laws were universal. In addition, Giere continues, it was the secularized version of Newton's characterization of science that prevailed in philosophy of science matters through a substantial part of the 20th century, as exemplified by the logical empiricists.

16. To this end, Keuzenkamp and Magnus (1995, p. 18) joke that the JRSS (*Journal of the Royal Statistical Society*) actually is the JSSR (*Journal of Statistically Significant Results*)!

17. Fisher (1973, p. 45) later backed away from this position: "No scientific worker has a fixed level of significance at which from year to year, and in all circumstances, he rejects hypotheses; he rather gives his mind to each particular case in the light of his evidence and his ideas." Such advice notwithstanding, the inviolability of the .05 significance level remains to this day.

18. At this juncture I offer my own mea culpas. I have searched for statistically significant results, for example, by experimenting with alternative model specifications, variable transformations, or by recasting one- and two-sided null hypotheses. And I have engaged in HARKing, on one occasion at the behest of a referee who asked "Where are the hypotheses?" in response to my original manuscript submission, which was completely free of them. Few are exempt from the inordinate burden of having to publish the "required" results.

19. That only 46 empirical articles were retained for analysis mirrors the rather stringent criteria necessary for inclusion in Gerber and Malhotra's (2008) "caliper" test for publication bias.

20. Compounding these publication bias deficiencies, in the significant difference model studies included in meta-analyses may suffer from marked construct validity ("comparing apples and oranges") and sample size (insufficient replications on the "same" topic) drawbacks. Or as Bangert-Drowns (1986, p. 388) has it, because studies involved in typical meta-analyses deploy only "roughly similar procedures," problems occur when averaging effects across independent and dependent variables which have been measured quite differently. Therefore, it frequently is unclear whether those studies selected for incorporation in meta-analyses are measuring the same constructs or relationships. This makes the results of conventional meta-analyses even more difficult to interpret and generalize (cf. Farley & Lehmann, 1986, pp. 15–16; Shadish et al., 2002, pp. 446–455). Note that because of the priority assigned to replication research, construct validity, sample size, and publication bias issues are minimal for meta-analyses undertaken in the significant sameness paradigm.

APPENDIX TO CHAPTER 2

An Empirical Regularity Not to Be Proud Of: Inadequate Statistical Power in the Social and Management Sciences

In the Neyman–Pearson variant of the significant difference paradigm, the researcher proposes two hypotheses. One is the null hypothesis, H_0, which the investigator is hoping to reject (nullify) in favor of the other, the alternative (research) hypothesis, H_A. Unfortunately, two kinds of mistakes can happen in deciding between H_0 and H_A. There is the erroneous rejection of H_0 called a Type I error, α. And there is the erroneous acceptance of H_0 called a Type II error, β. Statistical power, then, is defined as $1\text{-}\beta$, or the probability of rejecting a false null hypothesis.

When statistical power in a research literature is low or marginal significance tests are prone to yielding null results, and therefore may be viewed as less worthy of being published. At the same time, and acting in the opposite direction, when power is low

> the veracity of even statistically significant results may be questioned, because the probability of rejecting a true null hypothesis may then be only slightly smaller than the probability of rejecting the null hypothesis when the alternative is true. . . . Thus, a substantial proportion of published significant results may be Type 1 errors. (Rossi, 1990, p. 647)

As can be inferred from the above, low power produces an inconsistent and questionable empirical literature wherein the results of some studies are statistically significant while those of others are not. This last point is vital and therefore receives explicit consideration in Section 3.5 of the book, a more suitable context for its discussion.

The power of a statistical test, Jacob Cohen (1969, 1988, p. 4) notes, is predicated on the chosen level of significance, the effect magnitude in the population, and the size of the sample. Accordingly, Cohen (1988) facilitates the calculation of power exhibited in published work by supplying power tables with different significance levels, standardized effect magnitudes, and sample sizes for several commonly used statistical techniques. The latter include the *t* test, the statistical significance of Pearson's *r*, the statistical significance of the differences between correlation coefficients, the sign test, the test for differences between proportions (%'s), χ^2 (chi-square) tests, the *F* test in analysis of variance (ANOVA), analysis of covariance (ANCOVA), and multiple regression analysis, as well as power tests for multivariate analysis of variance (MANOVA) and multivariate analysis of covariance (MANCOVA). The reader is invited to refer to Cohen (1988) about the details involved in computing power levels for what he calls small, medium, and large effect magnitudes.

Cohen (1969, 1988) offers five guidelines when performing a power analysis: (1) the article as the primary unit of analysis, (2) a significance level of .05, (3) nondirectional tests, (4) only major statistical tests included, and (5) conventional definitions of small, medium, and large effect sizes. In addition, Cohen (1988, p. 56) recommends that the value .80 be used when there is no other rationale for selecting a desired power level.

Table 2A-1 displays the results of 16 retrospective power analyses of articles in leading journals (putatively reflecting best statistical practice) from a variety of behavioral and business disciplines. All of these studies followed Cohen's five recommendations.

Only 2 of the 16 studies displayed in Table 2A-1, both in marketing (Hubbard & Armstrong, 1991; Sawyer & Ball, 1981), are able to meet (in fact, surpass) Cohen's (1988) .80 benchmark for detecting medium effects in the population. This is probably because of the larger sample sizes enjoyed in survey, as opposed to experimental, research in marketing journals.

Across all 16 studies in Table 2A-1, the average probabilities of unearthing small, medium, and large effects are .24, .64, and .86. Minus the two marketing studies the corresponding figures are .22, .61, and .84. Consequently, if medium effect sizes are thought to be the norm in the social and management sciences, an investigator has only about a 60% chance of rejecting a false null hypothesis. This represents a substantial empirical regularity in the significant difference paradigm. Regrettably, it is not one to be proud of.

Table 2A-1 Statistical Power in the Social and Management Sciences

		Sample Sizes		*Effect Sizes*		
Discipline	*Investigators*	*Articles*	*Tests*[a]	*Small*	*Medium*	*Large*
Accounting	Lindsay (1993b)	43	1,871	.16	.59	.83
	Borkowski, Welsh, & Zhang (2001)	96	1,782	.23	.71	.93
Communication	Chase & Tucker (1975)	46	1,298	.18	.52	.79
	Chase & Baran (1976)	48	701	.34	.76	.91
Education	Brewer (1972)	47	373	.14	.58	.78
Management	Mazen, Hemmasi, & Lewis (1987)	84	7,215	.31	.77	.91
	Mazen, Kellog, & Hemmasi (1987)	44	3,665	.23	.59	.83
MIS[b]	Baroudi and Orlikowski (1989)	57	149	.19	.60	.83
Management and psychology	Mone, Mueller, & Mauland (1996)	210	26,471	.27	.74	.92
Marketing	Sawyer & Ball (1981)	23	475	.41	.89	.98

(Continued)

Table 2A-1 (Continued)

Discipline	Investigators	*Samples Sizes*		*Effect Sizes*		
		Articles	*Tests*[a]	*Small*	*Medium*	*Large*
	Hubbard & Armstrong (1991)	14	92	.39	.90	.96
Psychology	J. Cohen (1962)	70	2,088	.18	.48	.83
	Chase & Chase (1976)	121	3,373	.25	.67	.86
	Sedlmeier & Gigerenzer (1989)	54	—	.21	.50	.84
	Rossi (1990)	221	6,155	.17	.57	.83
Speech pathology	Kroll & Chase (1975)	62	1,037	.16	.44	.73
Average				.24	.64	.86

[a]The very high number of tests in some of the studies often is because they include all individual correlation coefficients in a correlation matrix.

[b]MIS is Management Information Systems.

PHILOSOPHICAL ORIENTATION—SIGNIFICANT SAMENESS

> Management research places much less emphasis on empirical regularities than we should expect, and that is required, for scholarship that ultimately concerns itself with the real world. (Helfat, 2007, p. 185)
>
> We must not settle for pretend knowledge. (Wells, 2001, p. 497)

3.1 Introduction

Those espousing significant sameness understand that knowledge does not emanate from the rote application of statistical rituals. Allied with this recognition, and portrayed therefore in the first part of Section 3.2, is the crucially important refutation of the myth that the *p*-value is an "objective" measure of evidence in the generation of knowledge. The second part of Section 3.2 offers a more realistic, and accordingly much messier, conception of socially produced knowledge from the significant sameness vantage. It shows that knowledge arises from the conduct of many studies (replications), by many people, over an extended period of time, which may (or may not) win the backing of the scientific community. This section also outlines the role of confidence intervals (CIs) in the acquisition of facts which, in turn, provide the impetus for the creation of theory. The balance of Section 3.2 illustrates the superiority of overlapping CIs versus reliance on *p*-values as a measure of replication success.

Following in Section 3.3 is a discussion of the model of science informing the significant sameness approach, a postpositivist theory called *critical realism*. This model emphasizes *abductive*, as opposed to hypothetico-deductive, reasoning. It is a model accurately reflecting how science progresses.

Additionally, as Section 3.4 reveals, negative results are valued in this paradigm. This is because they mark the boundary conditions of an empirical regularity's expanse. In doing so they can spur theory building by explaining why a limit to a generalization exists. In this same spirit, Section 3.5 makes the case that null results with adequate statistical power are as deserving of publication as their non-null counterparts. Comments summarizing the chapter are made in Section 3.6.

3.2 Conception of Knowledge

The significant sameness paradigm sees the development of knowledge as cumbersome because data rarely speak for themselves (Bamber, Christensen, & Gaver, 2000; Fay, 1996, p. 204). It is ingenuous to view scientific facts as being created by rejecting the null hypothesis in what Gigerenzer (2004, p. 587) calls "mindless statistics." Statistical significance testing is mostly window dressing. To see this, one would think that the ubiquity of such testing presupposes its indispensability in empirical work. Yet it is remarkable how unpersuasive the results of statistical significance tests are; in everyday practice they are not taken seriously (Guttman, 1985, p. 5). For example, Summers (1991, p. 130) challenges his readers to come up with a hypothesis in economics that has fallen into disrepute over the outcome of a statistical test. Likewise, Ziliak and McCloskey (2008, p. 120) are unaware of any advance in economics since World War II that has turned on a test of statistical significance. And Guttman (1977, p. 92) asserts that "no one has yet published a scientific law in the social sciences which was developed, sharpened, or effectively substantiated on the basis of tests of significance." Concerns about the ineffectiveness of statistical significance tests at changing the minds of scholars are found also in Keuzenkamp (2000, p. 164), Lindsay and Ehrenberg (1993, p. 218), and Spanos (1986, p. 660). But if the results of significance tests fail to convince scientists about the veracity of a finding, why use them?

Moreover, because of its revered status among social and management scientists, it is of the utmost importance to contest Fisher's (1973, p. 46) allegation that the *p*-value is an objective measure of evidence against H_0. Drawing on some of my previous work with Murray Lindsay (Hubbard & Lindsay, 2008), Section 3.2.1 shows that several arguments can be marshaled against Fisher's claim.

3.2.1 The P-Value Is Not an Objective Measure of Evidence

P-Values Exaggerate the Evidence Against the Null Hypothesis, H_0

I begin with what is a most telling indictment of the *p*-value as a plausible inferential index, namely, its exaggeration of the evidence against H_0. This, in turn, makes "statistically significant results" relatively easy to attain.

Two-Sided Null Hypotheses. *P*-values exaggerate the evidence against two-sided (point null) hypotheses (Berger & Sellke, 1987), the kind tested all the time in the management and social sciences. A point null hypothesis is expressed as follows, $H_0 : \theta = \theta_0$ versus $H_A: \theta \neq \theta_0$, where θ_0 is a particular value of θ, usually zero. With this as background, using a Bayesian significance test for a normal mean, Berger and Sellke (1987, pp. 112–113) demonstrated that for *p*-values, that is, $Pr(x \mid H_0)$, of .05, .01, and .001, respectively, the *posterior probabilities* of the null, that is, $Pr(H_0 \mid x)$, for $n = 50$ are .52, .22, and .034. For $n = 100$ the numbers are .60, .27, and .045. It is clear that these discrepancies between p and $Pr(H_0 \mid x)$ are marked and raise strong doubts over the reasonableness of *p*-values as measures of evidence.

Berger and Delampady (1987) found similarly discrepant results between *p*-values versus posterior probabilities in both normal and binomial situations. This led them to suggest that the use of *p*-values be abandoned when testing precise (point null) hypotheses. Given this discussion, one must agree with Berger and Berry (1988) that the validity of empirical research based on moderately small, including .05, *p*-values is open to challenge.

And besides, except for rare instances (cf. Wainer, 1999), it is impossible to defend in any epistemological sense the practice of point null—or *nil* as Jacob Cohen (1994, p. 1000) would have it—hypothesis testing. "Discovering" in the population that a difference between two means is not *precisely* zero, or that a correlation between two variables is not *precisely* zero, are trivial findings. It is hard to digest the idea that such findings are the lingua franca of empirical social and management science.

Taken literally, point null hypotheses of exactly zero differences between means or exactly zero correlations between variables do not exist in nature. In the real world point null hypotheses always are false, even if only to some small degree, such that large enough samples will lead to their rejection. Or as the celebrated statistician John Tukey (1991, p. 100) explained: "All we know about the world teaches us that the effects of A and B are always different—in some decimal place—for any A and B. Thus asking 'Are the effects different?' is foolish." This view is retold by Lyle Jones and John Tukey (2000, p. 413). But if the point null hypothesis always is false, what's the point of testing a point null hypothesis?

Frequency Distribution of P-Values. A number of studies (e.g., Berger, 2003; Hubbard & Bayarri, 2003; and especially Sellke, Bayarri, & Berger, 2001) commenting on a simulation of the frequency distribution characteristics of *p*-values are illuminating. The simulation is available as an applet at www.stat.duke.edu/~berger.

To illustrate its use, suppose we wish to carry out some tests on the efficacy of an advertising campaign (A-C) designed to increase the awareness among voters of some political candidate. The statistical significance test would be H_0 : A-C = 0 versus H_A : A-C ≠ 0. The simulation revolves around a long series of such tests on normal data (variance known) and records how often H_0 is true for *p*-values in specified ranges, say, approximately equal to .05 or .01. Devastatingly, this simulation of the behavior of *p*-values shows that even when "statistically significant" outcomes near the .05 and .01 levels are obtained, these findings often come from true null hypotheses of no effect or association. In particular, if it is assumed that 50% of the null hypotheses in the (A-C) tests are true, Sellke et al. (2001, p. 63) cautioned:

1. Of the subset of [A-C] tests for which the *p*-value is close to the .05 level, *at least* 22% (and generally about 50%) arise from true nulls.
2. Of the subset of [A-C] tests for which the *p*-value is close to the .01 level, *at least* 7% (and generally about 15%) arise from true nulls.

So a *p*-value of .05 may provide no evidence against H_0.

P-Values and Sample Size

Small Versus Large Samples. The sample size crucially determines statistical significance levels. As an example, Royall (1986) tells of well-known statisticians whose interpretations of *p*-values in small as opposed to large sample studies are completely at odds. Some statisticians maintain that a given *p*-value in a small sample study is stronger evidence against the null than the same *p*-value in a large-scale study, and vice versa. Seen in this light, a given *p*-value does not possess a fixed, objective meaning, being connected to sample size. Of profound influence, the larger the sample size, the easier the chances of being able to record statistically significant results.

Lindley's "Paradox". Lindley (1957) demonstrated that for any level of statistical significance, *p*, and for any nonzero prior probability of the null hypothesis, $Pr(H_0)$, a sample size can be estimated so that the posterior probability of the null hypothesis, $Pr(H_0 \mid x)$, is 1-*p*. In other words, a null hypothesis that is firmly *rejected* at the conventional .05 level in a Fisherian statistical significance test can nonetheless have 95% *support* from a Bayesian perspective. These diametrically opposed inferences constitute the paradox. As Johnstone (1986, p. 494) writes, the explanation for this conundrum is that regardless of how small the *p*-value, the likelihood ratio $Pr(x \mid H_0)/Pr(x \mid H_A)$ approaches infinity as the sample size gets larger. Therefore, for large *n*, a small *p*-value provides evidence in favor of H_0 instead of against it. The issue of the objectivity and usefulness of the *p*-value as a measure of evidence is hereby dealt a crippling blow.

P-Values and Effect Sizes

The suitability of the *p*-value as a reliable measure of evidence certainly must be called into question when it has little to say about the *effect size* reported in a study (Gelman & Stern, 2006). As it stands, a small sample study with a large effect can produce the same *p*-value as a large sample study with a small effect size. This is seen in Table 3-1, which shows Peter Freeman's (1993) hypothetical data on medical trials wherein all patients receive both treatments A and B and are asked to state their preferences.

The results of trial 1, with its 75% preference rate for A over B, would be taken to reveal a possibly huge endorsement of A's supremacy. Conversely, the results of trial 4, with a 50.1% preference for A, would be considered as enormous evidence that preferences for A versus B are essentially identical. Few investigators would regard the findings of these four trials as being equivalent, yet they all yield a *p*-value of .041. It is for such reasons that Freeman (1993, p. 1443) did an about-face on his opinions of the usefulness of *p*-values:

> This paper started life as an attempt to defend *p*-values. . . . I have, however, been led inexorably to the opposite conclusion, that the current use of *p*-values as the "main means" of assessing and reporting the results of clinical trials is indefensible.

In view of the above, Gibbons's (1986, p. 367) declaration, in an account titled "*P*-values," that "an investigator who can report only a *P*-value conveys the maximum amount of information contained in the sample" simply is wrong. Indeed, Berger, Boukai, and Wang (1997) note that the interpretation of *p*-values can be expected to change drastically from problem to problem.

Table 3-1 Hypothetical Data on Treatment Preferences

Trial	*No. Preferring A*	*No. Preferring B*	*% Preferring A*
1	15	5	75.0
2	114	86	57.0
3	1,046	954	52.3
4	1,001,455	998,555	50.1

Source: Adapted from P. R. Freeman (1993, p. 1446).

P-Values and Subjectivity

Another case of the deficiency of the *p*-value as an unprejudiced measure of evidence is apparent in the decision of whether to use a one-sided or a two-sided test of statistical significance (Goodman & Royall, 1988; Royall, 1997). While two-sided tests are the norm, sometimes researchers are told that they can halve the *p*-value if they anticipate a departure from the null hypothesis in a specific direction. Or as Goodman and Royall (1988) state it, despite the fact that the data are the same, the *p*-value is modified on the basis of the researcher's subjective impressions about the expected outcome of the study. They further mention that similar alterations of *p*-values take place where multiple comparisons are concerned.

P-Values Are Logically Flawed

The logical flaw is that the *p*-value does not compute the probability of the observed data under H_0, but this *as well as the probability of more extreme data.* Because of this, statistical significance tests are influenced by how the probability distribution is spread over unobserved outcomes in the sample space. Otherwise expressed, the *p*-value embodies not only the probability of what was observed, but also the probabilities of all the more extreme events that did not occur.

Numerous statisticians (e.g., Berger & Berry, 1988; Berger & Delampady, 1987; P. R. Freeman, 1993; Goodman, 1999; Royall, 1997; Schervish, 1996) allege that a valid measure of strength of evidence cannot include the probabilities of unobserved outcomes. Jeffreys (1939, p. 316) sums up this illogic about *p*-values as follows: "*What the use of P implies . . . is that a hypothesis that may be true may be rejected because it has not predicted observable results that have not occurred.* This seems a remarkable procedure." In this manner, McGrayne (2011, p. 56) records, Jeffreys believed that *p*-values "fundamentally distorted science."

Specification of an Alternative Hypothesis, H_A

Evidence Is Relative. In the case where an alternative hypothesis can be specified, the researcher is able to identify those findings as extreme or greater than the observed event. Therefore, Royall (1997) informs us, it is not low probability under A that makes an observation evidence against A. More properly, it is low probability under A when compared with the probability under a rival hypothesis B, so this makes it evidence against A versus B. This relativistic approach requires a weighing of the evidence between two competing hypotheses, a situation disallowed in Fisherian statistical significance tests. Fisher never saw the need for an alternative hypothesis.

In this context, consider Johnstone's (1986, p. 493) view that the law of likelihood is a better measure of evidence than *p*-values for assessing the believability of two (or more) vying hypotheses. In particular, if the likelihood ratio $Pr(x \mid H_0)/Pr(x \mid H_A)$ is larger than 1, then the evidence favors H_0 over H_A, and vice versa. Regrettably, Fisher's disjunction applies only to $Pr(x \mid H_0)$; it has nothing to say about $Pr(x \mid H_A)$. The *p*-value is a tail-area probability and not a likelihood ratio.

We're Interested in the Alternative (Research), Not the Null, Hypothesis. Making explicit an alternative hypothesis is not only a way of covering values more extreme than those observed on a null hypothesis. The alternative (research) hypothesis is the one investigators are concerned with. Berkson (1942, p. 326) saw this well ahead of others when critiquing Fisher's paradigm of null hypothesis testing:

> In the null hypothesis schema we are trying only to nullify something. . . . But ordinarily evidence does not take this form. With the corpus delicti in front of you, you do not say, "Here is evidence against the hypothesis that no one is dead." You say, "Evidently, someone has been murdered."

Statistical tests are more likely to be useful when they focus on the research hypothesis, rather than being preoccupied with rejection of the null hypothesis. Unfortunately, Fisher's statistical framework denies the existence of an alternative (research) hypothesis. In attempting to rectify this state of affairs, it is sometimes argued that Fisherian statistical significance testing has an implicit alternative (research) hypothesis that is simply the complement of the null. Yet as Hubbard and Bayarri (2003, p. 172) remind us, this argument is difficult to formalize. Questions arise concerning exactly what is the complement of an $N(0, 1)$ model. Is it the average differing from 0, the variance differing from 1, the model not being Normal? Fisher considered only the null model and wanted to see whether the data were congruent with it.

The above account demonstrates from a variety of positions that, contrary to Fisher's notification, adopted all too willingly by management and social scientists eager to establish their scholarly status, the *p*-value is anything but an objective and credible measure of evidence.[1] To repeat, objective, value-free measures of evidence are fallacies; subjective judgment *always* will be required in the statistical analysis and interpretation of data, be it classical (Fisher/Neyman–Pearson) or Bayesian in nature (see Berger & Berry, 1988; Birnbaum, 1962; Chatfield, 1985, 2002; Freedman, 1999; Johansson, 2011; Leamer, 1983; Lindsay, 1995; Perlman & Wu, 1999). Despite this, there are those in the social (e.g., Frick, 1996) and management (e.g., R. Kent, 2007, p. 38) sciences who persist in interpreting the *p*-value as an objective measure of evidence.

3.2.2 Knowledge Development—Significant Sameness

Scientific facts do not somehow arise from the outcomes of statistical significance tests. They are, rather, intellectually constructed assertions whose relation to the external world is neither immediate nor certain (Ravetz, 1971). Every research conclusion is the outcome of an imperfect process which is shaped by the formulation of the problem (following from its theoretical underpinnings); the data that are collected or made available (contexts); the method of analysis; and the skills, biases, and epistemology that influence the many decisions an investigator makes while carrying out the research and interpreting its output (Hubbard & Lindsay, 2013a, p. 1379). This is why methodological authorities continually warn that the results of a single study, no matter how well designed and statistically significant the outcomes, are virtually meaningless in the cultivation of high-level understanding and causal explanations of phenomena (Nelder, 1986, p. 112; Popper, 1959, p. 45; Yates, 1951, p. 33). Fisher (1966, p. 13), of course, was aware of this: "We thereby admit that no isolated experiment, however significant in itself, can suffice for the experimental demonstration of any natural phenomenon."

As Ravetz (1971) explains it in his rich work on the production of knowledge, scientific facts must surmount three hurdles: (1) a *social test of significance* (the scientific community views the result as deserving of additional research), (2) *empirical stability* (the result can be empirically replicated), and (3) *empirical invariance* (the result is capable of being generalized to applications other than those peculiar to its own construction). Therefore, although scientific representations are constructed socially, over time it is possible to differentiate between those representations which are or are not reasonably congruent with reality because of science's empirical nature (Giere, 1999). The significant sameness paradigm is altogether consistent with this *social* view of science. These last remarks are expanded upon below.

In contrast to the widely held view of those endorsing the significant difference paradigm, the attainment of scientific knowledge is not accomplished by solitary people (Bauer, 1994, p. 52; Eichenbaum, 1995,

p. 1620; Tukey, 1980, p. 23). Or as management theorists Pfeffer and Sutton (2006b, p. 46) declare: "Knowledge isn't generated by lone geniuses who magically produce brilliant new ideas in their gigantic brains. This is a dangerous fiction." Rather, the production of scientific knowledge in any field is a communal endeavor in search of the widest possible consensus of rational opinion (Hackett, 2005, p. 788; Ziman, 1978, p. 3). As such, knowledge creation *takes a great deal of time and the conduct of many, many studies.*

Locke's (2007) work is important here. He wishes to make the case for inductive theory building—not H-D—in social research, a case I obviously applaud.[2] In doing so Locke reinforces the view that this process takes a long period of time:

> The social sciences accepted the hypothetico-deductive model [and] . . . this meant that researchers often had to pretend that they had theories before they had a firm basis for any. This method makes for quick and often short-lived theories; in contrast, true inductive theorizing takes many years, even decades, and, I believe is far more likely to withstand the test of time. (p. 872)

As shown below, Locke offers three examples of the labyrinthine manner in which theories become inductively grounded to support his argument.

The first is Beck's (Beck, 1993; D. A. Clark & Beck, 1999) cognitive theory of depression, begun in 1956, which did not arrive full blown, but which meandered through "torturous paths" to reach its final form. The second is Bandura's (1986) social cognitive theory. Here, Locke (2007, p. 877) quotes Bandura (2005, p. 29) as saying:

> Theory building is for the long haul, not for the short winded. The formal version of the theory, that appears in print, is the distilled product of a lengthy interplay of empirically based inductive activity and conceptually based deductive activity.

The third is Locke's own research, with Latham, on goal setting theory, a theory ranked first in importance among 73 competitors by organizational behavior academicians (Miner, 2003). Locke (2007, p. 879) notes that it was only after 25 years of work, embracing some 400 studies by himself, Latham, and others, that they felt comfortable enough to develop a theory on goal setting (see Locke & Latham, 1990, 2002). Locke's experience meshes with Haig's (2013a, p. 137) estimate that it might take anywhere from 3 to 30 years for social scientists to construct explanatory theories to account for regularities/phenomena of interest.

Even in disciplines like physics, Ziman (1978, p. 40) relates, the time element in acquiring knowledge is unavoidable:

> [The researcher] learns how easy it is to persuade oneself of the validity of a model which later turns out to be false, and comes to realize that even in very strongly mathematical and well-defined scientific issues it may take a long time, much criticism and the death of many promising conjectures . . . before a reliable theory is well-based and thoroughly acceptable.

This explains why the content of undergraduate physics textbooks is approximately 90% true, while that of the primary physics journals is 90% false (Ziman, 1978, p. 40).

Assembling scientific knowledge calls for great *patience* (Gribbin, 2004, p. 462). From this aspect, theory development is better understood as a lengthy, arduous *process* (Faust, 1984, p. 131; Hull, 1988) rather than as a *product* (Kaplan, 1964, p. 409; Weick, 1995). Contemporary philosophy of science sees investigators beginning with very low-level theory, known to be defective and perhaps false, which undergoes active elaboration within a research network or program (Hubbard & Lindsay, 2013b, pp. 1395–1396; T. C. Jones & Dugdale, 2002). This is why talk of "confirmation" or "refutation" is meaningless since theories are, at best, only abstractions of reality (Suppe, 1977). It also validates Hacking's (1983, p. 15) position that accepting and rejecting theories plays a minor role in science.

In line with the above, as Kuhn (1970, p. 24) tells it, most researchers in the established sciences are not involved with inventing and testing new theories. Rather, they spend their careers engaged in "mopping up" or "normal" science activities. Which is to say they are occupied with the relatively prosaic jobs of extending the facts and predictions, along with the further articulation, of the dominant paradigm they labor within. The truth is that the reconstruction of important scientific findings seen in textbooks betrays the sense of how progress really is achieved because usually they fail to mention the many studies and false detours that led in the end to the acceptance of a theory by the scientific hierarchy (Blachowicz, 2009; Gower, 1997; T. C. Jones & Dugdale, 2002; Ladyman, 2002; Levy, 2010; Nash, 1963; Ravetz, 1971; Roberts, 1989). The untidiness inherent in scientific advance is captured perfectly by Livio (2013) in his recent book *Brilliant Blunders From Darwin to Einstein,* describing the monumental knowledge breakthroughs made by Charles Darwin, Albert Einstein, Fred Hoyle, Lord Kelvin, and Linus Pauling. Or as Rosenbaum (2002, p. 11) astutely summarizes this process:

> Scientific questions are not settled on a particular date by a single event, nor are they settled irrevocably. We speak of the weight of evidence. Eventually, this weight is such that critics can no longer lift it, or are too weary to try. Overwhelming evidence is evidence that overwhelms responsible critics.

Reiterating, only those raised in the significant difference school of thought see knowledge procurement as instantaneous, brought about by the unquestioning application of formal statistical protocols looking for $p \leq .05$ results in single-shot studies.

In practical terms, then, the question to be answered is: How, exactly, are scientific facts arrived at in the significant sameness paradigm? This time-consuming task requiring multiple studies (replications) is described below.

3.2.3 Point Estimates and Confidence Intervals (CIs)

The fixation with *p*-values deflects attention from gauging the size of the phenomenon under scrutiny, the latter being a crucial need in science (A. W. F. Edwards, 1992; Hubbard & Lindsay, 2008, 2013b; Lindsay, 1995; Tukey, 1969; Ziliak & McCloskey, 2008). How should these sizes be assessed? In the significant sameness paradigm, this is done in research programs (not by isolated authors) reporting sample statistics, effect sizes, and the confidence intervals (CIs) around them. Over time, through replication research, the CIs of additional (new) results about the phenomena in question can be compared with the increasingly robust baselines created by their myriad predecessors to see whether they are consistent with them (Hubbard & Lindsay, 2013b, p. 1394). In turn, these assembled magnitudes become the grist for the meta-analyst's mill (Eden, 2002, p. 842). The results of these more directed, fact-focused meta-analyses would permit the emergence of a rational consensus of opinion among researchers about the outcomes in any given area. This is totally at variance with the ethos of the governing significant difference paradigm, where editorial-reviewer insistence on novelty vitiates the attainment of cumulative knowledge, and hence the possibility of consensus, inviting and rewarding instead an anything-goes outlook with respect to empirical results (Andreski, 1972, p. 16; Pfeffer, 1993, pp. 612, 616; Simmons, Nelson, & Simonsohn, 2011, p. 1359). If a chemist claimed that she or he had performed a study in a "regular" environment showing that water freezes at 58°F, she or he would be laughed out of the academy. Likewise with an astronomer who writes of demonstrating that Newton's laws of motion and gravitation are false when applied to celestial bodies. But in an anything-goes world, with little in the way of credible yardsticks for judging the validity of results, all findings are equally admissible, provided that they come with the $p \leq .05$ seal of approval.

Why the use of CIs? To begin with, CIs supply all the information contained in a significance test and more (Natrella, 1960). For instance, CIs underline the desirability of estimation over testing. They indicate, also, the precision or reliability of the estimate via the width of the interval. Moreover, because they are couched in the same metric as the point estimate, CIs are easy to interpret and provide evidence on the substantive, as opposed to statistical, significance of a result. And while I do not condone this usage, a CI can be employed as a statistical significance test; a 95% CI not including the null value (usually zero) is equivalent to rejecting the hypothesis at the .05 level. In addition, the CI is a frequentist measure, that is, part of statistical orthodoxy.[3] From a pedagogical standpoint, therefore, the transition from emphasizing CIs rather than *p*-values ought to be a relatively straightforward one. Consequently, one is left wondering why CIs—a procedure that Tukey (1960) viewed as probably the most important among all types of statistical methods we know—are not routinely used, reported, and interpreted.[4]

Of singular interest from a significant sameness viewpoint, use of CIs promotes the acquisition of cumulative knowledge. It does so, in what Geoff Cumming and Sue Finch (2001), Bruce Thompson (2002), and Roger Kirk (2003) maintain is a largely unexplored but critical topic in the social sciences, by obligating the researcher to think meta-analytically about estimation, replication, and comparing intervals across studies. This is in keeping with the advocation that overlapping CIs be adopted as the criterion for a successful replication.[5] Overlapping CIs suggest credible estimates of the same population parameter(s). Fortunately, there have been some useful recent contributions in this area (see, e.g., Cumming, 2012; Cumming & Finch, 2001, 2005; Cumming & Maillardet, 2006; Fidler, Thomason, Cumming, Finch, & Leeman, 2005; Goldstein & Healy, 1995; Huberty & Lowman, 2000; Schenker & Gentleman, 2001; F. L. Schmidt, 1996; Smithson, 2003; B. Thompson, 2002; Tryon, 2001). So central, in fact, is the idea of overlapping CIs as a measure of replication success that its championing calls for greater explanation.

3.2.4 Overlapping CIs as a Definition of Replication "Success"

Significant Difference

A custom prevails in the social and management sciences of relying on the outcomes of significance tests to determine the success or failure of a replication. A replication success is defined as a result that was statistically significant ($p \leq .05$) in the initial study and continues to be so (in the same direction) in the follow-up (as criticized by, e.g., Bayarri & Mayoral, 2002; Humphreys, 1980; J. Miller, 2009; Ottenbacher, 1996; Rosenthal, 1990). This tradition is inimical to the development of empirical regularities because it ignores the pernicious influence of low statistical power, a condition endemic in the business and social sciences as referenced in Chapter 2. That is, two studies may each have similar quantitative relationships (slopes) or effect sizes, but a statistically significant coefficient is not found in the replication because statistical power is too low. Unfortunately, many researchers do not understand the link between power and the probability of obtaining a successful replication (Busche & Kennedy, 1984; Hubbard & Armstrong, 1994; Lindsay, 1993b; Ottenbacher, 1996; Tversky & Kahneman, 1971; Utts, 1991) because sampling variation and/or small *n* rarely is seen as a possible reason for disparate results (Gelman & Stern, 2006; Lindsay, 1993b; Ottenbacher, 1996).

Consider the following example from Frank Schmidt (1996) who found that a large study in the area of personnel selection using 1,428 subjects obtained a correlation of 0.22 between a single clerical test and job performance. Based on the median sample size of $n = 68$ found in the personnel psychology literature, Schmidt made 21 random draws (without replacement) from this larger study. Table 3-2 shows the correlation coefficients for the smaller studies. Only 8 of these (38%) reached statistical significance at the conventional .05 level. If observers had access to all 21 studies, they would probably say that these results are mixed, thus stimulating additional research to uncover likely explanations for the discrepancies found in the literature. But the bias against publishing negative results makes it highly unlikely that the discipline will have access to many of the other 13 studies reporting nonsignificance. As stated earlier, this leads to an inflated estimate of the population effect size. The average effect size for studies attaining statistical significance is 0.33, which is 50% larger than the real population effect size.

This example shows that the statistical significance test procedure is an unreliable criterion for certifying replication success.[6] Conflicting results are inevitable—even in situations where the only difference among studies is sampling error and/or differences in sample size. The consequences, however, are by no means trivial. The usual response is to search for additional moderator variables (interactions) to explain the "contradictory" findings. Yet these more elaborate models fare no better over a series of studies. A vicious cycle of proposing ever more complex models takes place until the research community abandons the area for lack of

Table 3-2 Random Draws ($N = 68$) Without Replacement From a Larger Study on Clerical Testing and Job Performance ($N = 1,428$) Possessing a Correlation of $r = 0.22$

		95% Confidence Interval	
Draw Number	*Correlation*	*Lower*	*Upper*
19	0.39*	0.19	0.59
5	0.38*	0.18	0.58
14	0.37*	0.16	0.58
8	0.36*	0.15	0.57
3	0.31*	0.09	0.53
16	0.29*	0.07	0.51
6	0.27*	0.05	0.49
17	0.26*	0.04	0.48
11	0.23	0	0.46
20	0.22	–0.01	0.45
21	0.21	–0.02	0.44
13	0.21	–0.02	0.44
9	0.20	–0.03	0.43
18	0.17	–0.06	0.40
7	0.15	–0.08	0.38
2	0.14	–0.09	0.37
15	0.14	–0.09	0.37
4	0.12	–0.12	0.36
12	0.11	–0.13	0.35
1	0.04	–0.20	0.28
10	0.02	–0.22	0.26

Source: Adapted from Schmidt, Frank L. (1996), "Statistical Significance Testing and Cumulative Knowledge in Psychology: Implications for the Training of Researchers," *Psychological Methods*, 1 (1), 121. Copyright © 1996 by the American Psychological Association. Adapted with permission.

*$p < .05$ (two-tailed).

progress. This is why Schmidt (1996, p. 120) correctly emphasizes that "significance testing in psychology and other social sciences has led to frequent serious errors in interpreting the meaning of data . . . that have systematically retarded the growth of cumulative knowledge."

A second way in which null hypothesis significance testing impedes the establishment of empirical generalizations and theory building is now addressed. Two studies, with both reaching statistical significance for the coefficient of interest (with similar signs), nevertheless may have quantitative relationships that are clearly at odds and/or display widely varying effect sizes. Such results would indicate that the relationship is different and/or that some other variable in need of identification is operating. Yet the focus on *p*-values will not necessarily provide this information, intimating once more that it is a poor criterion of replication success.

Gendall, Hoek, and Brennan's (1998) article is enlightening in this respect. They wanted to see if a one-dollar cash incentive would yield a higher response rate to a mail questionnaire compared with a no-incentive control group. Table 3-3 shows that after two reminders there was a 68.2% (386/566) response for the control group and a 76.6% (431/563) cooperation rate for the one-dollar group, the difference being statistically significant ($z = 3.11$; $p < .002$). Now review the hypothetical attempt to confirm the original finding. Table 3-3 conveys that in the replication the responses of the control and one-dollar groups were 32.7% (185/566) and 54.4% (306/563), respectively. Again, this difference is statistically significant ($z = 6.78$; $p < .0001$). Normally, a result like this is taken to be evidence of a successful replication. From a significant sameness perspective, however, the discussion in the next section will show that a profitable opportunity to expand our learning will be missed if one stops here.

Significant Sameness

A significant sameness outlook utilizing CIs produces very different conclusions for the two studies above. With respect to the Schmidt (1996) data, Table 3-2 shows that the 95% CI for each correlation coefficient overlaps with the other studies—even for the largest and smallest correlations. Overlapping CIs suggest that the studies are in agreement with one another, contrary to the false impression left by the traditional "nose counting" approach that only 8 studies support the hypothesis while 13 do not. In this manner, use of CIs offers the prospect of unifying an otherwise seemingly fragmented literature caused by adopting the *p*-value criterion of replication success. Significant sameness addresses

Table 3-3 Hypothetical Replication of Gendall et al.'s (1998) Study on Incentives and Mail Survey Response Rates

	Gendall et al. Study			*Replication*		
	Sample	*Responses*	*Proportion*	*Sample*	*Responses*	*Proportion*
No incentive	566	386	.682	566	185	.327
$1 incentive	563	431	.766	563	306	.544
Differences in proportions			.084			.217
$z = 3.11$; $p < .002$				$z = 6.78$; $p < .0001$		
Confidence intervals around differences in the no-incentive and $1 incentive groups						
Lower 95% CI:			.033			.162
Upper 95% CI:			.135			.272

Source: Adapted from Hubbard, Raymond and R. Murray Lindsay (2013a), "From Significant Difference to Significant Sameness: Proposing a Paradigm Shift in Business Research," *Journal of Business Research*, 66 (September), p.1383 with permission from Elsevier.

commonalities in data sets, the road to generalization. Seen in this light, it would be fascinating to examine how differently published articles would read had the attention been on CIs and not *p*-values (cf. C. Poole, 2001).

CIs, of course, cannot specify the value of the population effect size. This necessitates conducting a meta-analysis. Such an analysis of the data in Table 3-2 demonstrates that there is only one population value—0.22—and that the differences in sample correlations are due only to sampling error.

I return now to the Gendall et al. (1998) mail survey incentive study. While it is apparent that the use of a one-dollar incentive boosted the frequency of responses over the control group (as indicated by the significance test), the significant sameness approach reveals a *failure* to replicate. This happens in two ways. First, the proportion of responses is different in the control and one-dollar groups for the two investigations. This is observed in Table 3-3 by the non-overlapping 95% CIs for the Gendall et al. study (3.3%–13.5%) and the replication (16.2%–27.2%). Second, the percentage increase in cooperation rates for the control and one-dollar groups is markedly greater in the replication. In Gendall et al.'s study there was a 12% increase in replies (.084/.682), while this figure was 66% for the follow-up study. This much better relative performance for the latter occurred even though, in absolute terms, responses in the original study were far higher. These points signify that there are important distinctions between the two studies that need to be explored.

For instance, the impressively high average survey return rate of 72% found in Gendall et al. (1998) was for a New Zealand sample, while the corresponding figure for the American sample in the replication was only 44%. Perhaps this evidently greater willingness by New Zealanders to answer questionnaires is because they are less inundated by them than are Americans. Or possibly the respondent characteristics (beyond those of nationality) and/or content of the survey in the American replication were different from those in the original research. The crux of the matter is that the New Zealand and American investigations vary in ways that call for explicit attention. But this call will likely go unheeded because a "successful" replication, as defined by *p*-values, ignores these differences by erroneously declaring the two works to be equivalent.

In sum, the *p*-value is an unreliable criterion for deciding replication success and encouraging true learning. Significant sameness—does the same relationship (i.e., effect size, quantitative model) hold across many sets of data—must become the new yardstick of replication success. And the use of overlapping CIs around sample statistics and effect sizes is the vehicle for deciding this, a strategy not without its detractors.

Criticism of the Overlapping CI Criterion—and Rejoinder

Still another reason for promoting the use of overlapping CIs across different studies is to spotlight and/or sidestep the baneful effects of low statistical power common in traditional significance testing. Ball and Sawyer (2013, pp. 1389–1390), however, are critical of the use of overlapping CIs as the means for establishing significant sameness. Yet as the following discussion explains, in well-designed studies the proper interpretation of CIs does not result in misleading inferences (Hubbard & Lindsay, 2013b, pp. 1393–1394).

Observe, for instance, quadrant 1 in Figure 3-1 depicting overlapping CIs in a well-designed, high-powered test. Here, the results are informative: The parameters are from the same population. The situation in quadrant 2, showing overlapping CIs in a low-powered test, affords a less tidy interpretation. Under these conditions the investigator cannot infer that the results of the two studies are the same; all that can be concluded is that there is no evidence to show that they differ. But a low-powered study is *not* a well-designed one. This low power revealed in the (pronounced) width of the intervals signals the need for collecting additional data to supply more precise estimates of the parameter(s) of concern. Beyond this, a case can be made that low-powered research should not be published. In contradistinction, non-overlapping CIs in a low-powered test strongly suggests the existence of a real difference between the parameters of the two studies, despite neither being estimated reliably (quadrant 4).

Ball and Sawyer (2013, p. 1390) note further that when exceptionally high-powered tests are involved, usually via the use of enormous sample sizes (think data mining), the CIs of two investigations may not overlap even though their population parameters appear for all intents and purposes to be alike (quadrant 3). The validity of their criticism is acknowledged; when sample sizes become huge, CIs reduce to points.

Given the above, it must be said that in applying the concept of overlapping CIs as the criterion of a replication success, we are at pains to underline that this should not be done reflexively (Hubbard & Lindsay, 2013b, p. 1394). Although focusing on the precision of magnitudes' CIs is infinitely better than using *p*-values when it comes to weighing the plausibility of knowledge claims, no thoughtless application of a rule of thumb should ever substitute for the exercise of subjective judgment when making inferences from data. These inferences must be tempered by concerns such as whether a result looks "reasonable," that is, whether it is aligned with previous results and background knowledge (unless a

Figure 3-1 Interpreting Replication Outcomes: Confidence Intervals (CIs) and Statistical Power

	Statistical Power	
	High	**Low**
Overlapping CIs	[1] Informative: similar results across studies. Parameters are from the same population.	[2] Not informative: need to collect additional data.
Non-overlapping CIs	[3] Mixed: parameters may or may not be from different populations. Here, as always in science, background knowledge and subjective judgment are required to assess whether practical or theoretical differences exist.	[4] Informative: parameters are from different populations.

Source: Adapted from Hubbard, Raymond and R. Murray Lindsay (2013b), "*The Significant Difference Paradigm Promotes Bad Science,*" *Journal of Business Research*, *66* (September), p. 1394 with permission from Elsevier.

boundary condition has been reached), seems to have been arrived at competently, and so on. Science resists formalism, although the value of formal methods is recognized (Hubbard & Lindsay, 2013b, p. 1394).

3.3 Model of Science—Critical Realism

Whereas the significant difference model revolves around theory *testing*, the significant sameness paradigm concentrates on theory *development*. Of note, the idea of significant sameness is totally in accord with an increasingly influential postpositivist metatheory, attributed to Bhaskar (1978, 1979), called *critical realism*. This philosophy unfolded in part through the study of researcher practices and so is based on what scientists actually do (Haig, 2013b, p. 7; M. L. Smith, 2006). It is beginning to attract the attention of scholars in economics (Lawson, 1997, 2003), education (S. Clegg, 2005), geography (Yeung, 1997), management (Rousseau, Manning, & Denyer, 2008; Tsang & Kwan, 1999; Van de Ven, 2007), marketing (Easton, 2002; Hubbard & Lindsay, 2013a; S. D. Hunt, 2003), and sociology (Danermark, Ekström, Jakobsen, & Karlsson, 2002), among others.

While including aspects of both, critical realism offers an alternative philosophy to those found wanting—positivism/empiricism on the one hand and relativism/interpretivism on the other (Sayer, 2000, p. 2). That is, it provides a bridge across the philosophical divide between the quantitative and qualitative research camps (Lindsay & Hubbard, 2011). This is done by pointing to the importance of context while at the same time eschewing "naïve" realism. As such, the critical realist approach capitalizes on the virtues of both quantitative and qualitative research capabilities, seeing no dichotomy between them.

Briefly, this philosophy asserts that, first, the world exists independently of our knowledge of it. Second, science aims at developing genuine, but always imperfect and subject to change, knowledge about the world. That knowledge is produced socially, however, and as a consequence is theory laden, does not make it theory determined (Sayer, 2000, p. 47). Third, all theories concerning knowledge claims must be critically evaluated; knowledge is not immune to empirical check (Sayer, 1992, p. 5). It is via an exacting appraisal of competing theories that the scientific college, over time, is able to retain and improve on those which do a better job of approximating reality, while discarding those which do not. In short, not all knowledge is equally fallible (M. L. Smith, 2006).

In common with critical realism, the significant sameness model is philosophically grounded in abductive (explanatory) reasoning. Abduction—subsequently named inference to the best explanation (Harman, 1965) or retroduction (Lawson, 2003, pp. 145–146)—is a vital concept introduced by the American pragmatist philosopher Charles S. Peirce.[7] As pointed out in Charles Hartshorne and Paul Weiss's (1934) anthology of Peirce's work, "Abduction consists of studying facts and devising a theory to explain them" (1903, Vol. 5, p. 90). Like detective work, abduction is a method of inference for generating plausible (best) explanations for the facts we possess. It is to be understood that the facts referred to here are of the stubborn variety.[8]

Empirical regularities or facts can be predicted by theory (T→E) or can precede theory (E→T). The former avenue glorifies hypothetico-deductivism and the significant difference philosophy. The significant sameness model, however, emphasizes that by means of inductive enumeration the empirical regularities must come *first*. In this bottom-up (E→T) interpretation of research, the discovery of empirical generalizations *fuels* (high-level) theory development rather than vice versa (Ehrenberg, 1993b; Haig, 2013a, 2013b, 2014; Hubbard & Lindsay, 2013a, 2013b; Lynch, Alba, Krishna, Morwitz, & Gürhan-Canli, 2012). The rationale behind this is that the purpose of theory is to explain and systematize lower-level findings and generalizations (Rosenberg, 1986); consequently, before any theory explaining a process can be developed or tested, that process needs to be understood through the identification of repeatable facts, phenomena, or regularities. Fiske (1986, p. 75) considers this to be the first step to achieving progress (see also Ehrenberg & Bound, 1993; Hacking, 1983, ch. 9; Haig, 2014, ch. 2; Ladyman, 2002, p. 48). This is because, the rank and file of empirical social and management research notwithstanding, we are unlikely to be successful in trying to explain particular data sets, or other isolated events in individual studies, since they tend to be ephemeral and affected by idiosyncratic boundary conditions that are extremely difficult to ascertain. Put another way, much theory testing in these areas concerns entities and their interrelationships whose viability is far from established, a precarious exercise to say the least. This also accounts for the claim made repeatedly here that data seldom speak for themselves, as well as the attendant proviso that the results of single studies must be treated with circumspection.

In light of these handicaps it makes sense to uphold the overlooked view that fruitful theoretical interpretation typically occurs *after* a pattern (fact) has been empirically determined from the parsing of many data sets (studies). Thus, a preferable strategy is to anchor theory development around the discovery of repeatable facts or regularities in the behavior of phenomena; their relative stubbornness *demands* an explanation. Keuzenkamp (2000, p. 221) seconds this rationale, noting that facts can inspire economic theory. As does Ormerod (1997, p. 210), who recommends that theory building by economists should originate from the facts, and not from abstractions about how a rational world ought to operate. So, too, does Summers (1991, p. 140) when advising that economic theories should seek to explain regularities. Critical realists share this position. Lawson (2003, p. 146), for instance, acknowledges that finding causal mechanisms "presupposes" the existence of partial or demi-regularities. Danermark et al. (2002, p. 116) concur: "We claim that social scientific research is about identifying demi-regularities and from them trying to find explanations." Management (Rousseau et al., 2008, pp. 481, 487), marketing (Sharp & Wind, 2009, p. 122), and psychology (Haig, 2013a, p. 136, 2013b, p. 9) theorists harbor analogous thoughts. Finally, no less an authority than Sherlock Holmes reckoned, "It is a capital mistake to theorize in advance of the facts" (cited in N. L. Kerr, 1998, p. 201).

To bring to the fore this little-appreciated point, more Nobel prizes have been awarded to those discovering stubborn facts or empirical generalizations than have been granted for the construction of theories (Haig, 2005, p. 384). It also underlines Blaug's (1992, p. 134) conviction that scientific progress comes only from the maximization of facts, Hunter's (2001, p. 157) insistence that in the advancement of science, facts are at least as important as ideas, Ziman's (1978, p. 6) and Imrey's (1994, p. 65) desire that scientific knowledge should consist of firmly established facts and principles, Hambrick's (2007, p. 1349) proposal that we should be willing to start with the compilation of facts, Keuzenkamp's (2000, p. 22) urging that the goal of econometric inference should be discovering regularities that are simple and descriptively accurate, and Alba's (1999, p. 2) and Lehmann, McAlister, and Staelin's (2011, p. 157) calls to place greater stress on the obtainment of facts.

Following Ehrenberg (1993b), accounting for stubborn facts necessitates a sequential process of data-theory (or concrete-abstract in critical realist language) interactions over an expanding range of different conditions. At this stage much will be understood, for example, that some factors do not matter, whereas others do, providing for considerable familiarity with the phenomenon in question and for the possibility of linking the result with findings and/or theories in other areas. This process encourages further conjecture (T), requiring the testing of new implications (E). The cycle keeps repeating itself (i.e., E→T→E→T . . .), thereby leading to an increase in our depth of understanding of phenomena (cp. Eisenhardt, 1989; Eisenhardt & Graebner, 2007). As noted, it is the latter, rather than some one-to-one correspondence between theory and observations, that convinces researchers about the feasibility of theoretical explanations.

Furthermore, the detection of empirical regularities as a paramount driving force behind theory building not only applies to the management and social sciences, but has been of major importance in the physical sciences as well (R. Brown, 1963, p. 136). For example, Bernard Cohen (1985, p. 125) acknowledges that the detailed recordings of the Danish astronomer Tycho Brahe greatly influenced Newton's thinking and led eventually to the Newtonian revolution. Indeed, Wigner (1964, p. 995) maintained in his Nobel laureate address that recognizing the need to specify regularities within a certain range of conditions may be the greatest discovery in physics:

> It is often said that the objective of physics is the explanation of nature, or at least of inanimate nature. . . . It is clear that physics does not endeavor to explain nature. In fact, the greatest success of physics is due to a restriction of its objectives: *it only endeavors to explain the regularities in the behavior*

> *of objects.* This renunciation of the broader aim, and the *specification of the domain* for which an explanation can be sought, now appears to us as an obvious necessity. In fact, the specification of the explainable may have been the greatest discovery of physics so far. (emphasis added)

Given that the procurement of empirical regularities and their later generalization is a major influence on theory development, the widespread preoccupation in the business and behavioral literatures on theory testing, following the H–D model of science, makes little sense (Barwise, 1995, p. G32; Cook & Campbell, 1979, p. 24; Hubbard & Lindsay, 2002, p. 393; Toulmin, 2001, p. 213). Much more attention needs to be devoted to identifying empirical regularities, thus making the case that external validity considerations are as important as their internal cousins. Shadish, Cook, and Campbell (2002) share this position, writing that earlier statements by Donald Campbell and Julian Stanley (1966) that internal validity is the sine qua non of experimentation have been misinterpreted by readers for many years. In an effort to set the record straight, Shadish et al. announce, "Let us be clear: *Internal validity is not the sine qua non of all research*" (p. 98), and go on to say that "in this book, methods for studying external validity now receive the extensive attention that our past work gave to internal validity" (p. xvii).

Finally, while the importance of acquiring empirical generalizations has been emphasized to counteract its chronic neglect to date, this is not considered to be the foremost goal of science. Rather, the development of empirically grounded theory (explanation) and causal understanding of phenomena, along with practical applications of theory, are what is desired. However, the attainment of predictable regularities is seen as a critical means of achieving these goals within a framework of abductive inference that many (e.g., Haig, 2005, 2014; Hubbard & Lindsay, 2013a, 2013b; Ketokivi & Mantere, 2010; Lipton, 2004; Psillos, 1999; Rozeboom, 1997) testify is, in fact, an accurate description of how researchers actually behave in progressive science.[9] And it is why, in the final telling, for those researchers favoring a critical realist position, replications are essential to disciplinary advance (Koole & Lakens, 2012, p. 609).

3.4 The Role of "Negative" ($p > .05$) Results

In stark contrast to the publication bias against negative outcomes emblematic of the significant difference paradigm, such results are *welcomed,* provided they can be reproduced (Ehrenberg, 1975), when the emphasis is placed on significant sameness. This is because contradictory results can play an important role in identifying the *boundary conditions* of an empirical generalization's applicability (C. M. Christensen & Carlile, 2009; Cortina & Folger, 1998; Hubbard & Vetter, 1996; Lindsay & Ehrenberg, 1993).

Going against the grain of many researcher beliefs, theories do not produce universal covering laws of the kind suggested by Hempel (1965). In Hempel's world an event is explained by showing how it can be deduced from some universal or general law(s) conjoined with a set of initial conditions. But, as Ladyman (2002, p. 202) spells out, a critical requirement for the deduction to be valid is the empirical truth of the universal law(s). However, owing to Hume's "problem of induction"—just because all *observed* A have the property B does not entail that *all* A have the property B—we cannot guarantee the universality of laws.[10] As discussed earlier, there are no immutable laws in the sense that they exhibit absolute invariance. The positivists' goal of generating infallible knowledge is impossible.

For universal laws to arise would require a totally closed system in which an invariable sequence of "if A, then B" empirical events could be observed readily (Lawson, 2003, p. 5; Sayer, 1992, pp. 122–124). Because closure of this kind cannot be met in natural systems, other than in simplified experimental situations, even physical laws must be conditional. The idea that the natural sciences, physics included, deal with phenomena that are not context dependent is a myth (Holtzman, 1986, p. 348; Shadish et al., 2002, p. xv). Instead, "laws" are best viewed as *restricted* generalizations possessing extensive empirical support (Chalmers, 1999, p. 216; Giere, 1999, p. 93; S. Gordon, 1991, p. 34; Toulmin, 2001, p. 111; Turner, 1967, pp. 251–252).

When comparisons are made between the physical and social sciences, what frequently goes unsaid is that researchers in the former are able to apply theory in the real world because they work very hard at specifying the relevant conditions under which the effect will and will not come about. This is a laborious task that demands numerous carefully conducted studies to delineate the existence of boundary conditions affecting a result. Boyle's Law, for example, says that at a given temperature the volume and absolute pressure of a gas vary inversely. This law does not hold, however, when the density of a gas increases or when the temperature changes. In a similar vein, Snell's Law of Refraction does not hold for certain temperature ranges, light rays, or for transparent crystals (R. Brown, 1963). The point is, all major laws in the physical sciences—including those of Ampère, Charles, Coulomb, Faraday, Kepler, Newton, and Ohm—are similarly constrained (Giere, 1999, p. 90; Losee, 2001, pp. 191–192).[11]

While these empirical generalizations may be of a restricted nature, they have nevertheless been found to hold in sufficient instances as to render them of great practical value. Scott Gordon (1991, p. 603) elucidates on this:

> When necessary, a scientist will, without a qualm, use "Avogadro's number," which, though it has been computed from a limited set of specific cases, asserts that *all* gases, at equal temperature and pressure, contain 6.023×10^{23} molecules per gram molecular weight. In the *Handbook of Chemistry and Physics* there are literally hundreds of thousands of such universal numerical statements for particular elements and compounds: boiling points, melting points, solubilities, densities, X-ray diffraction angles, etc., most of which are not even given with ± qualifiers.

And should an anomalous result occur when using these generalizations, a clear message is sent that some boundary has been reached concerning its relevance.

The notion of limited empirical regularities heeds, and is explainable by, the tenets of critical realism, where Bhaskar (1978, p. 50) advises that the tangible impact of causal laws must be viewed as "tendencies" rather than inevitabilities. The reason for this is that the portrayal of laws as universal empirical regularities is both ontologically and epistemologically naïve. Social (and physical) reality is not transparent and cannot be drawn down to the realm of observable phenomena in closed systems. It is, rather, an ontologically "deep" stratified and differentiated open system composed of three separate domains: real, actual, and empirical (Bhaskar, 1978, p. 56; Sayer, 2000, pp. 11–12).

The *real* domain is the deepest, and here reside objects of inquiry (e.g., consumers) and structures of internally related objects (e.g., buyers and sellers) with causal powers or mechanisms capable of producing events in the world (e.g., exchange relationships). To invest causal powers to objects and structures is to comment on what they will or can do under appropriate circumstances because of their intrinsic natures or "ways of acting" (Bhaskar, 1978, p. 14). It is these causal mechanisms, working alone or in concert with others, if and when triggered, that generate events in the *actual* domain. And some of these events may be observed in the *empirical* domain. The fundamental aim of science is to develop causal explanations by finding or imagining necessary generative mechanisms to account for the nonrandom patterns of events that are observed empirically (Danermark et al., 2002; M. L. Smith, 2006; Sobh & Perry, 2006). Put succinctly by Little (1993, p. 185), "the central explanatory task for social scientists is to uncover causal mechanisms."

But since the causal mechanisms which objects and structures possess are only tendencies which may be reinforced, modified, or inhibited in complex interactions with other objects' and structures' mechanisms in an open social world, they may or may not reveal themselves empirically (Danermark et al., 2002, p. 163). Therefore, whatever empirical materializations happen will surely not be in the form of invariable configurations. The social world is not some highly aggregated deterministic model, but rather one in which causal relationships are contingent. This, parenthetically, forms the basis for Lawson's (2009, p. 765) criticism that insistence on mathematical deductivist modeling "is why modern economics has continually failed on its own terms."

Yet to deny the existence of universal empirical generalizations in no way prevents the occurrence of restricted ones. It is true, as the above examples show, that the chances of detecting regularities are higher in the more closed systems characteristic of the natural sciences (Sayer, 1992, pp. 121–123, 2000, pp. 14–15) than in the open social world. However, it is possible for social systems (e.g., social and work organizations, families, buyer-seller relationships, health care systems) to be quasi-closed, thereby permitting the discovery of approximate (Sayer, 1992, p. 124), demi- (Lawson, 1997, pp. 204–221, 2003, pp. 105–107), phenomenal (Little, 1993, p. 187), sufficient (Ormerod, 1997, p. 210), or stylized (Helfat, 2007, pp. 186–189; Keuzenkamp, 2000, pp. 220–221) regularities. Moreover, some of these are "often very stable" (Danermark et al., 2002, p. 165), "repeatable events" (Haig, 2013a, p. 143).

Further, the identification of boundary conditions provides a powerful heuristic leading to higher-order theoretical synthesis (Brinberg & McGrath, 1985; Lynch, 1999; Magnani, 2001; McGuire, 1983; Poincaré, 1908/2004; M. S. Poole & Van de Ven, 1989; Van de Ven, 2007). The history of science is replete with examples of how the recognition of anomalies has motivated scholars to refine theories or to develop new and better ones in the course of understanding why the boundaries apply (Ehrenberg, 1975; N. L. Kerr, 1998, p. 210; Trusted, 1979). For this reason, Clayton Christensen and Michael Raynor (2003b) state that negative results represent a triumph, not a failure.[12]

Historically, management and social scientists have been unconcerned with this ponderous task of ascertaining the scope and limits of a finding (Greenwald, Pratkanis, Leippe, & Baumgardner, 1986; Hunter, 2001; Tsang & Kwan, 1999; Wells, 2001)—the "hard slog" as Ehrenberg (2004, p. 41) put it. This is unfortunate because some (e.g., R. Brown, 1963, pp. 149, 154; Ehrenberg, 1993a, p. 385; Ehrenberg & Bound, 1993, p. 191; Gage, 1996, p. 14; Harvey, 1969, p. 111; Kincaid, 1996,

p. 3; H. A. Simon, 1990, p. 2) see no reason, in essence at least, why empirical generalizations of the same limited sort found in the physical sciences cannot be uncovered also in the social sciences.[13]

Of direct relevance on this matter is Hedges's (1987) intriguing analysis of the empirical cumulativeness—the degree of agreement among replicated studies—of research published in the physical and social sciences. Using the same statistical methods (weighted least squares, Birge ratios) yielding numerical indexes for purposes of comparison, Hedges was able to determine the consonance of experimental results in physics and psychology. The physics data consisted of 13 quantitative reviews of the mass and lifetime of stable (against strong decay) particles from the Particle Data group. Hedges's corresponding data for psychology came from 13 meta-analyses encompassing six sub-areas of the discipline: sex differences in spatial ability; sex differences in verbal ability and field articulation; the effects of open education on attitude toward school, mathematics achievement, reading achievement, and self-concept; the effects of desegregation on educational achievement; the validity of student ratings of college faculty; and the effects of teacher expectancy on IQ. No doubt surprising many people, Hedges announced that "the evidence presented here suggests that social science research may not be overwhelmingly less cumulative than research in the physical sciences" (p. 453). His conclusion is hugely encouraging.

In fact, there is additional vindication for the hopes aired by those believing in the possibility of discovering empirical regularities in the social and managerial sciences. As noted above, and discussed further below, regularities of varying strengths have been found in these fields.

As a case in point, the psychophysics area of psychology has Weber–Fechner's Law (Gescheider, 1976; Teigen, 2002; Turner, 1967), which states that $\Delta I/I = k$. Here, I is the intensity or size of an initial stimulus, ΔI is the amount of change in a stimulus required to produce a *just noticeable difference,* and k is an empirical constant. An example might be increasing the volume of music until it is perceived to be noticeably louder (ΔI) than the original volume (I) and recording k. Just like laws in the physical sciences, this one is of restricted applicability; it works quite well when applied to human sense modalities over the intermediate range of intensities, but breaks down at the extremes. The Flynn effect is another good example of an empirical generalization in psychology.[14] As Haig (2013a, p. 138) and Wikipedia (2015) describe, this effect reveals the striking regularity that IQ scores have increased by some three points each decade from about 1930 to the present in 20 countries, from areas as diverse as North America, Europe, Asia, and Australasia. Geography boasts the rank-size rule for the population size of cities in various countries. It is formulated as $p_i = p_1/i$, where p_i is the size of the population of the ith city when all cities are ordered 1, 2, 3, 4, . . . n in terms of descending population sizes, and p_1 is the population of the largest city. The rank-size rule follows a negatively sloping straight line when plotted on a log-log scale and holds decently for the United States (Uncles & Wright, 2004, p. 6). Political science offers the "iron law of oligarchies," positing that social organizations tend to form hierarchical orders (S. Gordon, 1991, pp. 35–36). Or see Berelson and Steiner's (1964), albeit lax, inventory of generalizations about human behavior found in the social sciences as a whole.

When it comes to the management disciplines, we have the Boston Consulting Group's Law of Experience, which predicts that as cumulative output doubles, unit costs are reduced by a fixed percentage. For example, if a company has, say, an 85% experience curve—where experience is defined as the combined effects of learning, volume, investment, and specialization—this means that each time cumulative production doubles, per unit costs drop to 85% of their preceding level. Grant (1998) observes that this "law" holds in a number of studies from the manufacture of bottle caps and refrigerators to insurance policies and long-distance phone calls. There is, too, the predictable manner in which innovations occur within companies (Christensen & Raynor, 2003a).

In the marketing area, regularity is seen in the profiles of innovators. Robertson (1971), for example, provides information from 21 independent studies spanning a wide range of product categories and populations describing these profiles. Innovators usually have higher income and education levels, possess a more venturesome personality, and are much more likely to be opinion leaders than non-innovators. Other candidates for empirical regularities in marketing include the retail gravitation effect, the market share–ROI (return on investment) relationship, and the 80–20 rule describing how the majority of sales in some industries (e.g., airlines, beer, bank revenues, college donations) come from a minority of the market (Hubbard & Lindsay, 2002; Kerin & Sethuraman, 1999; Sheth & Sisodia, 1999).

Andrew Ehrenberg's work is exemplary when it comes to establishing empirical generalizations within a marketing context, or elsewhere in the management and social sciences for that matter. For instance, Ehrenberg (1988) demonstrated the progress achieved in modeling brand purchase frequencies and repeat buying patterns using the negative binomial distribution up through the increasingly more general findings on individual purchase incidences and brand choice made possible by the comprehensive Dirichlet model. The appeal of this model is its ability to account, parsimoniously (only the market share of each brand is required as input), for many empirical patterns of buyer behavior, including those described by the duplication of purchase law and the double jeopardy (DJ) phenomenon (Hubbard & Vetter, 1996, p. 154). The DJ effect, which says that brands with smaller market shares not only are bought by fewer customers in a given time period (penetration level) but also are bought less often (average frequency of purchase), has undergone extensive replication confirming its applicability. For example, Ehrenberg and Bound (1993) note that DJ holds for over 50 different products (including convenience and shopping goods, differentiated and undifferentiated items, as well as tangible products and services). DJ also applies to different distribution channels, different countries, different time periods, and so on. They further tell of a few exceptions (boundary conditions) where DJ doesn't hold, or holds only partially. Dan Vetter and I submit that Ehrenberg's (e.g., Ehrenberg & Bound, 1993; Ehrenberg, Goodhardt, & Barwise, 1990) work "provides perhaps the quintessential example of the value of systematic replication and extension research in producing robust and generalizable results" (Hubbard & Vetter, 1996, p. 155).

Furthermore, economics has the Law of Demand, the inverse relationship between the price of a product and the quantity consumed at that price, which Blaug (1992, p. 139) hails as "one of the best corroborated statistical 'laws' of economics." Then there is Pareto's Law of Income Distribution stating that the latter was essentially the same in all countries (S. Gordon, 1991, p. 35; Strathern, 2001, pp. 212–213).

Still another example of the resilience of empirical relationships in economics is found in Thomas Piketty's (2014) improbable best-seller *Capital in the Twenty-First Century*. In this book Piketty analyzes over time the capital/income ratios (βs) exhibited by various countries. Written as $\beta = s/g$, this index is the ratio of a country's savings to growth rates. For instance, if $s = 10\%$ and $g = 2\%$, then $\beta = 5$. What this means is that if a country saves 10% of its national income each year and the annual growth rate of this same income is 2%, then in the long run the country will have amassed capital worth 5 years of national income. Clearly, as β gets larger we see a growing inequality in incomes favoring the "haves" over the "have-nots" and the potential havoc that could ensue as a result. Piketty (p. 26) shows that for Europe, aggregate private wealth was worth 6–7 years of national income from 1870 to 1910, fell to 2–3 years by 1950 (because of two world wars and the Great Depression), and rose uniformly to some 4–6 years in 2010. The pattern for the United States was about 3 years in 1770, 5 years in 1910, less than 4 years in 1950, and 4.5 years in

2010. As Piketty's gathering of the facts shows beyond doubt, the ratio of private wealth to income has increased steadily since 1950 and now rivals 19th century values. It may well eclipse those historical benchmarks.

Interestingly, the above discussion puts the lie to the objections of those of a more humanistic or qualitative persuasion that empirical generalizations cannot be detected in the social sciences. The real problem is that the dominant paradigm researchers operate within places no weight on their establishment. In addition, it would be remiss to leave the impression that qualitative researchers somehow are absolved of the responsibility for seeking valid generalizations. They are, in fact, just as culpable in this regard as their quantitative colleagues (Freese, 2007; Golden, 1995; Hubbard & Lindsay, 2002; J. A. Maxwell, 1992; Wells, 2001; Wilk, 2001). In any case it must be reminded that critical realists see no absolute split between quantitative and qualitative research methods. They accept fully the need for involving *both* perspectives in the discovery and understanding of regularities (Rousseau et al., 2008, pp. 486–487; Sobh & Perry, 2006, p. 1195; Van de Ven, 2007, p. 70).

Once empirical generalizations are identified, the goal of science is devising explanations for them. And this is where abductive reasoning comes into prominence because causal explanations, as opposed to causal descriptions, must be *invented* (Bunge, 1997). Most causal mechanisms, be they physical (e.g., gravity) or social (e.g., the market's "invisible hand") are hidden, and therefore must be conjectured or visualized. This is why Bunge (1997, p. 423) says: "Imagine what would have happened if Newton had abstained from positing unobservables, such as mass and gravitation, and from postulating laws, focusing instead on observable properties and events and their statistical correlations." The apparent discovery of the Higgs boson, or so-called God particle, is a very recent example of support for the existence of a heretofore hidden causal mechanism for understanding why matter has mass (Heilprin, 2012, p. 5A).

Moreover, causal explanations for some observed regularities exist in the business and social sciences. These explanations range from the embryonic to the fully fledged. An example of the former is the DJ phenomenon. To date, the causal mechanism(s) underlying this widespread occurrence remains little more than a simple description put forward originally by the sociologist McPhee (1963) that DJ will arise whenever competitive items (e.g., market shares, movies, politicians), through varying degrees of exposure to an audience, differ in their popularity (see Ehrenberg et al., 1990, p. 85, for a fuller telling; Habel & Lockshin, 2013, for recent developments). The Flynn effect portraying an approximately linear increase in IQs over a number of decades has several, some mutually supportive, explanations. Among potential candidates receiving ongoing attention are the impacts of improved nutrition, greater environmental complexity, better education, smaller family sizes, heterosis (the birth of genetically superior children from mixing the genes of parents), and increases in test-specific skills (Wikipedia, 2015). Via his fundamental inequality $r > g$ (where r denotes the average annual rate of return on capital, and g is the growth rate of the economy), Piketty (2014, pp. 25–26) supplies the rationale underlying the divergence in capital/income ratios mentioned earlier. As is traditionally the case, when $r > g$ then inherited wealth must necessarily outstrip national income, possibly to toxic levels. As a final example, also in economics, sophisticated (mathematical) competing mechanisms to account for downward-sloping demand curves have been available for some time. These include the income and substitution effects (Slutzky equation), the law of diminishing marginal utility and utility maximization, and revealed preference theory (see, e.g., Green, 1976, pp. 62–69, 81–83, 121–127; Phlips, 1974, pp. 16–26, 40–45).

It has been shown that negative results can play a valuable role in the significant sameness paradigm. This is especially so when they are backed up with adequate statistical power. Section 3.5 speaks further to this issue.

3.5 The Statistical Power of "Negative" ($p > .05$) Results

An investigation by Scott Armstrong and me is informative here (Hubbard & Armstrong, 1992). Based on a content analysis of 32 randomly selected annual issues each of the *Journal of Consumer Research* (*JCR*), *Journal of Marketing* (*JM*), and *Journal of Marketing Research* (*JMR*) over the period 1974–1989, we uncovered some 54 articles reporting negative (i.e., $p > .05$) results.[15] We inspected the statistical power found in these 54 papers.

Remember that the power of a statistical test (i.e., the probability of rejecting a false null hypothesis) depends on the selected level of significance (usually .05), the effect size in the population, and the sample size. Tests with insufficient statistical power are likely to yield null findings and thus be seen as being less entitled to publication. Conversely, statistically nonsignificant results buttressed by high power are potential contributions to knowledge (Fagley, 1985), suggesting, perhaps, that a boundary condition on some phenomenon has been met. This is because with high-powered negative results the probability of a Type II error (erroneous acceptance of the null hypothesis) must be low (Rossi, 1990, p. 646). Therefore, it is necessary to determine the statistical power of "insignificant" results.

We adopted Jacob Cohen's (1988) methods and recommendations, as outlined in the appendix to Chapter 2, in an attempt to calculate the statistical power of the 54 articles in our sample reporting nonsignificant outcomes. It was found that 11 of these could not be power-analyzed, 6 because they used methods for which power tests were unavailable and 5 because not enough information to do so was provided. The remaining 43 articles allowed power analyses of 410 statistical significance tests, or an average of 9.5 tests per article. Of these 43 papers, 19 were from *JCR,* 8 from *JM,* and 16 from *JMR.*

Table 3-4 portrays the frequency and cumulative percentage distributions of the average power of the 43 articles to detect small, medium, and large effect sizes in the population. Recall that Cohen (1988) advises that the value .80 be employed when there is no other basis for establishing a satisfactory power level. Following this advice, the mean power levels of these articles are seen to be high; the probabilities of detecting small, medium, and large effect magnitudes per article are .35, .89, and .99, respectively. Corresponding figures based on the 410 individual statistical significance tests are .36, .87, and .98.

Over the period 1974–1989, all three journals showed consistent power levels for discerning small, medium, and large effects. For *JCR* these figures were .40, .87, and .98; for *JM* they were .29, .86, and .99; for *JMR* they were .33, .92, and .99.

On average across all three journals, the 43 statistically nonsignificant articles showed reasonable probabilities of detecting even small effects. Fourteen of these (32.6%) revealed a 50-50 chance or better of doing so, and 2 were able to exceed the recommended level of .80. If medium effects typify marketing's literature, these articles, on average, had almost a 90% chance of distinguishing them. Ten of the papers (23.2%) did not reach the nominated .80 power yardstick (although 9 of these revealed power in the .60 to .79 range), 33% met or exceeded the .99 power level, and none were below .50. All 43 articles reached or outgained the 80% power benchmark to uncover large effect sizes. In fact, 32 (74.4%) of them had a .99 probability or more of rejecting a false null hypothesis (Table 3-4).

The average power levels of publications recording statistically nonsignificant findings were consistent over the 1974–1989 time period enclosing our work. Splitting these time periods by decades, for 1974–1979 the ability to detect small, medium, and large effects across all three journals was .29, .88, and .99. For 1980–1989, these values were .42, .90, and .99.

Table 3-4 Power of 43 Published Marketing Studies With Statistically Nonsignificant ($p > .05$) Results: 1974–1989

	Small Effects		*Medium Effects*		*Large Effects*	
Power	*Frequency*	*Cumulative %*	*Frequency*	*Cumulative %*	*Frequency*	*Cumulative %*
.99–			14	100.0	32	100.0
.95–.98			5	67.4	8	25.6
.90–.94			8	55.8	2	7.0
.80–.89	2	100.0	6	37.2	1	2.3
.70–.79	3	95.3	5	23.2		
.60–.69	5	88.3	4	11.6		
.50–.59	4	76.7	1	2.3		
.40–.49	0	67.5				
.30–.39	2	67.5				
.20–.29	14	62.8				
.10–.19	13	30.2				
Number of articles (43)	.35		.89		.99	
Number of tests (410)	.36		.87		.98	

Source: Adapted from Hubbard & Armstrong (1992, p. 133).

An argument might be made, of course, that one of the reasons these 43 articles with null results were published in the first place is precisely because of their overall high levels of statistical power, the imputation being that *unpublished* manuscripts with $p > .05$ findings were noticeably underpowered. It was not possible for us to evaluate directly the soundness of this argument because, to the best of our knowledge, the power levels of published and unpublished marketing (or other management or social science) papers with null results have never been examined.

It must be pointed out, however, that indirect evidence indicates that concerns about statistical power played a minor part in the publication decision. Authors of published works obtaining, or failing to obtain, statistically significant results do not seem to formally include power considerations in their research designs (J. Cohen, 1990). We found that none of the 54 articles with insignificant findings displayed power calculations (Hubbard & Armstrong, 1992, p. 134). Indeed, as mentioned earlier, 5 of these studies were published without supplying information needed to compute levels of statistical power. Sawyer and Ball (1981) questioned authors of empirical works published in five issues of *JMR* (November 1978 to November 1979) about how they decided on sample sizes. They reported that tangible calculation of statistical power is uncommon among researchers; only 4 of 28 respondents (14%) computed power before data collection, while 2 others did so afterward. Our content analysis of these same five *JMR* issues showed that none of the 59 articles using statistical significance tests told about power calculations, and only 4 alluded to the topic of inadequate sample sizes (Hubbard & Armstrong, 1992, p. 134).

Research papers with null results, but which show that they meet or exceed Cohen's (1988) advocated .80 power level, are capable of yielding helpful information. Sufficiently powered studies failing to reject H_0 can provide strong evidence of a negligible effect size in the population. On the other hand, statistically significant outcomes with high power, often realized with large samples, may mask trivial effect magnitudes. Seen in this light, null results can be useful. The publication of (adequately powered) null outcomes also can be expected to deter researchers from reentering blind alleys. Acknowledging the potential contribution to knowledge of "negative" findings, editorials in the *Journal of Clinical Neuropsychology* (Rourke & Costa, 1979), the *New England Journal of Medicine* (Angell, 1989), and the *Journal of Cerebral Blood Flow and Metabolism* (Dirnagl & Lauritzen, 2010) have confirmed an openness to publish well-designed manuscripts with null results. Social and managerial science journals can benefit by adopting a similar posture.

3.6 Conclusions

It has been shown from numerous perspectives that the *p*-value is not an objective measure of evidence against the null hypothesis, H_0. But even if it was, so what? That the differences between two means or correlations between two variables are not exactly zero hardly qualifies as valuable scientific knowledge. Yet this is what adherents of the significant difference school are all too willing to settle for.

More broadly, it must be said that the notion of the scientific method, in the sense that dutifully abiding by a prescribed set of methodological steps yields knowledge, is apocryphal. Bauer (1994, p. 128) clarifies this when writing that if all the sciences were united by application of the scientific method, then sociology and chemistry would differ only because of their subject matters.[16] Yet it is well known that sociology and chemistry are poles apart in terms of their scientific status. To repeat, it takes a great deal of time and many studies (replications focusing on the estimation of effect magnitudes, including their CIs and degree of overlap) by many people for the research authorities to filter the wheat from the chaff (Bauer, 1994, pp. 48–52). So from the above it cannot possibly be that following the rules of the so-called scientific method is what distinguishes science from non-science. Despite such observations, members of the significant difference paradigm cling resolutely to this illusion. Patrons of the significant sameness camp do not.

Among other things, this chapter has drawn attention to the importance of replication research in the quest for scientific knowledge. Chapter 4 underscores this claim.

Notes

1. Additional information on the fallibility of the *p*-value as a dependable measure of evidence—for example, its logical flaws and susceptibility to differences in interpretation due to alternative experimental designs—is available in Hubbard and Lindsay (2008, pp. 76, 78–80).

2. From the questions raised, for example, in Locke (2007, p. 879), it is clear that he and Latham were engaged in abductive (see Section 3.3) as well as inductive reasoning. Locke's not acknowledging this explicitly is understandable for there is some confusion among philosophers of science about the distinction between abductive and inductive inference. In particular, some philosophers view abduction as a form of induction (see, e.g., Ladyman, 2002, p. 219; Okasha, 2002, p. 30).

3. Introduced by Neyman (see, e.g., 1934, 1937), the construction of frequentist CIs is predicated on the use of random samples. There will be many (most?) occasions in the significant sameness paradigm where employing probability samples is not feasible (it is

important to emphasize that this stricture applies equally in the significant difference paradigm). Does this, then, prohibit the use of CIs? The answer is yes when adopting the dominant *statistical* model of generalization, where one is inferring from sample *to* population. When the less ambitious, but more practical, goal of *empirical* generalization—generalizing *across* many well-defined (sub)populations—favored by the significant sameness approach and its replication strategy is taken up, this is not a pressing concern. These issues are treated at length in Chapters 4 and 6.

4. Disturbingly, Estes (1997, pp. 330–331) allots a major part of this failure, based on several decades of editing journals, to the fact that a great many (psychology) researchers do not understand the theoretical and computational bases of such a measure. Evidence supports Estes's hunch. Responses from 473 authors of journal articles in psychology, behavioral neuroscience, and medicine caused Belia, Fidler, Williams, and Cumming (2005, p. 395) to conclude that many of them have "fundamental and severe misconceptions" about how CIs can be used to make inferences from data.

5. See also Wilkinson and the American Psychological Association Task Force on Statistical Inference (1999, p. 599).

6. See Hubbard and Lindsay (2013a, p. 1383, 2013b, p. 1394) and Kline (2004, p. 74) for further examples of the problems of using *p*-values to decide replication success.

7. Stigler (1999, p. 192), a well-known statistician, lists Peirce as one of the two greatest American scientific minds of the 19th century (physicist J. Willard Gibbs being the other).

8. In this connection, see Haig's (2014) excellent treatment of ATOM, or Abductive Theory Of Method.

9. For example, abduction has played a vital role in scientific advances in the past, as witnessed in Darwin's theory of evolution, Einstein's work on Brownian motion (Okasha, 2002, pp. 31–33), and Keynes's ideas in economics (Skidelsky, 2009, p. 58). It continues to do so, with Magnani (2001, p. 94) waxing that its contributions "have guaranteed the philosophical centrality of abduction in present-day cultural, scientific and technological developments."

10. This reprimand is captured by the well-known example that no number of empirical confirmations of white swans precludes the subsequent appearance of a black one (as in Australia).

11. This is why Starbuck's (2006, p. 167) recommendations in his *The Production of Knowledge* in the social sciences are off the mark:

> 1. Journals should refuse to publish studies that purport to contradict the baseline propositions. Since the propositions are known laws of nature, valid evidence cannot contradict them.
> 2. Journals should refuse to publish studies that do no more than reaffirm the baseline propositions. Known laws of nature need no more documentation.

In the first place, as just told, even laws in the hard sciences are circumstantial. Second, the management and social science literatures have little in the way of baselines against which to judge the authenticity of a new result. Hence the appeal to endorse a policy of significant sameness. Third, even if reasonable baselines existed, a result at odds with them may signal the discovery of a boundary condition, which itself can be extremely helpful.

12. It is not being suggested that every study with null or negative results be published or considered as meaningful. A methodology for judging when to accept a null hypothesis

needs to be formulated. See, in this respect, Cook, Gruder, Hennigan, and Flay (1979), Fagley (1985), Frick (1996), Hubbard and Armstrong (1992), Lindsay (1993b, 1994), Schimmack (2012), and especially Cortina and Folger (1998) and Hauck and Anderson (1986). This topic is revisited in Section 3.5.

13. There are those who disagree with such assertions. Among them are proponents of the Austrian/neo-Austrian school of economics who disdain quantitative estimates. If the latter have such fleeting shelf lives as the neo-Austrians seem to imply, then, of course, it is pointless to make them. Obviously, I do not share this belief, particularly as I call for *multiple* estimates (and their CIs) of the phenomenon at hand, rather than relying on the typical one-off result.

14. I am indebted to Brian Haig for suggesting this example.

15. Deciding whether an article did or did not reject the null hypothesis involved the use of Sterling's (1959, footnote 2, pp. 31–32) detailed criteria. For example, when a dominant hypothesis was apparent, this decision was relatively straightforward. When an article involved two or more variables and it wasn't clear which were the more important, if H_0 was not rejected for at least 50% of these variables, the study was classified as H_0 not rejected and vice versa. A similar logic was employed for works with multiple studies within a given article.

16. Shapin (1995, pp. 294, 306) drops comparable hints about the gullibility of such thinking among some members of the social sciences.

THE IMPORTANCE OF REPLICATION RESEARCH—SIGNIFICANT SAMENESS

> Replicability is almost universally accepted as the most important criterion of genuine scientific knowledge. (Rosenthal & Rosnow, 1984, p. 9)

> Modern scientists are doing too much trusting and not enough verifying—to the detriment of the whole of science, and of humanity. ("How Science Goes Wrong," October 19, 2013, p. 13)

4.1 Introduction

The quotations opening this chapter distill the value of replication research. This value underlies Steiger's (1990, p. 176) dictum that an ounce of replication is worth a ton of inferential statistics, an attitude shared by Moonsinghe, Khoury, and Janssens (2007). Indeed, it is generally acknowledged that replication research plays a vital role in protecting the integrity of the scientific enterprise.

In light of the above, this chapter explores in greater depth how such research operates in the acquisition of valid, reliable, and generalizable knowledge. To initiate this discussion, Section 4.2 provides a brief sketch of replication's part in this process. Then Section 4.3 introduces a sixfold typology of replications, while in the lion's share of this chapter Section 4.4 explicates their roles in knowledge development. In doing so the indispensability of replication research is established. The importance of researchers replicating their own work within a given paper, or so-called internal replications, is covered in Section 4.5. Concluding remarks are offered in Section 4.6.

4.2 A Succinct Overview of Replication's Role

By determining whether original results are repeatable, replication helps guard a literature against the uncritical absorption of specious empirical findings such as Type I errors, mistakes, and fraud (Giner–Sorolla, 2012, pp. 562–564; Hubbard & Armstrong, 1994, p. 234; Hubbard, Brodie, & Armstrong, 1992, pp. 2–5; Ioannidis, 2005b, p. 0101; Simmons, Nelson, & Simonsohn, 2011, p. 1359). As seen in Chapter 3, however, an even more important role played by replication research is the establishment of stubborn facts or empirical generalizations. These, in turn, prompt the search for explanatory theories to further their understanding. The quotations listed in Table 4-1 from scholars in various fields are a resounding testament to the absolute centrality replication research commands in the production of knowledge.

Table 4-1 The Centrality of Replication Research in Science

Author(s)	*Quotations*
Allen & Preiss (1993, p. 9)	Scientific knowledge comes from replication.
Amir & Sharon (1990, p. 51)	The basic claim . . . is that before an empirical result can serve as a basis for a psychological theory, it should fulfill at least two prerequisites: the test of reproducibility and the test of generalizability. The way to conduct both these tests is basically with replication studies.
Babbie (1992, p. 334)	Replication . . . is a general solution to problems of validity in social research.
Bakker, van Dijk, & Wicherts (2012, p. 547)	At the end of the day it is all about replication.
Berthon, Ewing, Pitt, & Berthon (2003, p. 511)	Replications are an important component of research in that they convert tentative belief into more reliable knowledge.
Blaug (1992, p. 38)	Replicable results [are] the hallmark of science.
Campbell & Jackson (1979, p. 6)	Replication research can and should be a primary factor in determining the validity, reliability and generalizability of research.
Carver (1978, p. 392)	Replication is the cornerstone of science.
Chatfield (2002, p. 4)	Thus statisticians, if they wish to be regarded as good scientists . . . need to give more emphasis to *collecting more than one data set* whenever possible, as that is the route to scientifically valid and generalizable results.
J. Cohen (1994, p. 1002)	But given the problems of statistical induction, we must finally rely, as have the older sciences, on replication.

Author(s)	*Quotations*
H. M. Collins (1985, p. 19)	Replicability . . . is the Supreme Court of the scientific system.
Cumming (2012, p. 120)	Replication is at the heart of science.
Darley (2000, p. 121)	Replication is at the heart of any science.
Dewald, Thursby, & Anderson (1986, p. 600)	The replication of research is an essential component of scientific methodology.
Easley, Madden, & Dunn (2000, p. 89)	We believe that replication is essential to the conduct of good science.
Frank & Saxe (2012, p. 600)	Replication is held as the gold standard for ensuring the reliability of published scientific literature.
Galak, LeBoeuf, Nelson, & Simmons (2012, p. 943)	An effect is not an effect unless it is replicable, and a science is not a science unless it conducts (and values) attempted replications.
Guttman (1977, p. 86)	But the essence of science is replication.
Guttman (1985, p. 9)	Substantive replication is required by science in order to help ensure objectivity. Different researchers should arrive at the same results.
Haig (2014, p. 52)	That phenomena detection accords replication pride of place among its research procedures is perhaps the strongest justification of the importance of replication in science.
J. K. Hartshorne & Schachner (2012, p. 1)	There are many considerations that go into determining research quality, but perhaps the most fundamental is replicability.
Honig, Lampel, Siegel, & Drnevich (2013, p. 18)	Certainly, the low regard that management scholars have for replication represents a major hurdle in systematically advancing knowledge in the field.

(Continued)

Table 4-1 (Continued)

Author(s)	*Quotations*
Hubbard & Armstrong (1994, p. 233)	Thus, replications and extensions play a valuable role in ensuring the integrity of a discipline's empirical results.
Hunter (2001, p. 157)	Replication studies are desperately needed in order to determine facts.
Ioannidis (2012, p. 645)	Efficient and unbiased replication mechanisms are essential for maintaining high levels of scientific credibility. [Absent these] In several fields of investigation . . . perpetuated and unchallenged fallacies may comprise the majority of the circulating evidence.
Ioannidis & Doucouliagos (2013, p. 999)	Reproducibility is essential for validating empirical research.
Iyengar (1991, p. 33)	Replication is a hallmark of the scientific method.
Jasny, Chin, Chong, & Vignieri (2011, p. 1225)	Replication—the confirmation of results and conclusions from one study obtained independently in another—is considered the scientific gold standard.
Kane (1984, p. 4)	Reproducibility remains the touchstone of the scientific method.
Kline (2004, p. 247)	Replication is a critical scientific activity, one not given its due in the behavioral sciences.
Koole & Lakens (2012, p. 608)	A scientific discipline that invests in replication research is therefore immunized to a large degree against flawed research practices.
Krueger (2001, p. 21)	If practicing scientists agree on anything, it is that evidence must be replicable.
Lamal (1990, p. 31)	Replication would seem to underlie the self-correction which is presumed to be another characteristic of the scientific method. Replication is necessary because our knowledge is corrigible.

Author(s)	*Quotations*
LeBel & Peters (2011, p. 375)	Across all scientific disciplines, close replication is the gold standard for corroborating the discovery of an empirical phenomenon.
Lindsay & Ehrenberg (1993, p. 227)	Successful replication is the bedrock of scientific knowledge.
Makel, Plucker, & Hegarty (2012, p. 541)	The path to better understanding psychological science goes through replicating important findings.
Mayer (1980, p. 170)	Neither originality, logical rigor, or any other criterion was ranked as "essential" by so many natural scientists as was replicability.
Mezias & Regnier (2007, p. 284)	Replication is at the heart of the normal science model.
J. Miller (2009, p. 617)	Scientific theories are built on replicable phenomena.
Mirowski & Sklivas (1991, p. 147)	It is a commonplace that the replicability of experiments is the single-most important attribute which distinguishes scientific from nonscientific research.
Mittelstaedt & Zorn (1984, p. 14)	That which isn't worth replicating isn't worth knowing.
Moonesinghe, Khoury, & Janssens (2007, p. 0218)	As part of the scientific enterprise, we know that replication . . . is the cornerstone of science and replication of findings is very important before any causal inference can be drawn.
Neuliep & Crandall (1993, p. 21)	Replication provides the substance of inductive force to support a . . . theory.
Nosek, Spies, & Motyl (2012, p. 617)	Replication is a means of increasing the confidence in the truth value of a claim. Its dismissal as a waste of space incentivizes novelty over truth.

(Continued)

Table 4-1 (Continued)

Author(s)	*Quotations*
Popper (1959, p. 45)	Only by such repetitions [of observations] can we convince ourselves that we are not dealing with a mere isolated "coincidence," but with events which, on account of their regularity and reproducibility, are in principle intersubjectively testable.
Reid, Soley, & Wimmer (1981, p. 3)	The importance of replication in scientific research has been acknowledged for several hundred years.
S. Schmidt (2009, p. 90)	Replication is one of the most important tools for the verification of facts within the empirical sciences.
Selltiz, Wrightsman, & Cook (1976, p. 63)	Any piece of research must be repeated by other investigators before its findings can be considered as reasonably well established.
Simon & Burstein (1985, p. 15)	[Replicability] gives credibility to the conclusions of scientific research.
Singh, Ang, & Leong (2003, p. 534)	For emerging fields of inquiry, particularly in the social sciences, replication represents an essential core practice.
Tukey (1969, p. 84)	Confirmation comes from repetition. Any attempt to avoid this statement leads to failure and more probably to destruction.
Uncles (2011, p. 579)	It is widely accepted that replication is central to normal scientific investigation.
Van de Ven & Johnson (2006, pp. 806–807)	A research finding, principle, or process is judged to be robust when it appears invariant (or in common) across at least two (and preferably more) independent contexts, models, or theories. A pluralist approach of comparing multiple plausible models of reality is therefore essential for developing objective scientific knowledge.
Ziman (1978, p. 56)	From the very beginning, the achievement of consensus between different observers concerning the results of repetitions of the same experiment is fundamental to the creation of any body of scientific knowledge.

Building on this, real science consists of the onerous, time-consuming job of seeking a network of interrelated generalizations in which increasingly higher-order generalizations are developed to account for lower-level phenomena (Hubbard & Lindsay, 2013b, p. 1396). This holds for basic facts (lower-level regularities), operations (progressing from specific operations, manipulations, and methods to higher-order constructs), theories (higher-level abstractions of related facts), and grand theories (linked theoretical structures). The validity (meaning) of our theoretical structures is based on these networks. And establishing such generalizations can *only* be achieved by undertaking *well-designed* replications so as to maximize our rate of learning. A typology of these kinds of replications is presented next.

4.3 A Typology of Replications

The various kinds of replications shown in Table 4-2 are taken from Tsang and Kwan's (1999, p. 766) two-dimensional framework. The first dimension consists of three sources of data which may be employed in a study. More specifically, the first of these notes whether the same data set was used. The second and third sources examine whether the data were collected from the same or a different population.

The second dimension of Table 4-2 deals with whether changes in research methods have been introduced in the investigation. When combined, these two dimensions result in six different types of replication studies that have unique significance: (1) checking of analysis, (2) reanalysis of the data, (3) exact replications, (4) conceptual extensions, (5) empirical generalizations, and (6) generalizations and extensions. This sixfold classification will be of assistance to replicating researchers when reporting

Table 4-2 A Typology of Replications

Source of Data	*Same Measurement and Analysis*	*Different Measurement and/or Analysis*
Same data set	Checking of analysis	Reanalysis of the data
Same population	Exact replications	Conceptual extensions
Different population	Empirical generalizations	Generalizations and extensions

Source: Tsang & Kwan (1999, p. 766).

how their work compares and contrasts with prior studies. Elucidation of the role of each of these kinds of replications in the attainment of knowledge follows.

4.4 Replication Research and the Acquisition of Knowledge

4.4.1 Checking of Analysis: Determining the Accuracy of Results

This is mostly taken to be independent reexaminations of original data sets, using the same methods of analysis, the objective being to ensure the soundness of published results by "checking" if they are error free (see Table 4-2). Wolins's (1962) early study did not bode well concerning the authenticity of published empirical findings. As part of a master's thesis, Wolins had his graduate student write to 37 authors whose articles had appeared in various (unlisted) American Psychological Association journals between 1959 and 1961 asking for copies of their raw data. While unfortunately not specific, Wolins (p. 657) said he encountered gross errors capable of influencing the conclusions of three of the seven usable data sets with which he was supplied.

Another example, from Feigenbaum and Levy (1993), draws attention to the dangers of taking published results at face value. They calculated a median divergence of 22% in the estimated regression coefficients in 36 econometric replications. More recently, Bakker and Wicherts (2011) examined the prevalence of the misreporting of statistical results in a random sample of high-impact (*Development and Psychopathology, Journal of Child Psychology and Psychiatry, Journal of Personality and Social Psychology*) and low-impact (*Journal of Applied Developmental Psychology, Journal of Black Psychology, Journal of Research in Reading*) psychology journals. Based on 281 empirical articles published in these journals during 2008, Bakker and Wicherts found that 18% of statistical results were reported incorrectly and that 15% of these articles contained at least one statistical conclusion which, upon recalculation, was found to be wrong (p. 666). Error rates generally were higher in the low-impact journals.

Replication research focused on checking the accuracy of published findings should be made easier when the data sets are publicly available. For those in the accounting, economics, and finance areas, for example, there are the Compustat and Center for Research in Security Prices (CRSP) data sets. In addition, more high-quality journals (e.g., *Journal of Applied Econometrics, Journal of Business and Economic Statistics*) have policies asking for computer programs and data from authors (R. G. Anderson &

Dewald, 1994). Other journals (e.g., *Demography, Industrial and Labor Relations, Journal of Econometrics, Journal of Environmental Economics and Management, Journal of Human Resources, Land Economics, Oxford Bulletin of Economics and Statistics, Public Choice, Quarterly Journal of Business and Economics*) have editorial standards that require authors to make data and computer programs available to other researchers, or solicit replication research (Fuess, 1996).

The study by Dewald, Thursby, and Anderson (1986) arising from the *Journal of Money, Credit and Banking* (*JMCB*) Data Storage and Evaluation Project typifies this kind of research. They collected data and computer programs from a sample of authors who had recently published articles in the *JMCB* or whose work currently was under review there. Merely 8 of the first 54 (14.8%) data sets examined by Dewald et al. were judged to be sufficiently complete to permit an effort at replication. Their diligence notwithstanding, rarely did they find the reanalyses to be simple. What Dewald et al. found was that transcription, coding, data transformation and programming errors were common in the data sets they received and that a few of these errors changed the nature of the outcomes. Given this scenario, one has genuine cause for concern about the accuracy of the results reported in the works of the 42% of the sample who, despite their obligations under the *JMCB* project to make available their data sets upon request, nevertheless refused to comply. The transgression continues. To illustrate this, McCullough, McGeary, and Harrison (2006) examined the *JMCB* online archive for the period 1996–2002 to see how many authors had conformed with the mandate to supply data and code to enable a replication of their results. They found that most authors did not fulfill this requirement and that of some 150 empirical articles only 15 (10%) could *potentially* be replicated. This discouraging news is echoed in McCullough (2007). It resonates further in Ioannidis and Doucouliagos's (2013, p. 997) view that the credibility of the empirical literature of economics is modest or even low.

Erroneous results can be influential, as an example from sociology attests. Employing data from a 1977–1978 Los Angeles sample, Weitzman (1985), in a book titled *The Divorce Revolution,* claimed a 73% decline in women's standard of living after divorce and a 42% rise in men's standard of living. These numbers, from both the men's and women's perspectives, were substantially higher than previous estimates. Following a thorough reconstruction of Weitzman's original data set, Richard Peterson (1996a) found evidence of a 27% decline in women's standard of living and a 10% increase in men's. While divorce was still shown to be harmful to women and beneficial to men, these new figures are not nearly as pronounced as those reported by Weitzman.

Yet in the 11 years between its debut and the publication of Peterson's (1996a) findings, Weitzman's (1985) book enjoyed the kind of publicity only dreamed of in academic circles. Peterson recounts its impact at some length:

> *The Divorce Revolution* received considerable attention in academic, legal, and popular publications. It was reviewed in at least 22 social science journals, 12 law reviews, and 10 national magazines and newspapers. The book received the American Sociological Association's 1986 Book Award for "Distinguished Contribution to Scholarship." . . . From 1986 to 1993, it was cited in 348 social science articles . . . and in more than 250 law review articles. . . . *The Divorce Revolution* was also discussed widely in the popular press: It was cited over 85 times in newspapers and over 25 times in national magazines from 1985 to 1993. Remarkably, *The Divorce Revolution* has also been cited in at least 24 legal cases in state Appellate and Supreme courts . . . and was cited once by the U.S. Supreme Court. (p. 529)

As of August 2013, this book has managed a striking 1,654 Google Scholar citations.

Peterson (1996b) adds that Weitzman's (1985) inaccurate findings have seriously marred policy discussions about no-fault divorce. Weitzman (1996) has subsequently conceded that her results were fallacious. Indeed, she was unable to replicate her own findings.[1]

Empirical results of late in other areas of sociology likewise fail to inspire confidence in their believability. Thus, Freese (2007, p. 165) cites several examples of the ready ease with which one can call into question the replicability of "articles published in prominent sociological forums."

4.4.2 Reanalysis of the Data: Determining Whether the Results Hold Up Using Different Analytical Methods

The research considered above used the same methods of analysis on the same data sets to check the accuracy of published findings. An alternative way of doing this is to reexamine the same data set with a *different* method of analysis to see whether study outcomes are sensitive to such changes (see Table 4-2). Another example from sociology/management concerning the legitimacy of the "Hawthorne effect," a concept that shouldered the "paradigmatic foundation of the social science of work" (Franke & Kaul, 1978, p. 623), underlines the value of this kind of approach. Using multiple regression techniques, Franke and Kaul (1978) and Franke (1980) discovered that three measured variables—managerial discipline, the economic adversity of the Great Depression, and time for rest—explained most of the variability in worker productivity at the Hawthorne plant in Chicago. Later regression analyses by Stephen Jones (1992) on an augmented Franke and Kaul data set also showed no evidence of a Hawthorne effect. These findings are completely at loggerheads with the story retold for decades, and based only on visual inspection of the original experimental data and anecdotal evidence, that unmeasured changes in improved social relationships among workers and management was responsible for increased output.

Ramifications of Sections 4.4.1 and 4.4.2 for the Integrity of the Management and Social Science Literatures

That erroneous results are published in scholarly journals and monographs harms a discipline's well-being. The problem isn't just that, devoid of a replication tradition, false results permeate the literature. The problem is that they *stay* there (Nosek, Spies, & Motyl, 2012, p. 620; Uncles & Kwok, 2013, p. 1399), where they can wreak havoc.[2] This is exacerbated when, invariably, some of these results find their way into textbooks. For once in this medium, such apocrypha take on the appearance of established truths, to be passed on unwittingly in the classroom each semester.

Distressingly, some published errors become classics in a field, going for years without detection. The so-called Hawthorne effect, just discussed, is a powerful reminder of this. Another is the management example concerning the truth of Frederick W. Taylor's account of the loading of pig iron at the Bethlehem Iron Co. in Pennsylvania. As Wrege and Perroni (1974) comment, for 60 years textbooks on management and industrial psychology have cited extensively Taylor's study as an example of the revolutionary nature of his contributions to scientific management. Yet after a careful reappraisal of the records themselves, Wrege and Perroni submit that this account (together with other parts of Taylor's works) has been accepted and propagated on faith, not fact.

Similarly, Armstrong (1996) indicates that use of the Boston Consulting Group's growth-share matrix, a business portfolio planning model which Kerin, Hartley, and Rudelius (2013, p. 35) say has been used in some form by more than 75% of the largest U.S. corporations, in fact misleads decision makers and is detrimental to company performance. In addition, management science techniques are said to be faulty in their recommendations to maximize market share rather than pursuing the goal of profit maximization (Armstrong & Collopy, 1996).

Or consider the case of management guru Michael Porter's (1980) five forces model of competitive strategy, a model "taught in nearly every business school strategy course worldwide" (Honig, Lampel, Siegel, & Drnevich, 2013, p. 135). Porter's framework has generated hundreds of millions of dollars from corporate and nonprofit sector clients and made him the most frequently cited academic in business and economics (Denning, 2012, p. 3). This, despite the fact that the model has no basis in fact or logic (Denning, 2012, p. 3) and could not offer practical advice to help prevent Porter's own Monitor Group consultancy from going bankrupt in 2012 (Honig et al., 2013, p. 135).

Instances of highly influential erroneous results that have languished in textbooks in the behavioral sciences also are not difficult to adduce. In psychology they include J. B. Watson's conditioning of Little Albert and Sir Cyril Burt's "twins" research. The former example concerns the behaviorist's famous experiment, carried out on his lone subject, Little Albert, in the winter of 1919–1920, on conditioned fear responses. Shrouded in contradictions and discrepancies, Samelson (1980, p. 621) argues that "Watson's study could not have become enshrined as the paradigm for human conditioning on the basis of its hard scientific evidence." Samelson reasoned, instead, that its appeal lay in the fact that the message at that time was one the audience was predisposed toward and suited their "desire to retell a good and convincing story" (p. 622).

The controversy debasing Cyril Burt's work on the heritability of intelligence stands as yet another monument to the damage occasioned by heedlessly accepting unchecked findings. Burt's studies of identical twins raised apart not only produced the most impressive estimate of the heritability of IQ, but crucially found no correlation between the socioeconomic backgrounds of the homes wherein the twins were reared (Tucker, 1994, p. 335; see also Samelson, 1980, pp. 622–624). So nature trumps nurture. This breakthrough, Arthur Jensen opined, "will most probably secure Burt's place in the history of science" (reported in Tucker, 1994, p. 335). Despite Burt's results being praised by a clutch of exalted scholars like Jensen, Raymond Cattell, Hans Eysenck, and Richard Herrnstein, the Princeton psychologist Leon Kamin declared them to be "not worthy of serious scientific attention" (Tucker, 1994, p. 339).[3] Kamin's assessment proved to be correct; Burt was accused of being a fraud.

In sociology, van Poppel and Day's (1996, p. 500) analysis of Dutch data contemporaneous with that of Emile Durkheim's French figures raises serious concerns about Durkheim's theory of suicide, a theory that some 100 years later remains "the customary starting point for both the sociological and the epidemiological analysis of suicide." Durkheim theorized that religious differences were responsible for the higher suicide rates of Protestants versus Catholics. However, van Poppel and Day showed that these differences are ascribable to nothing more ominous than the manner in which the two religions classified deaths.

Unhappily, there is much evidence to indicate that the sanctity of the business and social science literatures cannot be taken for granted. Note, further, that all of the examples given in this section, and there are many more besides, are poster children for the dangers inherent in overgeneralizing the findings of a single study raised in Section 2.2.5. One might think that this practice would be an obvious concern even to a casual reader. Yet it continues as a pillar of the significant difference paradigm.

4.4.3 Exact Replications: Determining Whether Results Are Reproducible

Tsang and Kwan (1999, p. 766) relate that exact replications follow the same basic procedures as those used in an earlier study on a new sample from the same population. This definition mirrors that used in my own work with Scott Armstrong, itself modifying the interpretations of Stephen Brown and Kenneth Coney (1976, p. 622) as well as Reid, Soley, and Wimmer (1981, p. 7), that

> a *replication* [is] a duplication of a previously published empirical study that is concerned with assessing whether similar findings can be obtained upon repeating the study. This definition covers what are variously referred to as "exact," "straight," or "direct" replications. Such works duplicate as closely as possible the research design used in the original study by employing the same variable definitions, settings, measurement instruments, analytical techniques, and so on. An example would be repeating the study with another sample drawn from the same population. (Hubbard & Armstrong, 1994, p. 236)

Therefore, Hubbard and Armstrong's (1994) description of what constitutes a replication parallels Lindsay and Ehrenberg's (1993, p. 217) adoption of the term a *close* replication. Common to Hubbard and Armstrong's, Lindsay and Ehrenberg's, and Tsang and Kwan's definitions of replication, close replication, and exact replication, respectively, is a conscious effort to ascertain whether a real effect (correlation, association, pattern, regularity, stubborn fact) exists. Alternatively expressed, these comparably labeled definitions are all concerned primarily with whether we are dealing with a reproducible or repeatable finding versus a "one-off" result. For this reason they are best performed at an early stage in the research process. Given that no known major factor was varied, a successful replication provides support for the internal validity of the original study

(Tsang & Kwan, 1999, p. 766). A failure to replicate, on the other hand, casts doubt on the robustness or reliability of the original finding. Moreover, such a failure may signal that additional work in the area could be fruitless or at least difficult.

An example of this genre, Tsang and Kwan (1999, p. 766) suggest, is Hinings and Lee's (1971) exact replication of the Aston study of organization structure. They faithfully retained the same measurement instruments and interviewing methods of the latter, and both investigations collected their samples from companies located in the English Midlands. By and large, Hinings and Lee's results conformed with those of the Aston study; strong positive correlations were found with respect to specialization, standardization, and formalization.

It is important to dispel the widely held belief that conducting exact or close replications largely is a waste of time. This attitude follows from the misinformed view that a well-designed and well-executed study is likely to replicate (Rosenbaum, 2001, p. 223). As such, an unsuccessful replication often becomes attributed to the incompetence of the replicating researcher rather than to the fact that the original finding might be suspect (Bamber, Christensen, & Gaver, 2000; Kuhn, 1970, p. 35). This is one reason for the bias against publishing negative results (Fanelli, 2012; Hubbard & Armstrong, 1992, 1997; Lindsay, 1994; Sterling, Rosenbaum, & Weinkam, 1995) mentioned in Chapter 2.

In questioning the merits of this outlook, it is useful to list the specific factors that could produce conflicting results between two or more studies (see Hubbard & Armstrong, 1994).

- Mistakes were made in one or more of the investigations. As a number of works (e.g., Bakker & Wicherts, 2011; Dewald et al., 1986; Evans, Nadjari, & Burchell, 1990; Feigenbaum & Levy, 1993; R. R. Peterson, 1996a, 1996b; Strasak, Zaman, Marinell, Pfeiffer, & Ulmer, 2007; Weitzman, 1996; Wolins, 1962) have shown, transcription, computer programming, statistical, and other errors are commonplace (Hubbard & Armstrong, 1994, p. 239). This prevalence of errors compromises the possibility of conducting independent replications, thereby preventing researchers from building on earlier work (Begley & Ellis, 2012, p. 532; Dewald et al., 1986, p. 600; Ioannidis, 2012, p. 646).
- The original result was due to chance (Hubbard & Armstrong, 1994, p. 238). At a minimum, if the .05 critical level of statistical significance is used routinely, then one can expect 5% of the rejections of true nulls to be Type I errors. In practice, however, many more tests are calculated than are reported in the average publication. Consequently, the experimentwise alpha (i.e., the probability of obtaining one or more false rejections of the

null hypothesis) is much higher than the nominal .05 level (Hubbard & Armstrong, 1992, p. 128; Hubbard & Vetter, 1996, p. 154; Hubbard, Vetter, & Little, 1998, p. 243; Lindsay, 1997).

- The replication and/or original study lacked sufficient statistical power, so that sampling error was at fault for producing discrepant results between the various studies (Hubbard & Armstrong, 1991, pp. 10–11). This is an ever present concern that was acknowledged in the appendix to Chapter 2 and will be noted further in Chapter 5.
- The meanings of the measured variables are not stable across studies. While the same measures were used, they were interpreted differently by participants. This is a real worry in the early stages of the research process.
- The original researcher "cheated" (see Andreski, 1972, pp. 109–110; Armstrong, 1983; Bailey, Hasselback, & Karcher, 2001; Bedeian, Taylor, & Miller, 2010; S. W. Davis & Ketz, 1991; Enders & Hoover, 2004, 2006; Fanelli, 2009, 2012; Honig & Bedi, 2012; Honig et al., 2013; John, Loewenstein, & Prelec, 2012; List, Bailey, Euzent, & Martin, 2001; Walsh, 2011). Research takes place in a publish-or-perish environment. By way of recollection, Section 2.4 told in some detail of how such a hypercompetitive atmosphere could lead investigators to engage in outright fraud and other pathological research behaviors. Moreover, as will be shown in Chapter 5, business and social science journals almost never publish exact or close replications. Therefore, in lieu of the high stakes for publishing, coupled with the low probability of being caught, researchers may be tempted to submit results whose "truth" value is indeterminate (Hackett, 2005; Sovacool, 2008).[4]
- Researcher bias played a role (Hubbard & Armstrong, 1994, p. 238; Rosenthal, 1966). Inadvertent researcher bias could contaminate the outcomes of the first study and/or the follow-up, thus leading to opposing results.
- The structures and causal mechanisms stipulated in the theory are not accurate (Tsang & Kwan, 1999, p. 769). Alternatively, in the replicated study, a different set of contingencies and/or obstructive mechanisms, leading to a discrepant set of events being recorded, are at play (Tsang & Kwan, 1999, p. 769).
- The original study's conclusions may not hold up over time (Hubbard & Armstrong, 1994, p. 239). This is likely to be more of a problem the greater the time lag between the initial study and the replication. Cook and Campbell (1979, p. 79) recognized this when commenting that generalizing across time is more difficult than generalizing across persons or settings.
- Editors may show a preference for publishing contradictory, rather than confirmatory, results because these may be seen as providing "novel" evidence on the topic at hand. Accordingly, Hubbard and Armstrong (1994, p. 239) conjecture that researchers might be more inclined to submit manuscripts for publication whose results conflict with earlier findings.[5]

- The replication failed because the conditions of the replication were inexact or not "close." Disparities unknown to the researcher existed between the replication and the original work. This might arise due to differences in the execution of the study (i.e., method) and/or the presence of a dissimilar set of contextual factors. Alternatively, some ignored variations between the two studies that were assumed to be of minor influence actually had a significant impact on the results. This is not all bad news, for, depending on the circumstances, it is possible that a search for which factor(s) caused a replication to fail can be "theoretically consequential" (Open Science Collaboration, 2012, p. 658).

With respect to this last bullet point, it is important to stress that an exact or close replication does not constitute an identical one. Or as Ziman (1978, p. 56) put it: "One cannot step twice into the same river." Many dissimilarities between two studies always will exist, for example, time periods, geographic areas, researchers, ideological dispositions, economic climates, sociocultural norms, organizational structures, and so on. Moreover, because of limited journal space the methods section in an article rarely describes all the details of the procedures employed (Hubbard & Armstrong, 1994, p. 241; Hubbard & Little, 1997, p. 41; Latham, Erez, & Locke, 1988, p. 754; Nosek et al., 2012, pp. 624–625). Of note, the crucial effects of "method specificity" were revealed in the Latham et al. (1988) investigation which was jointly designed by two antagonists to a debate concerning the effects of participation on goal commitment and performance. At the heart of the matter was the fact that the two disputants—Gary Latham and his coworkers and Miriam Erez and her colleagues—had obtained markedly different findings. Playing the role of mediator, Edwin Locke confessed his surprise at how a number of small differences between two studies can add up to yield contradictory results (Latham et al., 1988, p. 769). From this perspective the advice of Mellers, Hertwig, and Kahneman (2001), as well as Koole and Lakens (2012), for conducting what they call "adversarial collaborations" is helpful.

In summary, given all of the factors that can lead to conflicting findings, performing a successful close replication is far from guaranteed and should never be interpreted as trivial research. Dewald et al.'s (1986) experience in trying to reanalyze others' results in the *JMCB* project substantiates this viewpoint. They complained that such work rarely was straightforward and that in one case they were unable to replicate the results of an article even with the active assistance of the original author. For these reasons, Lindsay and Ehrenberg (1993, p. 220) are justified in calling the first successful replication a dramatic outcome.

Together with deciding whether a potentially stable result exists, in theory-testing situations a close replication often has more epistemic value than many realize because of the distinction between accommodation and prediction (Tsang & Kwan, 1999, pp. 768–769). Philosophers of science accord a theory more corroboration or epistemic support for making a successful prediction than for explaining already existing facts (Popper, 1963, ch. 1). In this connection, Tsang and Kwan (1999, p. 769) cite Maher's (1988) story of Dmitri Mendeleev. He constructed a theory of the periodic table to account for all of the 60 known elements in chemistry, resulting in the scientific community not being terribly impressed with his achievement. However, when Mendeleev used his theory to predict the existence of two unknown elements that were later independently confirmed, the Royal Society awarded him a Davy Medal.

As shown in Section 2.4, accommodation is a fact of life in the management and social sciences. To reiterate, researchers faced with null results are pressured to engage in behaviors geared toward the production of $p \leq .05$ findings. The problem here is that the results may be no more than by-products of the research method and/or due to chance, researcher biases, or cheating/fudging (Andreski, 1972, pp. 109–110; Bedeian et al., 2010; Fanelli, 2009; Greenwald, 1975; Hubbard & Armstrong, 1994; N. L. Kerr, 1998; Leamer, 1983; Lindsay, 1994, 1997; Simmons et al., 2011; Stroebe, Postmes, & Spears, 2012). Viewed against this backdrop, in a theory-testing situation it is not an overstatement to claim, as Tsang and Kwan (1999, p. 769; see also Ray & Valeriano, 2003, p. 82) do, that the first successful independent replication provides a theory with "a quantum leap in credibility."

As the discussion thus far in Section 4.4 has laid bare, there are huge benefits to be had simply by checking the integrity of the results of (even milestone) studies. And only replication research is capable of meeting this challenge.

After securing a few successful close replications, the value of further such research diminishes.[6] The real goal is to seek *consistency* of results in a *diverse* set of studies (Rosenbaum, 2001, pp. 223–224; Rosenthal, 1990, p. 5). Thus, it becomes necessary to introduce deliberate variations in some fairly major aspects of the inquiry to determine the range of conditions under which a result does or does not hold, or whether the finding is due to the particular manner in which the study was operationalized. This leads to a discussion of the part played by conceptual extensions in scientific advancement.

4.4.4 Conceptual Extensions: Determining Whether Results Hold Up When Constructs and Their Interrelationships Are Measured/Analyzed Differently

Tsang and Kwan (1999, p. 767) state that conceptual extensions are replications which use different procedures from those employed in the earlier study. These differences lie primarily in how theoretical constructs are measured and the way in which they interrelate with other constructs. In short, conceptual extensions are replications dealing with the topic of *construct validity* (Cronbach & Meehl, 1955). This is where the same theory holds for the replication and the original investigation (Tsang & Kwan, 1999, p. 767) but the operational details are varied in an effort to gain triangulation. Or as Anastasi (1982, p. 144) mentions, construct validation involves the collection of information from a number of sources which illuminate the development of the construct(s) of interest as well as its manifestations. This information on construct validity can only be gleaned by replications assessing a construct's convergent, discriminant, and nomological validity.[7] These different aspects of construct validity are discussed below.

Any study attempting to capture the essence of some higher-order construct—say, political apathy—must do so in a specific fashion. Given that there are numerous ways of measuring political apathy, the question arises as to whether the results of a study are due to the particular representation of the construct, that is, to method specificity. So it is necessary, employing conceptual extensions (or constructive replications), to see if a result is robust to changes in the specific instrument used (Brinberg & McGrath, 1985, p. 122). Note that this demonstration of *convergent validity* (D. T. Campbell & Fiske, 1959) basically revolves around the idea that two or more approaches to measuring the "same" hypothetical construct should correlate highly with each other and predict similar results. This is because the many specific ways in which an underlying construct *could* be measured all hope to embody its prototypical features.

As an example, consider Rumelt's (1974) well-known study on the positive impact of company diversification on economic performance. In demonstrating this connection, Rumelt adopted an ordinal measure of diversification when he divided organizations into four groups: single-business, dominant-business, related-diversified, and unrelated-diversified. Using a continuous measure of diversification, the Herfindahl index, Montgomery (1982) found an almost monotonic increase between this index and Rumelt's single-business through unrelated-diversified groups, thus providing evidence of convergent validity between the two different measures of diversification.

A further example, this time from psychology, also is instructive. Van Rooy, Rotton, and Burns (2006) looked at the convergent validity of three measures of aggressive driving—the Driver Anger Scale, Driver Vengeance Questionnaire, and Driving Behavior Inventory. The high correlations between these three scales of .87 to .95 showed strong evidence of said validity.

The rationale underlying convergent validity is based on employing measurement schemas that are maximally different (Iacobucci & Churchill, 2010, p. 259). However, if the method of data collection was the same in a series of replications, and it is suspected that different ways of gathering data may affect the outcome, then this needs to be addressed (Cook & Campbell, 1979).

In discussing research dealing with achievement motivation, for example, Selltiz, Wrightsman, and Cook (1976, p. 177) tell of a case where findings were affected by the manner in which the data were obtained. Pointedly, most studies measuring achievement motivation had used an instrument whereby a scoring system was applied to stories from responses to Thematic Apperception Test (TAT)–type pictures. Unfortunately, paper-and-pencil questionnaires for gauging achievement motivation, relying on respondents checking their levels of agreement and disagreement with an array of items, generally have shown a scant relationship with scores from TAT stories.

On the other hand there are positives to be taken in this connection. For example, Locke (2007, p. 878) admits that many commentators on the success of his laboratory studies on goal setting were skeptical of their transference to the "real world of work." The good news was that Gary Latham, who was to become Locke's lifelong research colleague, had replicated independently the importance of goal setting in field studies. These replications show that studies of goal setting are resilient to changes in the methods of data collection.[8]

Or see Card and Krueger's (1994, 1998, 2000) accounts on the effects of minimum wage laws on employment in fast-food chains in New Jersey and eastern Pennsylvania, as summarized by Rosenbaum (2001, p. 226). In 1992 New Jersey raised its minimum wage by some 20%, an initiative (higher price for labor) that might be expected to put people out of work. Using survey data, Card and Krueger (1994) found no evidence supporting this claim. In response to objections about their findings, Card and Krueger (1998, 2000) replicated their study utilizing payroll records from the U.S. Bureau of Labor Statistics and found similar results to their earlier effort. Their studies illustrate convergent validity between two different methods of data collection—primary and secondary—on the same construct (unemployment).

The principle of *discriminant validity* says that it is possible to differentiate empirically the target construct from other similar ones, as well as to specify how it is unrelated to these (Campbell & Fiske, 1959). That is, we are able to suggest which other variables/constructs are *correlated* with the construct of interest (and why), and how others should be *uncorrelated* with it (Kerlinger, 1973, p. 462). To illustrate this, Kerlinger (1970) showed that a scale to measure conservatism did, as predicted by theory, correlate substantively with the constructs of authoritarianism and rigidity, but not with measures of social desirability.

Or reflect on Saxe and Weitz's (1982) development of the Selling Orientation-Customer Orientation (SOCO) scale to determine the extent to which salespeople engage in customer-oriented selling. Consisting of 24 items, the SOCO scale indicates particular actions a salesperson might take when interacting with consumers. By happenstance, like Kerlinger (1970), Saxe and Weitz's investigation also involved respondents filling out a social desirability scale. The low correlations between the latter and the SOCO scores were taken as support for discriminant validity, that is, the responses of salespeople apparently were not biased by the need for approval (Saxe & Weitz, 1982, p. 345).

Finally, consider Shimp and Sharma's (1987) construction of an instrument called CETSCALE, designed to measure Consumers' Ethnocentric Tendencies regarding the purchase of foreign- as opposed to American-made products. At its core this 17-item scale assesses consumer ethnocentrism, that is, the degree to which buyers believe that it is inappropriate, unpatriotic, and even immoral to purchase goods from abroad because of the deleterious consequences for the domestic economy/society. Indications of CETSCALE's discriminant validity are seen in the positive correlations between the instrument and measures of dogmatism, politico-economic conservatism, and patriotism (Shimp & Sharma, 1987, p. 284).

Conceptual replications are key to establishing *nomological validity,* which is concerned chiefly with theory development. Or as Cronbach and Meehl (1955) emphasized, the very meaning of a theoretical entity/construct depends on its relationships with other such entities in what they described as a nomological, or theoretically intricate (Selltiz et al., 1976, p. 176), net. In particular, this form of validity focuses on the theoretical predictions emanating from such a network.

Jaworski and Kohli's (1993) high-impact article (based on citation counts) on the idea of *market orientation*, which refers to organization-wide marketing intelligence generation, dissemination, and responsiveness to it, unfortunately yields only partial evidence of nomological validity. While strong support for the hypothesized effects of a market orientation on employees' organizational commitment and *esprit de corps* were found, that for arguably the most compelling implication of all, business performance, was equivocal. On the other hand, Narver and Slater's (1990) equally authoritative piece (based also on citation counts) reported a substantial positive effect between market orientation and business profitability.[9]

Referring back to Saxe and Weitz's (1982) SOCO scale, tests concerning its nomological validity are confirmed in the predictions of the relationships between consumer-oriented selling and the ability of salespeople to help their customers and in the quality of customer-salesperson interactions. Saxe and Weitz admit, however, that these results are not strong. Similarly, evidence supporting the nomological validity of Shimp and Sharma's (1987, p. 280) CETSCALE, measuring the effects of consumer ethnocentrism on buyers' beliefs, attitudes, purchase intentions, and behaviors toward foreign-made products, was only moderately predictive.

Replication Research and Theory Development

As seen above, conceptual extensions play an integral role in theory development by showing evidence of a construct's convergent, discriminant, and nomological validities. This motif is sketched further in what follows.

Examine Figure 4-1. The left side of this figure consists of the intangible, abstract world which is not amenable to our sense modalities. It contains a network of unobservable hypothetical constructs (Cs), or mechanisms, as indicated by the single lines between them, some of which, however, are connected by double lines to the empirical world. In short, this network constitutes a theory (model) whose purpose is to describe, explain, predict, and possibly control the phenomena sometimes observed in the empirical world shown on the right side of Figure 4-1.

Note the central role played by replication research in Figure 4-1. First, it is replication research which is essential to the initial measurement, and further refinement, of the constructs themselves. Second, it is replication research which is responsible for monitoring the linkages (theoretical consistency) between these constructs. Third, it is replication research which judges the adequacy of this system of constructs for explaining *some* of what we see in the world around us. In summary, replication research is the unavoidable conduit for scientific advance.

Figure 4-1 Replication Research and Theory Development

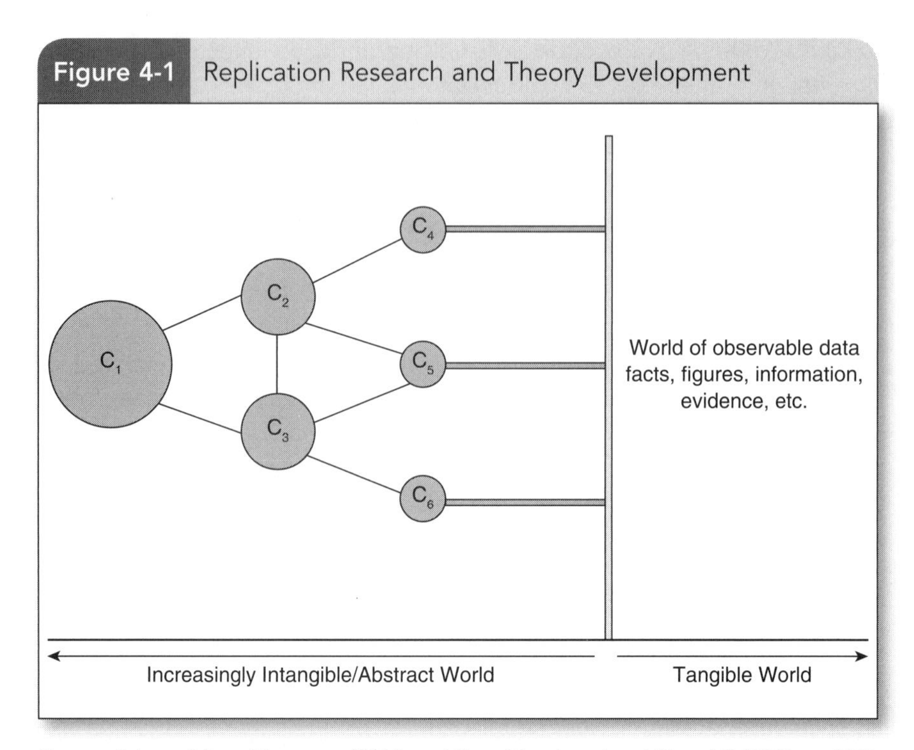

Source: Adapted from Margenau (1961, p. 16) and Iacobucci and Churchill (2010, p. 252). The Cs refer to hypothetical constructs within a theory (model).

Is Construct Validity Taken Seriously?

Despite the provision of examples of conceptual replications in this section, it would be deceiving to leave the impression that construct validation is a serious pursuit in the significant difference paradigm. It is not. Bolstering this charge is Peter's (1981) analysis of the construct validation practices in marketing following a review of more than 450 articles and research notes published in the *Journal of Marketing Research,* the mouthpiece for promoting assiduous methodological habits within the discipline no less, for the period 1973–1979. He found that only 12 of these works were concerned with validation issues. In prosecuting his case, Peter (p. 142) enunciated that few if any measures in marketing could satisfy demanding construct validation criteria.

Bearden, Netemeyer, and Mobley's (1993, p. 3) coverage of 127 multi-item measures in their *Handbook of Marketing Scales* echoes Peter's (1981) reservations when acknowledging that the inclusion of a scale in their book did not guarantee an acceptable level of quality. Bruner and Hensel (1994, p. xii) similarly lament that the vast majority of the authors of 588 marketing scales in their collection paid no attention to validity issues.

Finally, there seems to be no end of opportunities for constructive replication research in the strategic management area, where it is a rarity. Based on an investigation of data-based articles published in the *Academy of Management Journal, Administrative Science Quarterly, Management Science,* and *Strategic Management Journal* over the period 1998–2000, Boyd, Gove, and Hitt (2005, p. 244) concluded that few attempts to examine the convergent validity of constructs are reported. Worsening matters, Boyd et al. further told that most constructs were operationalized using only a single indicator. Indeed, Scandura and Williams's (2000, p. 1259) study, through a content analysis of articles in the *Academy of Management Journal, Administrative Science Quarterly,* and *Journal of Management* for the periods 1985–1987 and 1995–1997, reported that the data for the latter period indicate *lower* levels of internal validity, external validity, and construct validity in management research.

Further empirical grounds for this pessimism over the construct validity exhibited by measurement instruments in the significant difference school also can be invoked. Take Cote and Buckley's (1987) article. They examined the construct validation results of 70 published data sets using multitrait-multimethod (MTMM) matrices and spanning a large variety of social science disciplines, measurement methods, and constructs. By way of confirmatory factor analysis (CFA), Cote and Buckley discovered that, on average, the measures in these 70 data sets contained 41.7% trait (construct) variance, 26.3% method (systematic error) variance, and 32.0% random error variance. In other words, fully 58.3% of the variance in these constructs is due to measurement (systematic + random) error. Given such results, Cote and Buckley urged that more stringent validation of constructs is essential. But as we have seen, for the most part this is not taking place.

Bagozzi and Yi's (1991) findings likewise are bothersome. Working with four published data sets, they compared the results of three alternative ways—Campbell-Fiske procedure, CFA, and the direct product model—of analyzing MTMM matrices. Their reanalyses of the four data sets led them in two cases to results which were "markedly different . . . from those stated by the original authors," to an extent precluding any solid basis for establishing convergent and discriminant validity (p. 438).

4.4.5 Empirical Generalizations: Determining Whether Results Hold Up in New Domains

Assuming that the initial study is well designed, in a close replication one typically expects the same result to occur, subject to the reservations listed in Section 4.4.3. A failure to replicate is a surprise and the value of such a finding increases the closer the replication is to the original study, that is, increased doubt is cast on the reliability of the original findings. Conversely, in a replication focusing on the *external validity,* or empirical generalizability, of results, one is not nearly so sure of verifying the original findings. The goal is to see whether the same result holds across (sub) populations *despite* the differences in conditions that are present (or are introduced). What kinds of differences constitute an empirical generalization? According to Shadish et al. (2002, p. 83), these include differences in persons, settings, treatments, and outcomes.

Persons. The results of any study might be influenced by the characteristics of the respondents. Potential variables here include, from a multitude of others, age, sex, position, socioeconomic background, years of employment, personality dimensions, and reasons for being in the study. The latter is especially important in that those who wish to participate might (and often do) differ from those who refuse in ways which can affect the generalizability of the results (i.e., the nonresponse bias problem; see Blair & Zinkhan, 2006, on this issue). In addition, the use of students in experiments is the center of much controversy in terms of whether the results found for them generalize to other populations. This last point is noteworthy and as a consequence is looked at more closely in an appendix to this chapter.

Settings. In marketing, for example, this dimension might involve such things as the kind of industry, company size, culture (country), function, extent of interdependence among units, competitive environment, technology, market orientation, and strategy.

Treatments. Some relationships occur only within certain limits of values for independent variables; as such, any attempt at making a generalization needs to acknowledge the level of the stimulus (treatment variable) and how this affects the relationship (Shadish et al., 2002, pp. 49–50).

Outcomes. Cause-effect relationships may vary over different treatment outcomes. For example, the outcomes of an attempt to boost education performance among high school students could include the likes of improved grade point averages, more successful results on standardized tests, higher graduation rates, increased class attendance, decreased truancy, better employment opportunities, and so on. Shadish et al. (2002, p. 89) argue that consultation with stakeholders in the design of the study helps anticipate questions over the generalizability of research outcomes.

The plan is to show whether the changes in conditions that are introduced are irrelevant to the relationship of interest. In general, a result accrues a higher level of empirical generalization as the differences in conditions increase. However, one must deliberate carefully about the number of items to vary. If several factors are altered and the results do not replicate, we would be at a loss to explain this failure, requiring us to begin anew (Barlow & Hersen, 1984; Monroe, 1992a). In addition, even when conducting a replication, some factors should remain close for several studies in order to gauge their robustness (S. Schmidt, 2009, p. 99).

An example of a successful empirical generalization is seen in Netemeyer, Durvasula, and Lichtenstein's (1991) ability to extend to France, Germany, and Japan the results of Shimp and Sharma's (1987) 17-item measure of consumer ethnocentrism (CETSCALE) developed in the United States. Likewise, using a different set of products as well as an Australian sample, Clarke and Soutar (1982) nevertheless successfully generalized Kasulis, Lusch, and Stafford's (1979) American results with regard to consumer acquisition patterns of durable goods. Again, Hubbard and Little (1988) replicated Furse and Stewart's (1982) conclusion that promises of donations to charitable organizations fail to increase mail survey response rates over a no-incentive control group. They did so despite employing different samples (midwestern versus national) and selections of charitable institutions (unconstrained choice versus 10 options provided). Further, in Saxe and Weitz's (1982) account of the SOCO scale, salespeople themselves estimated their own degree of customer orientation. Noting, sensibly, that it would be more appropriate if buyers (not sellers) were the ones doing the evaluation of customer orientation, Michaels and Day (1985) replicated Saxe and Weitz's study from just such a perspective. In obtaining results similar to those of the latter, they reported that the SOCO scale appears to work as well with customers as with salespeople.

A couple of additional successful generalizations conclude this section. The first is Helmig, Spraul, and Tremp's (2012) replication of Jacobs and Glass's (2002) results demonstrating that media publicity constitutes a rare occurrence for most nonprofit organizations. Helmig et al.'s data were collected in Zurich, Switzerland, while Jacobs and Glass's were based on a New York City sample. The second case in point is Müller, Kroll, and Vogt's (2012) telling of how they were able to generalize from a number of previous studies, all employing rather artificial scenarios, the ruggedness of the "compromise effect"—the phenomenon that product choices situated between extreme alternatives (e.g., high and low prices) are viewed as more attractive and therefore more likely to be bought. They did so by introducing more realistic situations, for example, working with experienced consumers, real payments, binding choice conditions, and so on. All of the above examples illustrate the desirability of empirically generalizing results over variations in persons, settings, treatments, and outcomes. Indeed, because of its importance, the topic of empirical generalization receives further coverage in Chapter 6.

4.4.6 Generalizations and Extensions: Determining Whether Results Hold Up in New Domains and With New Methods of Measurement and/or Analysis

Described by Tsang and Kwan (1999) as popular, the sixth kind of replication is generalizations and extensions. These involve changes in both research procedures and populations under study. As Tsang and Kwan (p. 768) allude, however, this category is nettlesome because the broadness of the changes incorporated into a follow-up investigation make it difficult to classify whether the latter is a valid replication or a different paper altogether.

Following this line of inquiry, Murray Lindsay and I (Hubbard & Lindsay, 2002, p. 399) have argued that many generalizations and extensions in fact are mainstream studies dealing with theory testing. In other words, the data from such research focuses more on theory extensions than on extensions to previous *empirical* findings.

4.5 The Role of "Internal" Replications

Researchers wishing to determine the replicability of their findings can do so using "internal" analyses. While better than no "replication" at all, some internal analyses—like (double) cross-validation, the bootstrap, and the jackknife—involve the use of data from the sample in hand and yield exaggerated claims about the replicability of results (Thompson, 1994).

Going beyond this, aspects of both exact and conceptual replications can and should appear in a single, well-designed internal replication (cf. Barwise, 1995; Easley, Madden, & Dunn, 2000; Ehrenberg & Bound, 1993; Evanschitzky & Armstrong, 2013; Lindsay & Ehrenberg, 1993; Uncles & Kwok, 2013). For example, a single study might replicate the design of an earlier piece of research but, as advised in Section 4.4.4, also include a new measurement instrument (in addition to the original) to begin examining their convergent validity. To take another example, an investigator might collect data on, say, the supermarket food expenditure patterns exhibited by male versus female shoppers, by higher versus lower income groups, by people in different geographic areas, and so on. The basic point is that replication is not someone else's job; *every* researcher should consider incorporating replications in the design of his or her studies.

I sanction wholeheartedly as an integral component of the significant sameness model the policy of authors building internal replications into their studies and of replicating their own work in ensuing research. Having said this, a major word of restraint is in order: I value particularly the findings of external or *independent* replications which require a new sample. Researchers replicating their own work are more likely to confirm earlier results. In discussing this problem of "correlated replicators," Rosenthal and Rosnow (1984, p. 183) question the worth of 10 replications performed by the same researcher versus 10 replications each conducted by a different investigator.

Some journal editors, like McCabe (1984) and Monroe (1992a, 1992b), agree. In fact, McCabe stipulates that the editorial platform of the *Quarterly Journal of Business and Economics* is to reject the publication of replications carried out by the same author(s) because "clearly, it is the independence of the two works that makes replication valuable" (p. 79). Thompson (1994) concurs that external evaluations of the replicability of outcomes are inherently superior to their internal cousins. So do Cook and Campbell (1979, p. 79), Ioannidis and Doucouliagos (2013, p. 1001), Kmetz (2011, p. 185), and Koole and Lakens (2012, p. 609). As evidence of this confirmatory bias among correlated replicators, of the 266 replications found in Hubbard and Vetter's (1996) study of five management disciplines, some 51 involved authors replicating their own work. Of these 51, only 5 (9.8%) reported findings that conflicted with earlier results.[10] On the other hand, of the 215 extensions conducted by independent (different) researchers, 116 (54%) conflicted with earlier claims (Hubbard, 1995a, p. 458). Additional evidence of a verification bias among researchers replicating their own work, this time in psychology, is afforded by Makel, Plucker, and Hegarty (2012, p. 539). These authors discovered that only 3 out of 167 (1.8%) replications where at least one author was on both the initial and follow-up articles failed to replicate any of the original results.[11]

Fisher on Replication

As it happens, Fisher was a proponent of internal replications. According to Fisher Box (1978, p. 142), "Fisher had reason to emphasize, as a first principle of experimentation, the function of appropriate *replication* in providing an estimate of error." Indeed, Fisher Box (p. 142) states that the term *replication* was coined by Fisher: "The method adopted was replication, as Fisher called it; by his naming of what was already a common experimental practice, he called attention to its functional importance in experimentation." And he valued especially the part played by conceptual replications: "We may, by deliberately varying in each case some of the conditions of the experiment, achieve a wider inductive basis for our conclusions, without in any degree impairing their precision" (Fisher, 1966, p. 102).

From the foregoing, it is easy to imagine Fisher sympathizing with those asking for greater stress on replication. This being so, it is ironic that his encouragement of such work has been overlooked, while at the same time his widely misunderstood *p*-values flourish.

4.6 Conclusions

Most affiliates of the business and behavioral sciences have a very limited understanding of what replication research is about. All too often replication is viewed as being restricted to checking the accuracy of the results of prior studies and performing what have been called exact or close replications. Yet it has been shown with some choice examples how even this basic interpretation of replication research plays an essential role in scientific advancement.

Importantly, it was clarified that there is much more to replication research than this narrow (though exigent) perspective held by disciples of the significant difference paradigm. The truth of the matter is that replication research percolates every crevice of scientific discourse. As revealed at some length in Section 4.4, different kinds of replications provide the means necessary for judging the validity of our generalizations about phenomena of interest. Thus, exact replications allow us to appraise the internal validity of a study. They further enable the establishment of facts and the causal theories underlying them. Conceptual replications extend this development of causal theory by examining the validity of hypothetical constructs and their interrelationships. Specifically, conceptual extensions make possible the evaluation of a construct's convergent, discriminant, and nomological validity. What could be more important than this? Further, by permitting the investigation of whether the same (similar) findings hold up across (sub)populations, empirical generalizations assess the neglected topic of a study's external validity.

In sum, it is for good reason that replication research is said to lie at the heart of scientific progress.

It strains credulity, therefore, to find out in Chapter 5 that members of the significant difference paradigm in general are antagonistic toward conducting and publishing replications. This is unfortunate, for until the practice of replication research is accorded the respect commensurate with its role in knowledge development, the validity, generalizability, and usefulness of findings in the management and social science literatures must continue to be met with a jaundiced eye.

Notes

1. Unsurprisingly, Weitzman (1996) distances herself from the blame for her erroneous results. As Richard Peterson (1996b, p. 539) observes: "She claims that mistakes were made by computer experts, who botched the original analysis, did not properly check the data or analysis, incorrectly constructed the data files, and finally lost them." Peterson (1996a, p. 534) sees matters differently: "I have been unable to discover how Weitzman's results could have been obtained. The most likely explanation is that errors in her analysis of the data were responsible for producing her results."

2. This not only is true in the management and social sciences, but applies equally to findings in the medical and biological fields as well (W. W. Stewart & Feder, 1987). Illustrative of this is Glenn Begley and Lee Ellis's (2012, p. 532) account of how some nonreplicable landmark published articles on cancer treatments "had spawned an entire field, with hundreds of secondary publications that expanded on elements of the original observation, but did not actually seek to confirm or falsify its fundamental basis." There is, then, every incentive to check the repeatability of initial outcomes. See also note 4.

3. In displaying behavior of the most repugnant kind, it must be added that these same scientists (and others) were keen to stifle and discredit Kamin's observations.

4. Even when the odds of being caught ostensibly are higher, as in the biomedical sciences, Broad and Wade (1982), in their book *Betrayers of the Truth,* contend that scientific misconduct is rife. In making their case these authors investigated the area of biological testing, in which the Food and Drug Administration or the Environmental Protection Agency runs *external checks* on the validity of thousands of test results submitted by industry concerning the safety of new foods, drugs, and pesticides. Broad and Wade (p. 81) recited that the FDA and EPA "[turn] up bad science, outright error, and deliberate fraud by the bucketful." They further surmised that for every major case of fraud that becomes public, 100,000 other major and minor cases combined remain hidden in the scientific literature (p. 87).

Judson's (2004, p. 39) updated critique on this matter, in his book *The Great Betrayal: Fraud in Science,* is even more depressing:

> A scientist wanting to check another's work by direct repetition . . . encounters formidable obstacles as the enterprise is organized today. In a report in the *New England Journal of Medicine* in 1987, a committee of scientists and a philosopher analyzed a flagrant case of fraud that had just been uncovered at the University of California in San Diego. They pointed to the problem. "Replication, once an important element in science, is no longer an effective deterrent to fraud because the modern biomedical

> research system is structured to prevent replication—not to ensure it. It appears to be impossible to obtain funding for studies that are largely duplicative," they wrote. Academic credit . . . "tends to be given only for new findings."

Novel findings, recall, are a mainstay of the significant difference model. See also Sarasohn's (1993) *Science on Trial* as an earlier indictment of the behavior of some members of the biomedical research community. Deplorably, evidence of scientific malfeasance in this vital area runs on apace (see, e.g., Hotz, 2007; Martinson, Anderson, & de Vries, 2005; Titus, Wells, & Rhoades, 2008).

5. Bear in mind, however, the countervailing argument to this made earlier concerning allegations of researcher incompetence and the bias against publishing negative results.

6. For example, Rosenthal (1990) provides the following rule of thumb: The first close replication doubles our information, the fifth adds 20% to our information level, and the fiftieth replication adds only 2%.

7. The ideas of convergent and discriminant validation were introduced into the literature, via their multitrait-multimethod matrix, in Donald Campbell and Donald Fiske's (1959) seminal contribution.

8. These findings fit with Locke's (1986) extensive review of research in the fields of organizational behavior and human resource management showing that the results of laboratory studies generalize quite well to the field. Locke's conclusion is validated by Mitchell's (2012, p. 111) demonstration that effect sizes in laboratory and field settings in industrial-organization psychology are highly ($r = .89$) correlated.

9. According to data accessed from the Web of Science, Jaworski and Kohli (1993) has been cited some 540 times from 1993 to 2008, while Narver and Slater (1990) garnered 563 citations (Hubbard, Norman, & Parsa, 2010, p. 680).

10. Yet even here, in two cases the "conflicting" results were welcomed, pointing as they did to more harmonious race relations. Thus, the actual incidence of authors contradicting their own previous work is only 5.9%.

11. Although going against the grain, Makel et al. (2012, p. 539) also encountered a high proportion (64.6%) of successful replications among independent researchers.

APPENDIX TO CHAPTER 4

The Use of Student Samples in the Management and Social Sciences

The use of student samples is prevalent in the management and social sciences. For example, Aldag and Stearns's (1988, p. 256) content analysis of 226 empirical articles published during 1986 in five leading management journals demonstrated that, on average, 30.1% used students as test units. There was considerable variation in this figure among the journals themselves: *Academy of Management Journal* (13.7%), *Administrative Science Quarterly* (9.1%), *Journal of Applied Psychology* (23.8%), *Journal of Management* (28.6%), and *Organizational Behavior and Human Decision Processes* (70.2%). More recently, Collier and Bienstock's (2007) article, discussed further in Chapter 6, found that 25.7% of empirical studies published in the *Journal of Marketing* over the period 1999–2003 relied on student samples. For the *Journal of the Academy of Marketing Science* this number is 27.7%, while for the *Journal of Marketing Research* it rose to 60.5%.

One of marketing's most venerable periodicals, the *Journal of Consumer Research* (*JCR*), is particularly dependent on the use of student samples. Consider the data in Table 4A-1, which I compiled from a simple random sample of an annual issue of *JCR* for every year from 2001 to 2010. This procedure identified a total of 127 empirical studies and 145 sample types (some works used more than one kind of sample). The data reveal that only two (1.4%) sampling plans claim to be of a "representative" nature. A further 35 (24.1%) sampling plans use nonstudent convenience samples. Fully 103 (71.0%) sampling plans employed student samples, the vast majority being undergraduates. Indeed, it is not uncommon for the results of almost all empirical work published in *JCR* to be based on student opinions, as the April 2011 issue attests.

Lynch (1999), a highly respected academic consumer behavior and marketing scholar, defended the use of student samples in experimental (and presumably other) research. In this he was responding to Winer's (1999) apprehensions over the generalizability (external validity) of student sample results to those obtained from "real people" (other subpopulations). Lynch brushed off Winer's concerns on the grounds that the results from a sample of real people are no more or less generalizable than those from a group of students.

Table 4A-1 Types of Samples Employed in the *Journal of Consumer Research:* 2001–2010

	Undergraduate Students	*Graduate Students*	*"Students"*[a]	*Convenience*[b] *(Nonstudent) Samples*	*"Representative" Samples*	*Miscellaneous*[c]	*Total*
Number	82	6	15	35	2	5	145
%	56.6	4.1	10.3	24.1	1.4	3.4	100.0

[a]No distinction is made between undergraduate and graduate students.

[b]This includes 10 qualitative studies.

[c]Incorporates works which state "respondents" or "participants" (without further defining them) and citation analyses.

I sympathize with Winer's position. Whether student and real people attitudes/behaviors are similar or not ultimately is a matter for empirical inquiry. On this subject, Robert Peterson's (2001) results are sobering. His study was a second-order meta-analysis of 34 meta-analyses (cumulative sample size in excess of one million) involving students in social science research. Peterson concluded that it is imperative to replicate "research based on college student subjects with nonstudent subjects before attempting any generalizations" (p. 450). More recent empirical evidence confirms this view (Peterson & Merunka, 2014). Adding fuel to this fire, Henrich, Norenzayan, and Heine's (2010) contribution exposes the folly of trying to generalize the results of samples drawn from Western, Educated, Industrialized, Rich, and Democratic (WEIRD) societies—never mind U.S. undergraduates—to other contexts.

THE IMPORTANCE OF REPLICATION RESEARCH—SIGNIFICANT DIFFERENCE

> It is not the case that replications . . . are not important for the field. They are important. However, they do not belong in *JMR* [*Journal of Marketing Research*] if they take the place of high-impact articles. (Huber, 2007, p. 2)
>
> "Replication" should not be a dirty word. (Helfat, 2007, p. 188)

5.1 Introduction

This chapter evaluates the role of replication in the significant difference paradigm. I begin in Section 5.2 by reporting the proportion of empirical studies published in the management and social science literatures that is devoted to such work. In doing so it is made clear that in the significant difference paradigm this crucial scientific activity is something that receives much lip service but very little attention. This is followed in Section 5.3 by the provision of estimates on the critical issues as to whether replications typically support or contradict original results. Section 5.4 speaks to the timeliness of replication research because this has major bearings on its usefulness.

Thereafter, the largest segment of this chapter, Section 5.5, offers a number of reasons for the lack of replications found in print, with many of them indicative of widespread faulty thinking among constituents of the significant difference paradigm. Then, Section 5.6 presents figures on the publication incidence of commentaries, a proxy for the "openness" of the scientific community, in the management disciplines. Section 5.7, summarizing the key points discussed throughout, concludes the chapter.

5.2 The Publication Incidence of Replication Research in the Management and Social Sciences

As noted earlier, the core of the significant difference model is its focus on single works concerned with novel research. In this telling, the outcomes of significance tests in a one-off study are taken as satisfactorily dealing with the research question(s) at hand. With these now disposed of, the investigator moves on to another piece of original research.

Given this ethos it is not surprising to find that published replication research is uncommon. For example, Sterling (1959) did not discover a single replication in his inspection of four leading psychology journals listed in Section 2.4. Likewise, Bozarth and Roberts (1972) revealed that less than 1% of articles featured in three prominent counseling psychology journals between 1967 and 1970 were replications. Wilson, Smoke, and Martin (1973), meanwhile, reported that of 76 articles using significance testing featured in the *American Journal of Sociology, American*

Sociological Review, and *Social Forces* between July 1969 and June 1970, 4 (5.3%) were replications. In yet another survey, Kmetz (1998), cited by Kline (2004, p. 75), used an electronic database to investigate the frequency of papers specifically described as replications in some 13,000 works in the field of organizational science and 28,000 articles in economics. These frequencies were infinitesimal: .32% and .18%, respectively. When employing the entire history of the 100 psychology journals with the highest 5-year citation impact factors, Makel, Plucker, and Hegarty (2012, p. 537) ascertained the publication occurrence of replications in the field since 1900. Specifically, they discovered that 1.6% (5,051 of 321,411) of articles used the term *replication* in the text. A more probing analysis of 500 randomly selected articles from their database showed that only 1.1% qualified as either direct or conceptual replications. Finally, a similar investigation of the top 100 education journals ranked by 5-year impact factor found that merely 0.13% (221 of 164,589) were replications (Makel & Plucker, 2014).

Table 5-1 depicts the results of 13 studies estimating the proportion of replication research published in the management sciences.[1] Only 4 of the 13,355 empirical papers involved were classified as close replications, that is, replications trying to determine whether an original result is repeatable. A further 838 articles qualified as replications beginning the process of seeking to establish empirical generalizations (Hubbard & Vetter, 1996; Lindsay & Ehrenberg, 1993). All told, 842 papers, or 6.3%, were classified as replications.

The disciplines with the highest replication incidences are economics (10.7%), finance (10.0%), forecasting (9.4%), and accounting (8.6%).[2] Overlapping 95% confidence intervals (CIs) for these four areas suggest no real differences among the estimates, with the population value for each discipline being in the 6%–12% range. The advertising (6.0%) field publishes less replication research than economics and finance, but has overlapping CIs with accounting and forecasting. Management (5.3%) has overlapping CIs with both accounting and advertising. Bringing up the rear at 2.3%, marketing publishes the lowest percentage of replication research; its 95% CI does not overlap with those from any other business discipline. These figures suggest that conducting replication research is something honored more in the breach than in practice.

5.3 The Outcomes of Replication Research

The publication frequency of replications appears to be very low, and it shows in stark relief that the vast majority of empirical business and

Table 5-1 Estimates of the Frequency of Published Replications in the Management Sciences

Discipline and Authors	*Time Period*	*Number of Empirical Studies*	*Number of Replications*	*%*	*95% Confidence Interval*		
Accounting							
Hubbard & Vetter (1996)	1970–1991	373	32	8.6	5.7	–	11.5
Advertising							
Reid, Soley, & Wimmer (1981)[a]	1977–1979	501	30	6.0	3.9	–	8.1
Economics							
Fuess (1996)	1984–1995	275	62	22.5	17.6	–	27.4
Hubbard & Vetter (1992)	1965–1989	942	92	9.8	7.9	–	11.7
Hubbard & Vetter (1996)	1970–1991	980	82	8.4	6.7	–	10.1
Subtotal		2,197	236	10.7	9.4	–	12.0
Finance							
Hubbard & Vetter (1991)	1969–1989	555	55	9.9	7.4	–	12.4
Hubbard & Vetter (1996)	1970–1991	556	54	9.7	7.2	–	12.2
Hubbard & Vetter (1997)	1975–1994	1,423	144	10.1	8.5	–	11.7
Subtotal		2,534	253	10.0	8.8	–	11.2
Forecasting							
Evanschitzky & Armstrong (2010)	1996–2005	766	72	9.4	7.2	–	11.6

(Continued)

Table 5-1 (Continued)

Discipline and Authors	*Time Period*	*Number of Empirical Studies*	*Number of Replications*	*%*	*95% Confidence Interval*		
Management							
Hubbard & Vetter (1996)	1970–1991	1,222	65	5.3	4.0	–	6.6
Hubbard, Vetter, & Little (1998)	1976–1995	701	37	5.3	3.6	–	7.0
Subtotal		1,923	102	5.3	4.3	–	6.3
Marketing							
S. W. Brown & Coney (1976)[b]	1971–1975	465	13	2.8	1.3	–	4.3
Darley (2000)	1986–1995	970	22	2.3	1.3	–	3.3
Evanschitzky, Baumgarth, Hubbard, & Armstrong (2007)	1990–2004	1,389	16	1.2	0.6	–	1.8
Hubbard & Armstrong (1994)	1974–1989	835	20	2.4	1.4	–	3.4
Hubbard & Vetter (1996)	1970–1991	1,139	33	2.9	1.9	–	3.9
Zinkhan, Jones, Gardial, & Cox (1990)	1975–1984	263	13	4.9	2.3	–	7.5
Subtotal		5,061	117	2.3	1.9	–	2.7
Total		13,355	842	6.3	5.9	–	6.7

Source: Adapted from Hubbard, Raymond and R. Murray Lindsay (2013a), "From Significant Difference to Significant Sameness: Proposing a Paradigm Shift in Business Research," *Journal of Business Research*, 66 (September), p.1382 with permission from Elsevier.

[a]Replication estimate based on all, not just empirical, studies. It was not possible to determine the number of empirical papers only.

[b]Original replication estimate of 2.0% was based on all, not just empirical, studies. The present estimate of 2.8% is based only on empirical studies.

behavioral research consists of unverified single-shot studies. Nonetheless, one cannot necessarily conclude that an insufficient number of replications are being published. If replications routinely support original findings, then their rate of occurrence might be satisfactory. Therefore, it is useful to examine the outcomes of replication research. As shown, here the lessons are salutary.

A study by Joshua Hartshorne and Adena Schachner (2012), for example, gathered data from 49 psychologists in a cross-section of specialties about whether they had ever attempted to replicate a published study. They reported a total of 257 replication attempts. Hartshorne and Schachner were disappointed to find that, of these, "only 127 (49%) fully replicated the original findings" (p. 3). The remaining percentages were approximately evenly split between those partially replicating and those failing to replicate the original results. An important limitation of the Hartshorne and Schachner study, however, is that these percentage breakdowns of agreement on replication success or otherwise appear to apply only to *unpublished claimed attempts* at repeating a study and not to actual publications of replication research. This drawback is remedied in what follows.

Twelve of the 13 management surveys listed in Section 5.2 provided information to determine whether published replications supported, partially supported, or conflicted with the findings of the original study. This information, based on the conclusions reached by the replicating author(s) typically after inspecting *p*-values, is presented in Table 5-2. Overall, some 46% of replications conflict with their predecessors, 29% provide partial support, and only 25% confirm prior findings. In most fields the replications were more likely to contradict than corroborate earlier results. The percentage of replication failures, by discipline, is shown in parentheses: accounting (50%), advertising (40%), economics (52%), finance (57%), and marketing (46%).[3] Only forecasting (19%) and management (19%), for reasons unknown, buck this trend. These are disturbing results. They point to disciplines some distance removed from the possession of reliable knowledge.

So What About the Call for Evidence-Based Research?

A dearth of reliable knowledge, sadly, is why recent talk about creating evidence-based management and social science literatures (see, e.g., APA Presidential Task Force on Evidence-Based Practice, 2006; McHugh & Barlow, 2010; Pfeffer & Sutton, 2006a, 2006b; Rousseau, 2006; Tranfield, Denyer, & Smart, 2003) is premature. The research climate in

Table 5-2 Outcomes of Replications in the Management Sciences

Discipline and Authors	*Support*[a]	*Partial Support*[a]	*Conflict*[a]	*Total*
Accounting				
Hubbard & Vetter (1996)	6 (18.8)	10 (31.3)	16 (50.0)	32
Advertising				
Reid, Soley, & Wimmer (1981)	12 (40.0)	6 (20.0)	12 (40.0)	30
Economics				
Fuess (1996)	19 (30.6)	31 (50.0)	12 (19.4)	
Hubbard & Vetter (1992)	18 (19.6)	14 (15.2)	60 (65.2)	
Hubbard & Vetter (1996)	15 (18.3)	17 (20.7)	50 (61.0)	
Subtotal	52 (22.0)	62 (26.3)	122 (51.7)	236
Finance				
Hubbard & Vetter (1991)	11 (20.0)	11 (20.0)	33 (60.0)	
Hubbard & Vetter (1996)	12 (22.2)	11 (20.4)	31 (57.4)	
Hubbard & Vetter (1997)	23 (16.0)	42 (29.2)	79 (54.9)	
Subtotal	46 (18.2)	64 (25.3)	143 (56.5)	253
Forecasting				
Evanschitzky & Armstrong (2010)[b]	17 (36.2)	21 (44.7)	9 (19.1)	47
Management				
Hubbard & Vetter (1996)	32 (49.2)	24 (36.9)	9 (13.8)	
Hubbard, Vetter, & Little (1998)	11 (29.7)	16 (43.2)	10 (27.0)	
Subtotal	43 (42.2)	40 (39.2)	19 (18.6)	102

Discipline and Authors	*Support*[a]	*Partial Support*[a]	*Conflict*[a]	*Total*
Marketing				
Darley (2000)	7 (31.8)	4 (18.2)	11 (50.0)	
Evanschitzky, Baumgarth, Hubbard, & Armstrong (2007)	7 (43.8)	5 (31.3)	4 (25.0)	
Hubbard & Armstrong (1994)	3 (15.0)	5 (25.0)	12 (60.0)	
Hubbard & Vetter (1996)	7 (21.2)	11 (33.3)	15 (45.5)	
Zinkhan, Jones, Gardial, & Cox (1990)	3 (23.1)	4 (30.8)	6 (46.2)	
Subtotal	27 (26.0)	29 (27.9)	48 (46.2)	104
Total	203 (25.2)	232 (28.9)	369 (45.9)	804

[a]Values in parentheses are percentages. Assessment of outcomes is based on the conclusions reached by the authors conducting the replication.

[b]These determinations could be made for only 47 of the 72 replications listed in Table 5-1.

the significant difference paradigm blocks the development of benchmarks or facts for judging the plausibility of later findings based on earlier ones. This is through its demands for novel outcomes (something conceded by Pfeffer & Sutton, 2006a, p. 71, in their plea for evidence-based management scholarship) and disregard for replication research, the latter being the only way to establish facts and baselines.

Testimony to the absence of fiducial baselines for informing practice can be seen in Rynes, Giluk, and Brown's (2007, p. 989) survey of 85 editorial board members of leading management journals seeking input on the "five most fundamental findings from human resources research that all practicing managers should know." As Guest (2007, p. 1022) commented on this survey, results showing that only five topics were mentioned more than 10 times, with the top three receiving just over 20 mentions, hardly suggests academic consensus on what constitutes "fundamental findings."

A similar situation about the lack of well-replicated results needed to inform evidence-based practice exists in the field of marketing. To see this, consult Hanssens's (2009) edited book *Empirical Generalizations About*

Marketing Impact: What We Have Learned From Academic Research. This well-intentioned volume presents 83 "empirical generalizations." It is instructive to note, however, that 28 (33.7%) of these putative regularities are based on a single study. Generalizing on the back of a single investigation is better known as speculation. An additional 21 (25.3%) result from two studies (where it is by no means obvious that the second study is a genuine replication). In other words, a more accurate subtitle for this book would be *What We Are Only Just Beginning to Learn From Academic Marketing Research.* Comparable reservations apply to Wind and Sharp's (2009) special issue of the *Journal of Advertising Research,* "What We Know About Advertising: Lessons From Empirical Generalizations." This issue came with a boast of "21 Watertight Laws for Intelligent Advertising Decisions." Results like these simply corroborate Armstrong and Schultz's (1993) earlier accusation that marketing does not own a collection of "principles."

A push for evidence-based research in biomedicine preceded that in the management and social sciences. Worryingly, even in this area findings are suspect. Ioannidis (2005a), for example, showed that of 45 highly cited articles—more than 1,000 times between 1990 and 2003—published in major journals claiming effective interventions, 16% were contradicted by subsequent studies, 16% found effects that were stronger than later works, and a further 24% remain largely unchallenged. In sum, the veracity of 56% of these hugely influential studies is open to question. More generally, because of the "chase" for statistical significance and other assorted research biases, Ioannidis (2005b) pulls no punches when stating (backed by simulation studies) that most published medical findings are false. Moonesinghe, Khoury, and Janssens (2007) agree with him.

Empirical evidence by pharmaceutical companies attempting to replicate published scholarly research underscores Ioannidis's (2005b) misgivings. For example, researchers at Bayer HealthCare were able to replicate in-house only about 25% of 67 published studies in the fields of oncology, women's health, and cardiovascular diseases. It was noted also that most failures to replicate led to termination of the projects because support for the therapeutic hypothesis was insufficient to justify continued funding (Prinz, Schlange, & Asadullah, 2011, p. 713). In a similar vein, Begley and Ellis (2012, p. 532) tell of how investigators at the biotechnology firm Amgen could reproduce the outcomes of a mere 6 of 53 (11%) "landmark" studies targeting cancers, a result they called "shocking."

In closing, an emphasis on significant difference is anathema to knowledge cumulation. Ironically, a significant sameness orientation *defines* the very route to evidence-based practices, be they in the management, social, or biomedical sciences.

5.4 The Timeliness of Replication Research

All else being equal, timely efforts to replicate and extend the findings of initial works have greater value than those performed at later dates. Adapted from Hubbard and Vetter (1996, p. 155), Figure 5-1 conveys the relative utilities attached to successful and unsuccessful replication efforts conducted over the short and long terms.

Short-haul successful replications and extensions build belief in the integrity of early research outcomes. As an example, in accounting Jiambalvo's (1982) efforts at determining the accuracy and consistency of the performance evaluation criteria of CPA personnel were in line with Maher, Ramanathan, and Peterson's (1979) previous study in this area. In an example from the finance literature, Servaes (1991) corroborated Lang, Stultz, and Walking's (1989) results on the connection between takeover gains and the *q* ratios of targets and bidders. A third instance is Mitchell's (2012) ability to reproduce Craig Anderson, James Lindsay, and Brad Bushman's (1999) findings on the external validity of laboratory experiments in the area of industrial-organization psychology.[4]

Short-term, unsuccessful replications spread doubt on the findings of the original research (and perhaps also on those attempting to replicate it). Failed replications may indicate the need for more work in that area. An illustration is provided by Terpstra (1981), who hypothesized a negative relationship between the degree of methodological rigor and the reported success of organization development interventions. Woodman and Wayne's (1985) later study showed no support for this association. Similarly, Fishe and Wohar's (1990) findings in economics are at odds with Mankiw, Miron, and Weil's (1987) contention that the establishment of the Federal Reserve System in 1914 initiated a new stochastic structure of interest rates. To take another example in this category, in light of the staggering nature of the claims made by Bem (2011) to have demonstrated experimentally the existence of human precognition, that is, an ability to see events which have yet to occur, it is not surprising that speedy attempts at replication followed. As it happens, efforts to confirm Bem's results by Wagenmakers, Wetzels, Borsboom, and van der Maas (2011), Ritchie, Wiseman, and French (2012a, 2012b), and Galak, LeBoeuf, Nelson, and Simmons (2012) failed uniformly.

An unsuccessful replication carried out well after the original investigation may be of dubious worth. The attempted replication of Haire's (1950) "shopping list" study, conducted in the United States, about the reluctance of women to adopt instant coffee is a conspicuous example of this. Haire's work is hailed as a marketing classic, being included in various anthologies

Figure 5-1 The Relative Utilities of Replications Over Time

	Successful Replication/Extension Attempt?	
	Yes	**No**
Short run	Instills confidence in the robustness of initial results. Additional work required to appraise generalizability of results. Examples: *Accounting:* Jiambalvo (1982) corroborates Maher, Ramanathan, & Peterson (1979). *Finance:* Servaes (1991) supports Lang, Stultz, & Walking (1989). *Psychology:* Mitchell (2012) ratifies C. A. Anderson, Lindsay, & Bushman (1999).	Promotes doubt over the robustness of initial results. Additional work needed to resolve discrepancies: Examples: *Economics:* Fishe & Wohar (1990) contradict Mankiw, Miron, & Weil (1987). *Management:* Woodman & Wayne (1985) are at odds with Terpstra (1981). *Psychology:* Galak, LeBoeuf, Nelson, & Simmons (2012), Ritchie, Wiseman, & French (2012a), and Wagenmakers, Wetzels, Borsboom, & van der Maas (2011) cannot replicate Bem (2011).
Long run	Supplies cumulative evidence of the existence of a phenomenon as well as its scope and limits. Examples: *Management:* Bird & Fisher (1986) confirm Kirchner & Dunnette (1954). *Marketing:* Ehrenberg & Bound (1993) show the empirical stability of the Dirichlet model of buyer behavior.	Of dubious worth when the time lag is substantial. Earlier replications usually more informative. But can be useful when initial studies continue to be influential. Examples: *Marketing:* Webster & von Pechmann (1970), Arndt (1973), and G. S. Lane & Watson (1975) fail to replicate Haire (1950). *Psychology:* Shanks et al. (2013) refute Dijksterhuis & van Knippenberg (1998). Harris, Coburn, Rohrer, & Pashler (2013) disagree with Bargh, Gollwitzer, Lee-Chai, Barndollar, & Trötschel (2001).

Source: Adapted from Hubbard, Raymond and Daniel E. Vetter (1996), "An Empirical Comparison of Published Replication Research in Accounting, Economics, Finance, Management, and Marketing," Journal of Business Research, 35 (February), p.155 with permission from Elsevier.

such as Howard Thompson's (1981, pp. 382–393) *The Great Writings in Marketing,* lauding the virtues of projective techniques.[5] But as I inquired (Hubbard, 1994, p. 258), why did it take 20-odd years before a (by now, utterly predictable given the enormous changes in the socioeconomic environments) failure to replicate his results by Webster and von Pechmann (1970) in the United States, George Lane and Gayne Watson (1975) in Canada, and Arndt (1973) in Norway were published? Surely a work of this presumed stature merited earlier (and virtually cost-free, at that) replication. Additionally, I wondered why it took 28 years for James Anderson's (1978) paper stating that Haire's methods were flawed to appear in print.

A caveat is in order: An argument can be made that even relatively untimely failures to replicate the findings of previous research are valuable when the latter remain influential. To see this, consider the work done on priming in psychology. Sharing territory with subliminal perception, priming is the claimed ability to modify people's behavior on the basis of subtle external cues of which they are unaware.

Despite fears that the thriving area of priming research is poorly founded—Daniel Kahneman chastens, "I see a train wreck looming"—the results of work in this domain nonetheless have begun to infiltrate the options of policymakers wishing to nudge the populace ("How Science Goes Wrong," 2013, p. 26). Adding fuel to these concerns is the unreplicability of key studies on priming. These include the nine experiments in Shanks et al.'s (2013) investigation unable to reproduce Dijksterhuis and van Knippenberg's (1998) results, the latter with 534 Google Scholar citations, on priming intelligent behavior. In addition, there are Harris, Coburn, Rohrer, and Pashler's (2013) two failures to duplicate Bargh, Gollwitzer, Lee-Chai, Barndollar, and Trötschel (2001), which has amassed 1,236 citations, on high-performance-goal priming. The message under such circumstances is that efforts at replication are better late than never.

Long-term, ongoing, successful replication attempts obviously represent the most desirable picture. In this respect the versatility of the Dirichlet model to capture many empirical patterns of consumer behavior over several decades epitomizes how findings replicate over time (Ehrenberg & Bound, 1993; Ehrenberg, Goodhardt, & Barwise, 1990).

Dan Vetter and I (Hubbard & Vetter, 1996, p. 159) estimated that the average time lag between the publication of original and replication studies in the management sciences appears to be reasonable. This time lag, in years, by discipline is as follows: accounting (3.6), economics (4.2), finance (3.9), management (4.6), and marketing (5.6). Across all five disciplines, the average publication time lag is 4.3 years. While the timeliness of replications, on average, seems acceptable, the glaring problem remains that there are not nearly enough of them.

5.5 Why the Lack of Replication Research?

The sentiments propounded in Chapters 3 and 4 about the principle of replicability being the pivotal criterion for establishing scientific knowledge beg the question: Why is so little evidence of its application visible in the management and social science literatures? Together with Scott Armstrong (Hubbard & Armstrong, 1994, pp. 240–244), I have offered several reasons, expanded upon below, for this state of affairs.

5.5.1 Misinterpreting the *P*-Value as a Measure of the Replicability of Results

Many researchers erroneously interpret statistical significance as a measure of replicability (Goodman, 1992; Sohn, 1998; Tversky & Kahneman, 1971). Carver (1978, p. 385) refers to this as the "replicability or reliability fantasy," and he cites a textbook by a well-known psychometrician as guilty of subscribing to it. This fantasy continues to bewitch authors of statistics textbooks in the behavioral sciences (see, e.g., Heyes, Hardy, Humphreys, & Rookes, 1993, p. 35; Pagano, 2001, p. 228). Small wonder, then, that the same survey administered to members of the American Vocational Education Research Association (H. R. D. Gordon, 2001), the American Educational Research Association (Mittag & Thompson, 2000), and a national sample of Spanish psychology professors (Monterde-i-Bort, Frias-Navarro, & Pascual-Llobell, 2010) found too many of them agreeing that smaller *p*-values indicate a higher probability of replicable outcomes.

The results of Boos and Stefanski's (2011) article also raise problems with using the *p*-value as an indicator of the replicability of findings. In illustrating this, Boos and Stefanski comment:

> Recent articles have lamented the lack of reproducibility in statistically supported scientific findings, with examples in the health sciences particularly troubling. . . . For example, suppose that in a sample of journal articles, we find *p*-values mostly in the range of 0.005 to 0.05. [Based on their work] This suggests that the probability of *nonreplication* of published studies with *p*-values in the range of 0.005 to 0.05 is roughly 0.33. (p. 221; emphasis added)

Goodman's (1992, p. 877) estimate of the probability of a nonreplication of original results found to be statistically significantly different in the .005-.05 latitude is approximately .35.

Furthermore, there is a popular tendency to misinterpret the complement of the *p*-value as a direct index of replicability. Oakes (1986, p. 80), for instance, showed that 42 of 70 (60%) British academic psychologists believed that an experimental outcome that is significant at the .01 level has a .99 probability of being statistically significant if the study were replicated. Many individuals in Gigerenzer, Krauss, and Vitouch's (2004, p. 395) survey of German psychologists also misinterpreted this 1–*p* criterion as the probability of replication success. The percentage doing so was reported for three groups: students who had passed at least one statistics course in which significance testing was taught (41%), faculty not teaching statistics (49%), and, most disquieting of all, faculty *teaching* statistics (37%). Such widespread misconceptions about statistical significance and replicability have no doubt undermined the perceived need for actually *performing* them.

If academic researchers misinterpret *p*-values as indices of replicability, it stands to reason that this mistake also is made by writers of popular press offerings. In reviewing the book *Meaningful Marketing: 100 Data-Proven Truths and 402 Practical Ideas for Selling More With Less Effort* (Hall & Stamp, 2003), for example, Sharp (2004, p. 105) states that the authors commit the classic error of confusing statistical significance with replicability, citing them as follows: "With statistics, we can quantify the likelihood that what we're observing is a reproducible and reliable truth versus a coincidental one-time random event." Unfortunately, this is not the case.

The gist of this matter is straightforward: There is no formal warrant connecting *p*-values with the replicability of results. Acknowledging this explains why Cumming's (2008, p. 286) simulation study concluded that the *p*-value is a "dramatically" vague and extremely unreliable measure of replication success. Consistent with a significant sameness orientation, Cumming recommends the adoption of CIs over *p*-values.

5.5.2 Information Needed to Perform a Replication Is Difficult to Obtain

Because of limited journal space, the initial article may lack specifics about the procedures used (e.g., measurement instruments, analytical methods, sampling approaches) to allow for a replication or extension to take place. Alternatively, if a replication is attempted, then some of the procedures involved will be only approximations of those employed in the first study. This could easily lead to contradictory findings between the two works.

To perform an accurate replication or extension may therefore mean that the would-be replicator must contact the author(s) of the original study for additional information, which could include requests for the provision of the raw data or a copy of the questionnaire used. How are requests such as this normally handled? This section, based primarily on Hubbard and Little (1997), reveals that many management and social science researchers can be stubbornly uncooperative. Responses to lesser requests are examined first.

Lesser Requests

Lesser requests do not involve asking an author for the data. In this context, Madden, Franz, and Mittelstaedt (1979) contacted the authors of 60 papers from the 1975–1977 American Marketing Association and Association for Consumer Research annual conference proceedings. They were asked if they would be prepared to share research instruments and other details of their work to allow replication, but Madden et al. were careful to stress that no request was being made for the raw data. Nevertheless, half of the authors refused to give any information.

In another study, Eaton (1984) asked 45 authors of articles in psychology journals for missing summary statistics needed to perform a meta-analysis. Under these conditions, 20 (44%) of the authors did not comply. Gasparikova-Krasnec and Ging (1987), on the other hand, were more successful. They queried 72 authors from a random sample of psychology journals for information not included in their articles so that they could conduct a replication. Eighteen (25%) refused assistance (see Table 5-3).

Requests for Raw Data

All of the results in this section are summarized in Table 5-3. Recall from Section 4.4.1 that an early study by Wolins (1962) involved contacting 37 authors of articles published in psychology journals requesting copies of their raw data for purposes of replication. It does not paint an atmosphere of open scholarly exchange. Of the 32 authors who responded, 21 said that their data were lost or had been accidentally destroyed. For the remaining 11, 2 demanded control over anything that might later be published with them. Data from only 9 authors were received, and for 2 of these replications could not be performed because the information did not arrive in a timely enough fashion to be included in Wolins's study. In total, 28 (76%) authors failed to respond.

Following up on Wolins's article, Craig and Reese (1973) chose one issue from each of four psychology journals (*Journal of Comparative and*

Table 5-3 How Authors Respond to Requests for Information[a]

	Lesser Requests			*Requests for Raw Data*				
Response Categories	*Madden, Franz, & Mittelstaedt (1979)*	*Eaton (1984)*	*Gasparikova-Krasnec & Ging (1987)*	*Wolins (1962)*	*Craig & Reese (1973)*	*Reid, Rotfeld, & Wimmer (1982)*	*Dewald, Thursby, & Anderson (1986)*[b]	*Wicherts, Borsboom, Kats, & Molenaar (2006)*
Number of requests	60	45	72	37	53	99	154	141
Nonresponses	14 (23.3)	11 (24.4)	11 (15.3)	5 (13.5)	8 (15.1)	36 (36.4)	37 (24.0)	20 (14.0)
Responded but refused to cooperate[c]	16 (26.7)	9 (20.0)	7 (9.7)	23 (62.2)	25 (47.2)	14 (14.1)	27 (17.5)	83 (59.0)
Total failure to cooperate[d]	30 (50.0)	20 (44.4)	18 (25.0)	28 (75.7)	33 (62.3)	50 (50.5)	64 (41.6)	103 (73.0)

Sources: Adapted from Hubbard, Raymond and Eldon L. Little (1997), "Share and Share Alike? A Review of Empirical Evidence Concerning Information Sharing Among Researchers," *Management Research News*, *20*, p.44.

Wicherts, Jelte M., Denny Borsboom, Judith Kats, and Dylan Molenaar (2006), "The Poor Availability of Psychological Research Data for Reanalysis," *American Psychologist*, *61* (October), p.727. Copyright © 2006 by the American Psychological Association. Adapted with permission.

[a]Numbers in parentheses are percentages.

[b]These results are based on the average responses of the three groups involved in this study.

[c]Includes researchers who would not release their data, the data were lost or destroyed, were proprietary, and so on.

[d]Made up of nonrespondents plus respondents refusing to cooperate.

Physiological Psychology, Journal of Verbal Learning and Verbal Behavior, Journal of Personality and Social Psychology, and *Journal of Educational Psychology*) and asked the lead author of each article for copies of his or her raw data for reanalysis as part of a master's thesis. Cooperation rates were better than those found by Wolins. From a mailing of 53 requests, 45 replies were obtained, 20 of which included the raw data or a summary of them. An additional 7 were willing to send materials if Craig and Reese could be more precise about what kinds of data they wanted, 9 simply refused to share, 5 commented that the requested information was unavailable (e.g., in storage), while 4 said that it had been lost or destroyed. Total failure to accede to requests in this case was 62% (33). Craig and Reese make no mention of whether they reexamined any of the supplied data sets.

Reid, Rotfeld, and Wimmer (1982) sent questionnaires to all first authors of empirical articles published in 1978–1979 issues of the *Journal of Advertising, Journal of Advertising Research, Journal of Consumer Research, Journal of Marketing,* and *Journal of Marketing Research*, asking for information, including raw data, needed to replicate their works. Ninety-nine requests were involved in their study. Of these, 49 responses noted that the materials were available, with some leaving it at this and others providing varying amounts of data with their replies.[6] Fourteen authors said that the requested materials were unavailable, because they either had been lost or discarded (11) or were proprietary (3). Thirty-six researchers did not respond to the double-mailing requests. Total failure to oblige solicitations was 51% (50).

One of the most worrisome accounts of the obstacles confronting fellow researchers as they try to collect information from their peers to enable replication is that of Dewald, Thursby, and Anderson (1986), mentioned earlier. They write that in July 1982 the *Journal of Money, Credit and Banking* (*JMCB*), with funding from the National Science Foundation (NSF), inaugurated the *JMCB* Data Storage and Evaluation Project. As a fundamental aspect of this project, *JMCB* changed editorial policy by asking researchers for copies of their data and computer programs so that these could be accessed by others upon request. Also, this modification in policy was implemented retroactively and was applicable to all works published in *JMCB* after 1980.

Dewald et al. (1986) placed authors into three groups. The first group was made up of those whose work was published before the onset of the *JMCB* project in July 1982, and who therefore were unaware that they would be contacted later for copies of their materials. The second group comprised authors whose manuscripts as of July 1982 had been accepted for publication in *JMCB* but were yet to appear in print. Authors of manuscripts submitted to *JMCB* as of July 1982, but were still under review, constituted the third group.

Of 62 requests made to those in the first group, 22 authors complied by submitting programs or data, 20 did not respond to repeated requests, and 20 said they could not provide the information because it was lost or discarded (14), was proprietary (2), or was conveniently available from published sources (4). Dewald et al. (1986) were concerned that so many authors were not able to supply programs and data, particularly when for some the requests came as little as 6 months after their articles were published. In total, the nonsubmission rate for group one (nonrespondents plus those responding but unable to honor the request) was 65%.

As might be expected, the nonsubmission rates for groups two and three were lower than group one's: 6 of 27 (22%) and 18 of 65 (28%), in turn. But even here cases existed in which data already were lost or destroyed. Puzzlingly, for group three no fewer than 16 (25%) authors would not comply with two appeals for information from the very journal currently viewing their manuscripts for publication. Total failure to oblige requests across all three groups was 42% (64).

Dewald et al. (1986, p. 588) despaired over the lack of collaboration in exchanging information, warning that

> for other researchers, however, private interest prevailed and our request was either refused or ignored. We note that NSF Policy Number 754.2 *requires* that computer programs and data which have been produced with the assistance of NSF grants be made available to other researchers either by publication, duplication, or loan to the researcher. Investigators have the first right of publication, but the NSF rule requires that the programs and data be made available to others. It appears that this policy is seldom enforced and that investigators either are unaware of the policy or unafraid of the penalties for failure to comply with it.

Analogous problems of individuals flouting the rule of sharing data with others when their research projects are supported by tax dollars are found elsewhere (see, e.g., Ceci & Walker, 1983).

The studies recounted above all involved requests for information from colleagues largely before the onset of user-friendly electronic data transference technologies. Given the fact that "nowadays, one would think that sending data is but a few mouse clicks away," Wicherts, Borsboom, Kats, and Molenaar (2006, p. 727) were optimistic that in this task they would fare much better than their precursors. As it happens, they were sorely disappointed.

Wicherts et al. (2006, p. 726) e-mailed the corresponding authors of 141 empirical articles published in the last two 2004 issues of four major American Psychological Association (APA) journals: *Developmental Psychology, Journal of Consulting and Clinical Psychology, Journal of Experimental Psychology: Learning, Memory, and Cognition,* and *Journal of Personality and Social Psychology.* They asked for copies of the raw data from these articles so that they could be reanalyzed with a view to assessing the robustness of the findings to the presence of possible outliers.

Because these 141 authors had been published in APA journals, Wicherts et al. (2006, p. 726) had every reason to trust that all of them had signed the Certification of Compliance With APA Ethical Principles, which dictates the sharing of data for reanalyses. This injunction from the APA (2001, p. 396) is clear enough:

> After research results are published, psychologists do not withhold the data on which their conclusions are based from other competent professionals who seek to verify the substantive claims through *reanalysis* and who intend to use such data only for that purpose, provided that the confidentiality of the participants can be protected and unless legal rights concerning proprietary data preclude their release. (emphasis added)

Nevertheless, as exhibited in Table 5-3, following more than 400 e-mail requests over a period of 6 months, Wicherts et al. (2006, p. 727) found that 103 authors (73%) refused to honor them, a figure almost identical to that found by Wolins (1962) some 50 years earlier.[7]

All told, the data featured in Table 5-3 show social science and business research communities to be largely unhelpful regarding information sharing. When it comes to lesser requests, the average failure to cooperate is 38.4%; for requests for raw data, it is 57.4%. These numbers suggest that Ioannidis's (2012, p. 646) castigation of an academic establishment in which "the vast majority of the collected data, protocols, and analyses are not available and/or disappear soon after or even before publication" perhaps is overblown. But neither is it without foundation.

Especially disconcerting, information is available relating the willingness to share data with the verity of the statistical significance of published outcomes. More specifically, Wicherts, Bakker, and Molenaar (2011) recalculated the accuracy of t, F, and χ^2 test statistics claiming significance at $p < .05$ extracted from 49 papers appearing in the *Journal of Personality and Social Psychology* and *Journal of Experimental*

Psychology: Learning, Memory, and Cognition in 2004. Corresponding authors of 21 (42.9%) of these articles had earlier shared some data with Wicherts et al. (2006). Interestingly, Wicherts et al. (2011, p. 4) encountered 10 cases from 7 of the 49 (14.3%) articles in which recalculated *p*-values were above .05, even though they featured in print as $p < .05$ results. Every one of these instances of misreported findings came from authors who had refused to share data.

Of course, it is only natural for researchers who have spent much time and effort in constructing a database to want to have exclusive rights over it for some specified period of time. More so if it came from primary data sources. The tricky issue, obviously, is establishing how long this moratorium should last. In answering this, Mayer (1980, p. 176) advises that "one or two years, combined with the usual publication lag for the initial paper should give them sufficient time to exploit their data mine ahead of others." In this same context, Sieber (1991, p. 9)—see also Hauser (1987)—notes that since 1985 the NSF has called for social science researchers to deposit data and associated materials in a designated public archive within one year following a project's completion. It would seem that additional work on determining appropriate time limits on the sharing of data sets would be time well spent.

Ceci and Walker (1983) have proposed that a change in emphasis from a Cartesian to a Baconian conception of science would help in bringing about data sharing among researchers. By way of analogy, Ceci and Walker write, the Cartesian investigator sees science like the running of a race, where one's colleagues are direct competitors. The goal is to win at all costs. In contrast, the Baconian interpretation views science as a cooperative enterprise focused on the improvement of human welfare. Plainly, cultivating a Baconian perspective in the management and social sciences would be beneficial. Unfortunately, the research publication system at hand in the significant difference paradigm, with its publish-or-perish mores, favors a Cartesian rather than a Baconian philosophy. The significant sameness paradigm, in opposition, epitomizes the latter view.

5.5.3 Replication Attempts Suffer From Low Statistical Power

It has been suggested that because of a belief in the "law of small numbers" replication research might be statistically underpowered and hence less worthy of publication (Sawyer & Ball, 1981; Tversky & Kahneman, 1971). Considering the preponderance of low-powered *non-replication* research published in the social and management sciences, documented in Chapter 2, this query makes sense.

The topic of the statistical power witnessed in replication (and original) studies has been investigated, in the marketing area at least, by Scott Armstrong and me (Hubbard & Armstrong, 1991, 1994). We followed standard procedures for conducting a power study described by Jacob Cohen (1988) and outlined earlier in the appendix to Chapter 2 and in Section 3.5.

Of the 20 replications in our study (Hubbard & Armstrong, 1994, p. 240), 5 could not be power-analyzed because they involved techniques like factor analysis for which power tests do not exist, did not use significance tests, or provided insufficient information to calculate power levels. Six of the 20 original studies were excluded for the same reasons.

Adopting Cohen's (1988, p. 56) criterion of .80 as a satisfactory power benchmark, the average power of both the 14 original and 15 replicated studies is impressive (Table 5-4). The mean power to detect small, medium, and large effect sizes in the original articles is .39, .90, and .96; for the replications the corresponding values are .36, .90, and .99. When power comparisons are made on the basis of the individual number of statistical significance tests carried out in the original articles (92), the average power to distinguish small, medium, and large effects is .41, .94, and .98, respectively, while for the replications (102) these figures are .38, .91, and .99 (Table 5-4).

If medium effect sizes are typical of the empirical marketing literature, Hubbard and Armstrong (1991, p. 11) report, all but 2 of the 15 replications exceeded the advocated power baseline of .80. All 15 did so if large effect sizes in the population are posited. Conversely, 1 of the 14 original articles failed to meet the 80% chance of detecting a medium effect; it had only a 50% chance of identifying a large effect.

As Table 5-4 makes clear, irrespective of whether replications support, partially support, or conflict with earlier findings, the mean power of an article to isolate small, medium, and large effects is consistently high. The same conclusion applies to the results from individual significance tests (see the bottom half of Table 5-4).

Whether our findings with respect to the statistical power of replications (and original works) are generalizable is something which can be determined only by further empirical research. But they are certainly encouraging. Indeed, it would be remiss not to mention that the power shown in our 14 original marketing articles to discover small (.39), medium (.90), and large (.96) effects is indistinguishable from their counterparts—.41, .89, and .98—found in Sawyer and Ball's (1981) 23 original marketing articles (see Table 2A-1).

Our doubtful speculation (Hubbard & Armstrong, 1992, p. 133) in Section 3.5 that the reason marketing articles with null results were

Table 5-4 Average Statistical Power of Original and Replicated Marketing Studies

Category	*Sample Sizes*	*Effect Sizes*		
		Small	*Medium*	*Large*
Number of original articles	14	.39	.90	.96
Number of replications	15	.36	.90	.99
Number of statistical tests in original articles	92	.41	.94	.98
Number of statistical tests in replications	102	.38	.91	.99
Articles (replications)				
Support	2	.50	.99	>.99
Partial support	4	.33	.94	>.99
Conflict	9	.35	.87	.98
Statistical tests (replications)				
Support	4	.54	.98	>.99
Partial support	27	.29	.91	>.99
Conflict	71	.41	.91	.99

Source: Adapted from Hubbard & Armstrong (1991, Table 6).

published is because they display high power holds analogously in the present circumstances. That is, we are skeptical that the replications were published based on their high power, while unpublished replications were rejected for their lack of it. We acknowledged, however, that it was not possible to assess directly the merits of this argument because, as far as we know, the power levels of published and unpublished replications have yet to be compared.

5.5.4 Ample Replication Research Is Published in Lower-Tier Journals

I have belabored that a trademark of the significant difference paradigm is the sacredness of engaging in so-called original or creative research. The results described in Section 5.2 showing the paucity of replications are based primarily on content analyses of the leading journals in the management and social sciences, where originality may be thought to be especially prized. The possibility exists, therefore, that a trickle-down effect is at play whereby journals in the lower tiers publish far more replication research than those of the first rank. As it turns out, there is no empirical support for this thesis. For example, Dan Vetter and I (Hubbard & Vetter, 1997) examined the publication frequency of replication research in a random sample of 1,423 empirical papers from three tiers of the finance journal hierarchy for the period 1975–1994. The journals in each tier are as follows: first tier (*Journal of Business, Journal of Finance, Journal of Financial Economics, Journal of Financial and Quantitative Analysis*), second tier (*Financial Analysts Journal, Financial Management, Journal of Banking and Finance, Journal of Portfolio Management*), third tier (*Financial Review, Journal of Business Finance and Accounting, Journal of Financial Research, Journal of Money, Credit and Banking*). We found the percentages of empirical research allocated to this work to be 10.4% for the first-tier journals, 7.5% for those in the second tier, and 12.5% for the third tier.

A matching pattern of results arose for replication frequencies found in the first (*Academy of Management Journal, Administrative Science Quarterly, Strategic Management Journal*), second (*California Management Review, Journal of Management Studies, Sloan Management Review*), and third (*Human Resource Management, Journal of General Management, Journal of Management*) tiers of journals in the management field for 1976–1995 (Hubbard, Vetter, & Little, 1998). Following an analysis of 701 empirical articles, the percentage of replication research published in these three tiers was 5.3%, 2.8%, and 6.9%, respectively. Consequently, we see little attention paid to this form of research in both leading journals and those of lesser prestige.

In addition, replications published in lower-tier journals are less likely to gain recognition. As an illustration of this, Sterling, Rosenbaum, & Weinkam (1995) cited an article by Begg and Berlin (1988) in which a positive result reported in a leading journal continued to influence medical practice long after later publications in less reputable outlets challenged its reliability.

5.5.5 Replications Are of Little Importance

Some editors, reviewers, and authors may question the value or importance of replications. For example, the argument goes, if the original study apparently is well designed, why bother to replicate it? Alternatively, if the study was poorly designed, a replication would be pointless. Moreover, it could be said that an exact replication of a prior study cannot discern whether the manipulated and dependent measures used were construct valid (Hubbard & Armstrong, 1994, p. 242). Even those things that an exact replication might indicate, such as errors in data analysis or experimenter effects in the original study, could easily leave the reader with more questions than answers. As an example, suppose that Smith replicates Jones's previous work and does not reach the same results. This could end up with Smith and Jones trading accusations as to who is responsible for the contradictory outcomes. Jones could state that Smith did not perform an accurate replication, made recording mistakes, introduced bias, and so on. Smith could respond that Jones made errors (recall Section 4.4.3).

The difficulties involved in determining whether the burden of proof lies with Smith or Jones (or both) for a failure to replicate undoubtedly will compound when an attempt at empirical generalization is performed. Besides those mentioned above, additional reasons for conflicting results could involve differences due to alterations in manipulated or measured variables, the inclusion of new constructs and their interactions, and the non-generalizability of findings to other (sub)populations, time periods, contexts, and so on (Hubbard & Armstrong, 1994, p. 242). Faced with such manifold problems, editors, reviewers, and researchers may decide that little is to be gained in trying to resolve the issue(s). This conclusion is exacerbated in light of what follows.

5.5.6 Original Works Are Not Worth Replicating

The literature presented in Hubbard and Armstrong (1994, pp. 242–243) inspiring replication attempts encompassed a diverse set of topics such as subliminal advertising, the double jeopardy effect, shopping lists, the effects of music on marketplace behavior, contributions to charities and mail survey response rates, and racial attitudes. These kinds of works often attract attention because they found surprising and/or controversial results. Some became classics, and many speak to concerns which have powerful implications for marketing theory and practice. By definition, most published articles will *not* make this kind of impact.

However, the all-important requirement to publish in a publish-or-perish landscape can be expected to deflect attention from the detailed examination of important marketing issues to those short-term ones that are more likely to appear in print. If this is true, members of the discipline may be of the opinion that much empirical research is just not important, valuable, or interesting enough to justify the time and costs that replications entail.

By far the most irrefutable data—among academicians at least—backing the thesis that much published work is not worthy of attention, never mind replication, is offered by citation counts. This topic is explored further in Chapter 8. Suffice it to say here that most research goes uncited.

On the other hand, there is evidence to suggest that there are enough papers in the literature deserving of replication. For example, Armstrong and Hubbard (1992) asked marketing academicians and practitioners to rate the importance of 20 data-based studies on consumer behavior. When considering simultaneously the criteria of importance, surprising hypotheses, and statistical significance, 4 (20%) of the studies were deemed to be of value. Our survey of reviewers on the editorial board of the *Journal of Consumer Research* discovered that approximately 35% of articles published in that journal were judged to be important. In this connection Monroe (1992b) and Nosek, Spies, and Motyl (2012, p. 622) suggest as a touchstone for research worthy of replication articles which have been cited frequently. More specifically, Makel et al. (2012, p. 541) offer, by their own admission, the arbitrary heuristic that an article which has been cited 100 times merits replication.

5.5.7 Peer Review Protects the Soundness of the Empirical Literature

Replication research may be thought to be superfluous because peer review will identify mistakes, if any, in submitted manuscripts. Clearly, this is an unreasonable assumption. To be sure, peer review may catch an error in, say, mathematical operations. But it is expecting too much to suppose that this process will uncover errors in statistical analyses, even if these are of an elementary kind. So the hope that peer review safeguards the validity and reliability of empirical results is unfounded.

In addition, viewing replication research solely as a means of checking the accuracy of published findings is an important, but much too narrow, definition of its role. As demonstrated in Chapter 4, replication permeates the scientific enterprise.

5.5.8 Editorial-Reviewer Bias Against Publishing Replications

Survey results strongly support the charge of an editorial-reviewer bias against the publication of replication research. For example, Steven Kerr, James Tolliver, and Doretta Petree's (1977, pp. 138, 140) survey of 429 editors and review board members of 19 management and social science journals found evidence of this bias, even when it is acknowledged that the replications had been done competently. Rowney and Zenisek (1980) administered a version of the Kerr et al. questionnaire to 268 reviewers of Canadian psychology journals and obtained similar results. Neuliep and Crandall's (1990) survey of 288 past and present editors of social science journals also offers convincing evidence of a bias against publishing replications.

In a later study involving responses from 80 social science journal reviewers, Neuliep and Crandall (1993, p. 27) said that several of them spurned replications as a waste of time and journal space. These reviewer apprehensions are consistent with Uncles's (2011, p. 579) reading that it is common to hear that "surely it [replication research] isn't worth the journals [*sic*] space." Again, Madden, Easley, and Dunn's (1995) analysis of the replies from 107 natural and social science journal editors found that while the former generally approve replication as a necessary part of research, the latter were much more lukewarm concerning its importance. They further testified to a similar lack of enthusiasm over the role of replication among the editors of 16 advertising, communication, and marketing journals. Finally, Easley, Madden, and Gray's (2013, p. 1459) follow-up study, based on feedback from 56 editors of natural and social science journals, concluded that the latter think replications aren't creative and unfairly displace original and important works.

Most of the editorial-reviewer bias against publishing replication research is attributable to the preferences shown for reserving limited journal space for original work. Past editors of the *International Journal of Research in Marketing,* Stremersch and Lehmann (2007, p. 1–2), personify this attitude: "The journal will not be very receptive to work that merely replicates well-established findings. . . . [W]e will favour truly new ideas and methods."[8] So does Eliot Smith, a former editor of the *Journal of Personality and Social Psychology* (*JPSP*), who published an inordinately controversial article by Bem (2011) supporting the idea of human precognition mentioned earlier in this chapter. Yet Smith did not even send out for peer review a manuscript submitted to *JPSP* by Ritchie et al. (2012a) which failed to replicate Bem's findings. As reported by Aldhous (2011) in *New Scientist,* Smith's rationale is that *JPSP* has a goal of promoting only the best original research and "does not publish replication studies, whether successful or unsuccessful." Further, Smith acknowledged the desirability of publishing replications, but "the question is where. There are hundreds of journals in psychology." The not-too-subtle implication is that while replications should appear in public somewhere, they are not to sully the pages of leading journals such as *JPSP*. The anti-scientific import of these remarks notwithstanding, it was documented in

Section 5.5.4 that lower-tier journals are no more likely to feature replication research than their more illustrious kin. Every journal seems to be in the business of publishing only unprecedented research.

The main fear among many editors and reviewers, and indeed researchers at large, appears to be that a greater readiness to entertain the publication of replications will appropriate too much scarce journal space, thus leaving less room for creative or novel inquiries. Yet as Nosek et al. (2012, p. 617) demur, "replication is a means of increasing the confidence in the truth value of a claim. Its dismissal as a waste of space incentivizes novelty over truth." Never mind that a novel result that cannot be replicated by definition is not creative (Makel & Plucker, 2014). And in any case, the assumption that replications hamper the publication of original research invites the question as to whether such an argument holds water. In other words, does the current (past) publication frequency of replications impede the visibility of unique works? Fortunately, this is an empirical question.

Do Replications Jeopardize the Publication of Original Research?

In answering this question, turn your attention to Table 5-5, adapted from Hubbard and Vetter's (1996) investigation of the publication incidence of replication research in five business disciplines: accounting, economics, finance, management, and marketing. Our study was based on a content analysis of a 25% (50% for marketing) simple random sample of annual issues of 18 business periodicals from these five fields for the period 1970–1991. It incorporated a total of 4,270 empirical articles and notes, of which 266 (6.2%) were replications. Finance (9.7%) is the discipline recording the highest frequency of such work, with marketing (2.9%) at the low end.

Of special concern to the present discussion is the last column of Table 5-5, which shows the percentage of research journal space, in pages, taken up by replications. The numbers are trifling. To see this, in comparison with responses from a polling of 1,292 psychologists recommending that 20% of journal space *should* be dedicated to replications (Fuchs, Jenny, & Fiedler, 2012, p. 641), on average, replications in the five business areas *actually* consume a mere 3.1% of this space. The breakdown by disciplines is as follows: accounting (4.5%), economics (2.9%), finance (4.4%), management (3.6%), and marketing (1.3%). Worries that replications threaten the publication of original research cannot be justified. On the contrary, the data in Table 5-5 illustrate quite vividly that there is abundant room for a major expansion of published replication research in business journals. The same could be said of periodicals in the social sciences.

Replication Research Is Unimaginative and Uninteresting: Rebuttal

Another important component of the editorial-reviewer bias against publishing replications is that such work is said to be deficient in terms of both imagination and interest to readers. Following these lines of thought, as revealed by Neuliep and Crandall (1993, p. 26) in particular, conducting replications is widely seen as an inferior type of research that is unlikely to earn recognition for those involved (S. W. Brown & Coney, 1976; Easley, Madden, & Dunn, 2000; Gilbert, 2012; Hubbard & Armstrong, 1994; Koole & Lakens, 2012; Madden et al., 1995; Mittelstaedt & Zorn, 1984; Monroe, 1992b). In magnifying this view Kane (1984, p. 3),

Table 5-5 Journal Space Devoted to Replication Research in Five Business Disciplines: 1970–1991

Discipline[a]	*Number of Empirical Studies*	*Number of Replications*	*Percentage of Replications*	*Percentage of Journal Space for Replications*[b]
Accounting	373	32	8.6	4.5
Economics	980	82	8.4	2.9
Finance	556	54	9.7	4.4
Management	1,222	65	5.3	3.6
Marketing	1,139	33	2.9	1.3
Total	4,270	266	6.2	3.1

Source: Adapted from Hubbard, Raymond and Daniel E. Vetter (1996), "An Empirical Comparison of Published Replication Research in Accounting, Economics, Finance, Management, and Marketing," *Journal of Business Research,* 35 (February), pp. 158, 162 with permission from Elsevier.

[a]The accounting sample included 22 issues each of *The Accounting Review* and the *Journal of Accounting Research* and 13 issues of the *Journal of Accounting and Economics.* The economics sample comprised 22 issues each of the *American Economic Review, Quarterly Journal of Economics,* and *Review of Economics and Statistics* as well as 33 issues of the *Journal of Political Economy.* The finance sample consisted of 22 issues each of the *Journal of Finance, Journal of Financial and Quantitative Analysis,* and *Journal of Money, Credit and Banking,* while for the *Journal of Financial Economics* it was 18 issues. Twenty-two issues each of the *Academy of Management Journal* and *Administrative Science Quarterly,* in conjunction with 30 issues of the *Journal of Applied Psychology* and 33 issues of *Organizational Behavior and Human Decision Processes,* made up the management sample. Lastly, the marketing sample was composed of 44 issues each of the *Journal of Marketing* and *Journal of Marketing Research* and 37 issues of the *Journal of Consumer Research.*

[b]Refers to the amount of journal research space, in pages, allocated to replication research.

an exponent of the practice of replication, complains that those engaged in this work often are accused of intellectual mediocrity, lacking in creativity, time-wasting, and perhaps harboring a bullying spirit. Repeating his previous thoughts on this matter (D. W. Stewart, 2000, p. 688), I encourage you to contemplate the remarks of David Stewart (2002, p. 4), a former editor of the *Journal of Marketing*, who in a farewell address titled "Getting Published: Reflections of an Old Editor" offered the following words of advice to its readership:

> Replications are not compelling. . . . A replication that works has a "so what?" character. A replication that does not work raises questions about why. Replications may fail for many reasons, and most of these reasons are not interesting.

Factor in Huber's (2007, p. 2) opinion in an introductory quotation to this chapter and the general tenor among some former editors of marketing journals about the place of replication research in the pursuit of knowledge leaves little to the imagination. Such attitudes, coming from editors and reviewers—the gatekeepers of a discipline's vitality and future direction—are quite astounding. They betray, in an off-handed fashion to boot, a complete misunderstanding of the value of replication in the research mission. They also put into relief why Helfat (2007, p. 188) had to be on the defensive when countering that *replication* is not a dirty word.

Yet crucially, replications can be creative and make an impact (Hubbard & Lindsay, 2013b, p. 1396). To see this we showed how Kinney and Wempe's (2002; hereafter KW) replication of Balakrishnan, Linsmeier, and Venkatachalam (1996; hereafter BLV) offers a rewarding instance of how a single, imaginative study that incorporates aspects of both exact/close and conceptual extension replications (see Chapter 4) can promote learning by putting recognizable bounds on earlier findings. Unlike BLV, KW found that just-in-time (JIT) adopters, on average, were able to substantively outperform their non-adopting industry counterparts in terms of return on assets. BLV's results were much more confined, with improved return on assets witnessed only for those adopters with diffuse customer bases. This outcome led BLV to conclude that firms with concentrated customer bases, that is, powerful customers, were not able to benefit from the JIT-related cost savings. In an effort to closely replicate BLV's findings, KW took great pains to ascertain the between-study differences to see if these influenced the outcomes. Only when the subsamples were stratified by company size were KW able to closely replicate BLV's findings.

KW's replication yielded two major contributions. First, based on only two studies there is now quite commanding evidence indicating that the performance advantage does not extend to companies below a particular size threshold. For firms above this threshold (i.e., medium to large size), KW reported that an advantage exists. Furthermore, KW's analysis showed that company size has a nonlinear effect insofar as beyond a certain threshold there is no noticeable effect across firms. Second, KW were unable to replicate BLV's results based on stratifying the samples according to customer concentration. This finding, together with the exact/close replication built around stratifying samples by company size, casts doubt on the validity of BLV's explanation. In turn, this led KW to propose alternative rationales for the findings, that is, making a potential theoretical advance.

In addition, KW made efforts to place bounds on the applicability of their chief findings by performing a number of internal replications. These suggested that the JIT payoff is concentrated among the earliest (pre-1990) adopters, consistent with a first mover advantage. Also, KW looked at how long the advantage lasts and saw that it declined in years four and five. By year six, JIT adopters had no upper hand, probably because this practice was now being employed by rival companies. Finally, KW eliminated some competing explanations for their results by discovering that other potentially differentiating factors did not affect the results. It is difficult to deny that KW's study not only is inventive, but also makes an impression; and it certainly deserved being published in accounting's flagship journal.

Howard et al. (2009) offer a second example of the innovation and originality made possible when conducting replication research. They were concerned with determining whether effect sizes obtained in an emergent literature on implementation intentions (cf. Gollwitzer, 1999)—whereby adherence to even simple plans, like subjects recording on a piece of paper exactly when and where they were going to write the first page of their autobiography (Howard et al., 2009, p. 151), can result in marked impacts on behavior—were trustworthy. In pursuing this aim they performed three experiments leading to an affirmative response.

What makes Howard et al.'s (2009) article noteworthy are the extraordinary lengths they went to in order to approximate what Popper (1963) calls severe tests for corroborating the existence of a phenomenon. For example, various members of their research team (eight in all) held contradictory positions about the efficacy of implementation intentions—from robust, to exaggerated, to absent—thus minimizing possible experimenter bias in the *planning* of the studies. Pains also were taken to counterbalance any systematic effect the instructor may have had in biasing outcomes while *carrying out* the studies. Finally, in *analyzing* the data from no fewer than three different angles—null hypothesis significance testing, meta-analysis, Bayesian—Howard et al. make the strongest case imaginable for the credibility of their findings. Can anyone truthfully say that the design, handling, and analysis of these experimental replications did not call for the highest standard of researcher skills and ingenuity?

Despite the above examples, faced with deeply held editorial-reviewer prejudices, together with the choking publish-or-perish atmosphere descriptive of academe, researchers are better off sticking with "original" work and leaving replications to others. Undertaking replication research is almost guaranteed to thwart, if not destroy, career advancement. Sadly, views like these constitute a quintessential example of the fallacy of composition, also known as the tragedy of the commons (Hardin, 1968): What's good for the individual (researcher) is not necessarily good for the group (scientific community).[9] In short, such views help to institutionalize "disastrous science" (Hunter, 2001, p. 157).

5.6 The Publication Frequency of Critical Commentary

Ostensibly, science is a public, transparent enterprise motivated by the fundamental desire to create knowledge. In this way, the habitual practice of replication affords the most powerful assessment of the integrity and generalizability of a discipline's empirical literature. Nonetheless, it would be a mistake to neglect the important role of the research comment and rejoinder in this same endeavor. Between them, the comment and rejoinder foster a scholarly dialogue and exchange of ideas, elicit a diversity of opinion, and provide a forum for publicly thrashing out the merits of research results. It is no doubt for such reasons that in marketing, for example, the AMA Task Force on the Development of Marketing Thought (1988, p. 7), concerned about the usefulness of academic research, appealed for greater emphasis on this medium of scholarly expression in both the *Journal of Marketing* and *Journal of Marketing Research.*

Yet the evidence on the publication of commentaries in marketing journals points unreservedly in the opposite direction. Based on an annual random sample of issues from the *Journal of Consumer Research, Journal of Marketing,* and *Journal of Marketing Research,* Scott Armstrong and I (Hubbard & Armstrong, 1994) found that, as a percentage of all published work, the appearance of commentaries declined from 1974–1979

(12.9%) to 1980–1989 (8.4%). Decreases in the publication frequency of commentaries also are seen in economics (based on random samples of the *American Economic Review, Journal of Political Economy,* and *Review of Economics and Statistics*) between 1965–1975 (25.2%) and 1976–1989 (17.0%; Hubbard & Vetter, 1992); in finance (based on random samples of the *Journal of Finance, Journal of Financial Economics, Journal of Financial and Quantitative Analysis,* and *Journal of Money, Credit and Banking*) between 1969–1982 (15.3%) and 1983–1989 (4.7%; Hubbard & Vetter, 1991); and in management (based on random samples of the *Academy of Management Journal, Administrative Science Quarterly, Journal of Applied Psychology,* and *Organizational Behavior and Human Decision Processes*) between 1970–1979 (6.0%) and 1980–1989 (2.6%; Hubbard, 1990). Of course, these data do not speak to the publication rate of commentaries in more recent years. Consequently I updated this number for the *Journal of Marketing* by inspecting the contents of a randomly drawn annual issue for the 10-year period 2000–2009. Of the 85 research reports involved, none were reflections on published work. So much for the recommendation of the AMA Task Force.

5.7 Conclusions

Under the auspices of the significant difference paradigm, this chapter has shown that the empirical literature of the management and social sciences consists almost solely of unsubstantiated, one-off results. Very few replications of any kind are found in these areas. Compounding this troublesome state of affairs, of the few replications that are published in the business disciplines, almost one half, on average, are said to contradict their predecessors.

Several reasons were offered to account for the paucity of replications in the extant business and behavioral science literatures. Some of these are symptoms of researcher misunderstandings and occasionally ignorance. The major culprit, however, is an editorial-reviewer bias which, with few exceptions, is at best indifferent, and at worst hostile, toward that most critical axiom of scientific practice—establishing the replicability of results.

Furthermore, the lack of replication research, coupled with a dwindling publication of research comments and rejoinders, means that the scholarly environment characterizing the significant difference paradigm is becoming increasingly private in orientation, and not the open, vibrant world that science is supposed to inhabit. It is not the kind of world conducive to the emergence of empirical regularities, strong theory, and knowledge that can inform management practice and help prescribe effective social policies.

Notes

1. The studies listed in Table 5-1 involved, by discipline, content-analyses from among the following publications. Accounting: *The Accounting Review, Journal of Accounting and Economics, Journal of Accounting Research.* Advertising: *Current Issues in Advertising and Research, Journal of Advertising, Journal of Advertising Research, Journal of Broadcasting, Journal of Communication, Journal of Consumer Research, Journal of Marketing, Journal of Marketing Research, Journalism Quarterly, Proceedings of the American Academy of Advertising, Proceedings of the American Marketing Association, Proceedings of the Association for Consumer Research, Public Opinion Quarterly.* Economics: *American Economic Review, Journal of Political Economy, Quarterly Journal of Business and Economics, Quarterly Journal of Economics, Review of Economics and Statistics.* Finance: *Financial Analysts Journal, Financial Management, Financial Review, Journal of Banking and Finance, Journal of Business, Journal of Business Finance and Accounting, Journal of Finance, Journal of Financial Economics, Journal of Financial and Quantitative Analysis, Journal of Financial Research, Journal of Money, Credit and Banking, Journal of Portfolio Management.* Forecasting: *International Journal of Forecasting, Journal of Forecasting.* Management: *Academy of Management Journal, Administrative Science Quarterly, California Management Review, Human Resource Management, Journal of Applied Psychology, Journal of General Management, Journal of Management, Journal of Management Studies, Organizational Behavior and Human Decision Processes.* Marketing: *Journal of Consumer Research, Journal of Macromarketing, Journal of Marketing, Journal of Marketing Research.*

2. A digression regarding Fuess's (1996) higher replication estimate of 22.5% for the economics discipline is necessary. His figure is based solely on the incidence of replication research published in the *Quarterly Journal of Business and Economics*, a journal whose editorial policy is unique insofar as it actively solicits such work. Excluding the Fuess study, the replication estimate in economics is 9.1%, ±1.2%.

3. Coincidentally, these numbers are in general agreement with the 47% of respondents to an online survey of 1,292 psychologists from 42 countries indicating that they had doubts about the replicability of findings in their area (Fuchs, Jenny, & Fiedler, 2012, p. 641).

4. It must be added that Mitchell (2012) obtained less impressive results on the generalizability of effects from laboratory experiments in social psychology research.

5. Projective techniques are employed when it is thought that the motivation for human behavior is largely unconscious. Such methods, developed mostly in clinical psychology, include word association tests, sentence completion, and storytelling (which is what Haire used).

6. In common with Craig and Reese (1973), no indication is given as to whether these 49 responses would permit a replication, and no information on any reanalyses of these data sets is offered.

7. The most recent APA (2010, p. 12) *Publication Manual* continues to advocate the importance of investigators retaining and sharing data:

> Researchers must make their data available to the editor at any time during the review and publication process if questions arise with respect to the accuracy of the report. Refusal to do so can lead to rejection of the submitted manuscript without further consideration. In a similar vein, once an article is published, researchers must make their data available to permit other qualified professionals to confirm the analyses and results (APA Ethics Code Standard 8.14a, Sharing Research Data for Verification). Authors are expected to retain raw data for a minimum of five years after publication of the research. Other information related to the research (e.g., instructions, treatment manuals, software, details of procedures, code for mathematical models reported in journal articles) should be kept for the same period; such information is necessary if others are to attempt replication and should be provided to qualified researchers on request (APA Ethics Code Standard 6.01, Documentation of Professional and Scientific Work and Maintenance of Records).

Whether this latest appeal has any more impact than those preceding it is an open question.

8. It is clear at this stage that Stremersch and Lehmann (2008) are way off base. As shown in this book there are precious few examples of well-established findings in marketing, or elsewhere in the business and behavioral sciences for that matter. Moreover, Chapter 8 reveals that little work in these same areas truly is original.

9. Still another twist in the editorial distortion of the research publication process must be raised. This involves increasing evidence of editors manipulating the impact factor of their journals by coercing authors to cite articles published in them more frequently in their (revised) submissions (Honig, Lampel, Siegel, & Drnevich, 2013). Since replication research is widely ignored, if not demeaned, by editors, such work is highly unlikely to feature in these "requests," thus further dampening their citation rates.

CONCEPTION OF GENERALIZATION/EXTERNAL VALIDITY

> We have not put much emphasis on random sampling for external validity, primarily because it is so rarely feasible in experiments. (Shadish, Cook, & Campbell, 2002, p. 91)

> In the last analysis, external validity—like construct validity—is a matter of replication. (Cook & Campbell, 1979, p. 78)

6.1 Introduction

Scientists are interested chiefly in results possessing some degree of generality. Suppositions concerning generalizability or external validity—the degree to which a causal relationship is maintained over changes in persons, settings, treatments, outcomes (Shadish et al., 2002, p. 83), and times (Cook & Campbell, 1979, p. 37)—are at odds between those with significant difference and significant sameness outlooks. This is an issue with far-reaching implications. It is tackled in this chapter.

Specifically, Section 6.2 adumbrates the conception of generalization shared by those of a significant difference frame of mind. This view embraces the notion of *statistical* generalization, with its dependence on random sampling, or the so-called representative model of generalization. Despite the recipes found in almost all statistics and methods texts for implementing this approach, it is shown that the representative model is difficult to put into practice and can, surprisingly, even impede knowledge development.

While recommending it in principle, Section 6.3 explains that those advocating significant sameness are familiar with the stringent limitations of the above approach. Emphasis in this paradigm therefore is on *empirical* generalization, that is, attempting to project results across many small, clearly identified (sub)populations. More often than not this process will involve the use of nonrandom, especially purposive, samples. This is not an undue concern; major scientific advances witnessed over the centuries took place without the benefit of randomization.

A brief summary of the contents of this chapter are offered in Section 6.4. It crystallizes the pragmatism, not to say superiority, of the significant sameness over the significant difference philosophy of generalization.

6.2 Significant Difference

In the social and management sciences, external validity has become equated with statistical generalization (Hubbard & Lindsay, 2013a; Lee & Baskerville, 2003), or what Cook and Campbell (1979) call the *representative* model of generalization. The importance of representativeness was brought to the attention of social scientists by survey researchers for whom generalizing *to* a population with known probability of error was vital (Mook, 1983, p. 384). As a consequence, authors of popular and multiple-edition methodology texts (e.g., Babbie, 1992, p. 197; Henry, 1990, p. 11; Iacobucci & Churchill, 2010, p. 285; Kerlinger, 1973, pp. 118–119; Selltiz, Wrightsman, & Cook, 1976, pp. 515–516; J. L. Simon & Burstein, 1985, p. 107) prescribe routinely that samples should be as representative as possible of the target population. This has led to an emphasis on random (representative) sampling, calculating sample averages, and performing tests of statistical significance.

Representativeness or statistical generalization is critical in what Deming (1975, p. 147) calls *enumerative* studies, whose objective is to describe or estimate some specific population parameter on the basis of drawing a random sample from the appropriate population members. Hence the focus on generalizing from sample *to* population. An example might be "What is the proportion of empirical articles published in leading academic journals which use statistical significance testing?" In this kind of study, canvassing the entire population yields the "correct" answer.

By way of contrast, in what are referred to variously as *analytic* or comparative studies, the goal is to make predictions about some process or cause system, or changes to it (Deming, 1975, p. 147). An example in this case might be attempting to answer the question "Will increasing advertising expenditures raise sales revenues for product A in company B?" In such circumstances representativeness (sought via random/probability sampling) often is impractical and even harmful to knowledge development.

To see this it is necessary to review Hahn and Meeker's (1993) elaboration of Deming's (1975) ideas. These statisticians urge us to remember that statistical generalization is premised on compliance with two fundamental requirements (Hahn & Meeker, 1993, p. 3). First, the population must be finite, well defined, and unchanging. Second, a random sampling mechanism must be involved in the choice of units. This is because "randomisation will suffice to guarantee the validity of the test of significance, by which the result of the experiment is to be judged" (Fisher, 1971, p. 21).

Fulfilling these prerequisites means that the researcher is able to define the "relevant" population and specify a *sampling frame,* that is, a listing of population elements from which the random sample will be drawn. When presented in textbooks this procedure seems simple enough. In reality, Hahn and Meeker (1993, p. 4) warn that for analytic (and even enumerative) studies, things are far different because

> we define an analytic study to be a study in which one is *not* dealing with a finite, identifiable, unchanging collection of units, and, thus is concerned with a process, rather than a population. . . . [I]n our experience, the great majority of applications encountered in practice [including most designed experiments, p. 10] . . . involve analytic, rather than enumerative studies. Moreover, it is inherently more complex to draw conclusions from analytic than from enumerative studies; analytic studies require the critical (and often unverifiable) added assumption that the process about which one wishes to make inferences is statistically identical to that from which the sample was selected.

Honoring these demands is riddled with difficulties in social situations. So as Hahn and Meeker (1993, p. 5) declare, random sampling rarely is a viable option in analytic cases owing to the fact that "one is no longer dealing with 'an aggregate of identifiable units,' [so] there is no relevant frame from which one can take a random sample." This is in line with fellow statisticians' experiences, like Chatfield's (2002, p. 4) and Kass's (2011, p. 8), that genuine random samples are few and far between in the non-experimental (and even experimental) sciences, but nevertheless are analyzed habitually as if they are. It is also a view consistent with the psychologist Mook's (1983, p. 381) assertion that "nobody does it," that is, selects a random sample from the population of interest, and Gigerenzer's (2004, p. 599) observation that seldom are experimental subjects randomly sampled from a specific population. This drawback is further highlighted in the first quotation beginning this chapter.

Information from Shaver and Norton (1980a, p. 5, 1980b, p. 11) corroborates the above opinions. They estimated that of empirical work published mostly in the 1970s in the *American Educational Research Journal, Social Education,* and *Theory and Research in Social Education,* fully 81%, 87%, and 82%, respectively, involved the use of nonrandom samples.

More recent evidence from the field of marketing, displayed in Table 6-1, affirms the prevalence of nonrandom samples. This evidence is based on Collier and Bienstock's (2007) content analysis of 481 data-based studies published in the *Journal of Marketing, Journal of Marketing Research,* and *Journal of the Academy of Marketing Science* for the 5-year period 1999–2003. Overall, 86.9% of these studies employed nonrandom samples, with 83.6%, because of the preponderance of student

Table 6-1 Sampling Procedures Used in Three Marketing Journals: 1999–2003

	JM		*JMR*		*JAMS*		*Total*	
Sampling Procedure	*Number*	*%*	*Number*	*%*	*Number*	*%*	*Number*	*%*
Nonrandom	105	77.2	217	93.1	96	85.7	418	86.9
Convenience	96	70.6	213	91.4	93	83.0	402	83.6
Judgment	5	3.7	1	0.4	2	1.8	8	1.7
Snowball	3	2.2	1	0.4	1	0.9	5	1.0
Quota	1	0.7	2	0.9	–	–	3	0.6
Random	31	22.8	16	6.9	16	14.3	63	13.1
Simple	30	22.1	8	3.4	13	11.6	51	10.6
Stratified	1	0.7	3	1.3	2	1.8	6	1.2
Systematic	–	–	–	–	1	0.9	1	0.2
Probability	–	–	4	1.7	–	–	4	0.8
Cannot determine	–	–	1	0.4	–	–	1	0.2
Total	136		233		112		481	

Source: Adapted from Collier & Bienstock (2007, p. 172).

Note: JM = Journal of Marketing, JMR = Journal of Marketing Research, JAMS = Journal of the Academy of Marketing Science.

samples (Collier & Bienstock, 2007, p. 170) being of a convenience variety. The percentage of nonrandom samples by journals is *Journal of Marketing* (77.2%), *Journal of Marketing Research* (93.1%), and *Journal of the Academy of Marketing Science* (85.7%). Table 6-1 shows a detailed breakdown of the types of sampling plans employed by researchers.

Not only is the prospect of random *sampling* from a clearly defined frame a precarious option, it should be added that even random *assignment* can be complicated. As Shadish et al. (2002, pp. 276–278) argue, random assignment may be neither feasible nor desirable. First, randomized field experiments can take years to set up and are thus untimely when quick answers are sought. Second, when a reasonable amount of evidence is available concerning the beneficial effects of a treatment, a randomized experiment may be superfluous. Third, randomly assigning subjects to dangerous treatments (e.g., smoking) is unethical. Fourth, some randomized studies may be done prematurely and therefore waste resources. Fifth, the loss of test units through experimental mortality is worrisome. For these reasons and others, Shadish et al. counsel, "it is easy to forget the many practical problems that can arise in implementing randomized experiments" (p. 277). Typically, however, random assignment is easier to implement than random sampling, especially when the test units are students.

Regardless of the telling distinction between enumerative and analytic studies, this topic "has successfully eluded most books and most teaching" (Deming, 1975, p. 147). And despite its "central role in applications," this same distinction had to be reiterated almost 20 years later by Hahn and Meeker (1993, p. 3). Even in the present day, statistics and methodology textbooks continue to ignore this issue.

The following incomplete list of examples illustrates how drawing a random sample in practice can pose major headaches:

- It may be difficult to define the "relevant" population. For example, based on an inspection of six volumes of the *American Educational Research Journal* covering the period 1968–1977, Shaver and Norton (1980b, pp. 11–12) were disappointed to find that few articles contained an explicit definition/description of the population from which the sample came. Their overall impression was that these definitions and descriptions were cursory and incidental. Shaver and Norton (1980a, p. 5) came to this same conclusion with regard to research published during the 1970s in the periodicals *Theory and Research in Social Education* and *Social Education.*
- It is often the case that sampling frames are unavailable. Henry (1990, p. 85), in his book *Practical Sampling,* writes that obtaining lists of the general U.S. population is difficult, if not impossible. It is a concern raised

also by Blankenship and Breen (1993, p. 169). This is one of the reasons favoring the use of cluster samples, where sampling frames are needed only for those units in the randomly selected clusters (subgroups) used in the investigation. Matters get worse. Assessing external validity necessitates the sampling of settings, treatments, and observations in addition to people. But as Shadish et al. (2002, p. 344) comment, locating comprehensive lists for these is extraordinarily problematic.

- If available, sampling lists may be obsolete or otherwise inaccurate. Such lists vary in quality.
- Rarely is there a one-to-one correspondence between the population and the sampling frame, thus precluding attainment of a true random sample. Use of the telephone directory as a sampling frame is a clear case in point. The two circles in Figure 6-1 depict the population and the sampling frame, the degree of overlap revealing their congruence. Yet the area labeled A in the figure shows people who are still listed in the "current" phone book but no longer belong to the population due to death or leaving the region. Conversely, the area labeled B contains people who belong to the population but are not included in the phone directory because they don't own a phone, are new to the region and so have yet to be listed, have unlisted numbers, or communicate via cell phone.
- Nonresponse bias, generally in the form of not-at-homes and refusals to cooperate (Iacobucci & Churchill, 2010, p. 331), can be a scourge in survey research. This bias compromises procurement of a random sample.
- Owing to exorbitant costs, random sampling may not be feasible. Therefore, no company test markets its products on the basis of a random sample of variously sized metropolitan areas. Here, judgment or purposive sampling is the only realistic alternative. So Des Moines, Iowa, is chosen subjectively as a test market on the grounds that it is "representative" of consumers in the Midwest. Following this logic, San Antonio, Texas, is

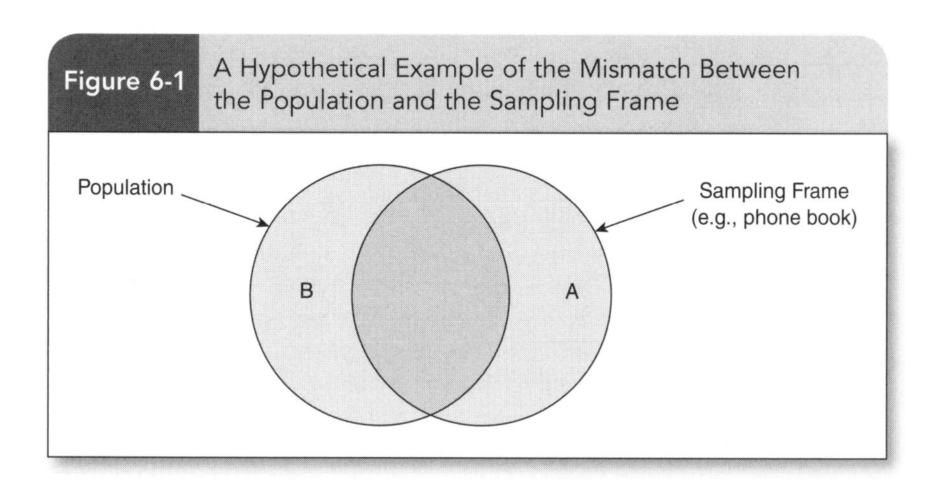

Figure 6-1 A Hypothetical Example of the Mismatch Between the Population and the Sampling Frame

believed to "mirror" the Southwest, and Baltimore, Maryland, is held to be "typical" of the Northeast.

In a similar vein, much-needed everyday information is accepted without blinking on the basis of nonrandom samples. Take the Consumer Price Index as an example. Iacobucci and Churchill (2010, p. 287) alert us to the fact that

> most people believe the Consumer Price Index (CPI) reflects prices everywhere in the U.S. Actually, the CPI is called the Consumer Price Index for Urban Wage Earners and Clerical Workers, and it samples a little over 50 cities, selected on the bases of judgment and political pressure.

Resource constraints likewise dictate the use of subjective judgment in the *snowball* sampling of "special" populations. If the target population consists of "multi-millionaires," a survey of the, say, Des Moines, Iowa, metropolitan area will turn up some of these, but this is a grossly wasteful process. Better to locate some multi-millionaires nonrandomly and have them identify others for the study.

- Classifying random sampling as representative sampling is something of a misnomer. Use of random sampling does not guarantee representative results.[1] If a random sample estimate of a population parameter comes from toward or in the tails of a distribution, then it will be markedly unrepresentative. Pivotally, of course, use of random samples legitimizes the calculation of the sampling error associated with an estimate of a population parameter. And this is why random sampling, if viable, is the road to follow.
- The prerequisite that the population be unchanging also is of concern. Sampling units do change as, for instance, Anastasi (1982, pp. 339–341) describes in connection with fluctuating IQ scores. Also, test units can change in response to experimental "maturation." Further views on the changing nature of populations are raised by Henry (1990, p. 85).

A host of factors conspire to prevent the selection of an authentic random sample. This explains why Hahn and Meeker (1993, p. 10) are comfortable with their determination: "It seems safe to say that in applications the simple textbook case of an enumerative study in which . . . from the target population . . . one has a random sample . . . is the exception, rather than the rule." It also underlies Deming's (1975, p. 151) somber prognosis that

> little advancement in the teaching of statistics is possible, and little hope for statistical methods to be useful in the frightful problems that face man today, *until the literature and classroom be rid of terms so deadening to scientific enquiry as null hypothesis, population (in place of frame), true value, level of significance for comparison of treatments, representative sample*. (emphasis added)

Yet from what I have seen, today's statistics and methodology textbooks show that nothing much has changed in the past 40 years when it comes to mentioning as an aside, much less itemizing, the difficulties involved in obtaining genuine random samples.

Putting things into perspective, the representative model of statistical generalization often is impractical/unworkable. Most studies in the management and social sciences cannot meet the cardinal assumptions that samples be selected randomly from well-defined, finite, and unchanging populations. This is no cavil, but a severe blow to those of a significant difference persuasion; without random sampling, traditional methods of statistical inference collapse. In Fisher's (1947, p. 435; cited in Lindsay, 1995, p. 37) words, "tests of significance would be worthless." Given dire consequences like these, it makes no sense to ground a conception of generalization on precepts which all too often cannot be upheld.

Compounding the problems associated with the suitability of the representative model of generalization is the fact that it is an *illusory* notion when applied to predictive studies. The reason for this is that assessments of external validity must feature variations that are included in the study together with those which are not (Denrell, 2003; Hubbard & Lindsay, 2013a, p. 1384; Shadish et al., 2002, p. 83). To be secure in making wide-ranging generalizations normally would obligate sampling from a "super-population" or universe composed of all likely, say, buyers and sellers (e.g., past, present, and future), settings, manipulations, stimulus levels, time periods, national and organizational cultures, researchers, methods and measures, and so on (Leviton, 2001). It is obvious that no individual study has this reach and therefore provides only meager insight for appraising the external validity of a result (Leviton, 2001, p. 5197).

Finally, and counterintuitively, not only is representativeness an overly ambitious goal, but efforts to obtain it can reduce the precision of the results in terms of their application or generalizability to specific contexts. In other words, when making a prediction it is necessary to entertain the various conditions (subpopulations) which might impact a result and address them *explicitly.*

An example suggested by Deming (1975, p. 150), and embellished here, clarifies this position. Suppose we are interested in estimating the average yield per acre of two varieties of seed corn, A and B, based on a simple random sample of those countries around the world amenable to the growing of corn. Imagine this survey finds that A has a 15% greater yield than B—a result which is statistically significant. Such a result supports the generalization that the *aggregate* measures of central tendency differ systematically (i.e., not by chance, assuming a true null hypothesis) in the overall population. It indicates that, *on average*, the yield per acre of corn seed A exceeds B's across the world. However, this inference, in Deming's (p. 150) take, is "nigh useless" for predicting, say, the situation existing in certain parts of rural Ukraine specializing in corn production with farmers with more than 20 years of crop-growing experience. Otherwise expressed, the aggregate result is scattered over so many conditions, such as country, soil types, rainfall, temperatures, angle of inclination of the land, degree of weeding, fertilizer usage, age of farming equipment, and farmer education levels, that no practical inference can be drawn because it does not pinpoint *when* or *where* the relationship of a 15% greater yield for A over B may or may not be expected to hold.

Continuing along these lines, turn to David Kent and Rodney Hayward's (2007) informative article "When Averages Hide Individual Differences in Clinical Trials." In most randomized clinical trials, patients are assigned to experimental and control groups, and the efficacy (or otherwise) of the treatment is evaluated by conducting a statistical significance test on the differences in the *average* scores between the two groups. But as Kent and Hayward point out, "because many factors other than the treatment affect a patient's outcome, determining the best treatment for a particular patient is fundamentally different from determining which treatment is best on average" (p. 60). They go on to say that reporting a single number (the mean) is misleading:

> Indeed, some patients may benefit substantially from a treatment even when the overall results from a trial are negative. Or a treatment with benefit on average may be extremely unlikely to help most patients, while being more likely to harm than help some others. But unless the trial investigators *analyze their data looking for these* [risk-stratified] *subgroups, the physicians cannot know whether they exist.* (p. 62; emphasis added)

Unfortunately, after reviewing the results of 108 clinical trials in four major (unlisted) journals in 2001, Kent and Hayward found that only one followed a risk-stratified subgroup approach. They noted further that they had not found many since.

The examples above by Deming (1975), Hahn and Meeker (1993), and Kent and Hayward (2007) show that much is wrong in the acquisition and analysis of data in the significant difference paradigm. They also help to account for the seemingly paradoxical result that "a true statement about averages across a population may be false for every subset of that population, including every individual in that population" (Starbuck, 1993, p. 889).

6.3 Significant Sameness

Cook and Campbell (1979) believe that a strong case can be made that external validity is enhanced more by several heterogeneous small studies than by one or two large ones employing representative (probability) sampling:

> With the latter one runs the risk of having heterogeneous treatment, measures that are not as reliable as they could be, and measures that do not reflect the unique nature of the treatment at different sites. Many small-scale experiments with local control and choice of measures is in many ways preferable to giant national experiments with a promised standardization that is neither feasible nor even desirable. (p. 80)

Of major importance here is the use of purposive or judgment samples (Deming, 1975, p. 151; Hahn & Meeker, 1993, p. 7; Hubbard & Lindsay, 2013a, p. 1384) married with a replication strategy which attempts to generalize results *across* multiple narrow, well-defined (sub)populations. This is in accord with McKelvey's (1978, p. 1438) reckoning that "solid findings about a narrower population are better than marginal findings of questionable generalizability to a broadly defined population." Then these multiple studies must be examined to see whether the same (similar) result holds, which defines how empirical (predictive) generalizations are developed.

A key concern in the design of persuasive observational (nonrandomized) studies is to replace the *control* that is so evident in laboratory experiments with a careful *choice* of environment, one that permits the treatment effects to be seen with great clarity (Rosenbaum, 2001, p. 223).[2] This can be accomplished using judgment samples in a replication research program.

Note, however, that the above account describes how much scientific headway actually is realized. Shadish et al. (2002, pp. 353–354) outline five principles of generalized causal inferences that scientists use every day in their work. First is the notion of *surface similarity,* meaning scientists generalize by rating the likenesses among phenomena. Second is the *ruling out of irrelevancies,* whereby scientists generalize by deciding on those features of persons, settings, treatments, and outcomes which are irrelevant because they do not change a generalization. Third is *making discriminations,* or finding limits to a generalization. Fourth is *interpolation and extrapolation,* whereby scientists generalize within the range of sampled persons, settings, treatments, and results, as well as those beyond them, respectively. Fifth is *causal explanation,* or scientists generalizing by developing and testing explanatory theories. Take note that none of these five principles involve formal (random/probability) sampling techniques.

In cautionary mode, Shadish et al. (2002, p. 356)

> remind the reader that these purposive sampling methods are not backed by a statistical logic that justifies formal generalizations, even though they are more practical than formal probability sampling. Still, they are by far the most widely used sampling methods for generalized causal inference problems.

That reliance on nonrandom samples is pervasive in the sciences is not a concern. Indeed, entire disciplines such as archaeology, dendrochronology, and paleontology, not to mention police forensic work, are sustained on the basis of convenience or accidental, let alone purposive, samples.

As Beveridge (1951, p. 101, cited in Flyvbjerg, 2001, p. 75), Deming (1975, p. 149), Ehrenberg (1983, p. 249), and Rubin (2007, p. 20) point out, much of our knowledge of causal effects has been created through the use of small, nonrandom samples in analytic studies (e.g., hospitals and clinics). For instance, the statistician Freedman (1999) describes how the employment of nonrandom samples was instrumental in drawing the causal links between contaminated water and the spread of cholera, and between cigarette smoking and lung cancer, to name but two examples. Importantly, he concurs that "most of what we know about causation . . . is derived from observational studies" (Freedman, 1999, p. 255).

Fisher—the father of randomization—also used nonrandom samples. Specifically, his work at Rothamsted Experimental Station was based on convenience samples. To be sure, Fisher used random assignation, but the samples themselves were strictly of the convenience kind.[3]

In the management field, Bamberger and Pratt (2010) similarly advise readers to be mindful that much scientific headway has been, and continues to be, achieved through the use of nonrandom samples. That people need to be reminded today that colossal advances in scientific knowledge accrued over the centuries without benefit of Fisher's randomization and statistical testing is deeply troubling, and says much about the monopolistic grip exercised by the significant difference paradigm in the social and management sciences. It also speaks poorly of our dominant conception of science itself.

6.4 Conclusions

This chapter has shown that replication research is invaluable when generalizing results on the basis of nonrandom samples. Random sampling is required in order to employ legitimately formal methods of statistical inference. For this reason, working with random samples is considered a methodological necessity by members of the significant difference school, though we have seen that this is problematic to implement in reality. But even when accomplished, it is extraordinarily difficult to identify examples of knowledge breakthroughs in the behavioral and business disciplines made possible by the application of random sampling and statistical significance testing. The fact is, use of nonrandom samples predominates in actual scientific life. Yet this is not a concern if a replication strategy is followed.

When considering matters related to the external validity of findings, then, the lesson is clear. It makes more sense to adopt the significant sameness notion of empirical generalization over that of statistical generalization prescribed by the significant difference philosophy.

Notes

1. In fact, nonrandom samples can produce more accurate estimates than those based on randomization. This was demonstrated empirically by "Student" (William S. Gosset) in his 1938 posthumously published article, rebutting his mentor, Fisher, on balanced versus random arrangements of field plots.

2. This would include the matching of units in experimental and control groups on covariates of interest. Rubin (2007) provides a useful discussion of this approach in observational study contexts.

3. Fisher's views on the nature of probability and random sampling are controversial and sufficient to warrant an appendix to this chapter.

APPENDIX TO CHAPTER 6

Fisher's Views on Probability and Random Sampling

Fisher's ideas on these topics are contentious. Much of his work involved practical situations in which time and costs permitted the gathering of only limited experimental data and random sampling from a fixed, well-defined population (à la Neyman–Pearson) was not possible. Hence Fisher's concerns with small sample theory and inference in the single case. Statistical inference for Fisher (1922, p. 311) is

> accomplished by constructing a hypothetical infinite population, of which the actual data are regarded as constituting a random sample. The law of distribution of this hypothetical population is specified by relatively few parameters, which are sufficient to describe it exhaustively in respect of all qualities under discussion.

In other words, the data at hand—on which the tests of significance are to be carried out—are conceived as simply one of the data sets that could have been, and in this particular instance were judged to be, produced by hypothetical infinite repetitions of the experiment. Here is Fisher's (1922, p. 313) account:

> It should be noted that there is no falsehood in interpreting any set of independent measurements as a random sample from an infinite population; for any such set of numbers are a random sample from the totality of numbers produced by the same matrix of causal conditions: the hypothetical population which we are studying is an aspect of the totality of the effects of these conditions, of whatever nature they may be. The postulate of randomness thus resolves itself into the question, "Of what population is this a random sample?" which must frequently be asked by every practical statistician. . . . [And so we interpret] the available observations as a sample from a hypothetical infinite population.

Or as Fisher (1959, p. 78) later proclaimed, "The only populations that can be referred to in a test of significance have no objective reality, being exclusively the product of the statistician's imagination through the hypotheses which he has decided to test."

The idea of a hypothetical infinite population as an imaginary construct was, however, baffling to philosophers of science (Hacking, 1965; Hogben, 1957) and statisticians (Inman, 1994; Kendall, 1943) alike. Zabell (1992, p. 374) used an interesting analogy in attempting to square Fisher's views on probability:

> In Fisher's writings, probability often seemed to live a curious Jekyll and Hyde existence; for much of the time, probability leads a quiet and respectable life as an objective frequency (Dr. Jekyll), but it occasionally transforms before our very eyes into a rational degree of belief or even a psychological mental state (Mr. Hyde, of course).

In seconding Zabell's (1992) evaluation, Efron (1998) characterizes Fisher's philosophy as a series of compromises between the Bayesian and frequentist viewpoints. Startlingly, these interpretations arise despite Fisher's (1970, p. 9) passionate renunciation of inverse probability (Bayesianism). According to some (e.g., Gigerenzer, 1993; D. A. Lane, 1980; E. L. Lehmann, 1993), Fisher struggled throughout his life over the meaning of probability.

CONTRASTS OVER STATISTICAL ISSUES

> I believe that the almost universal reliance on merely refuting the null hypothesis as the standard method for corroborating substantive theories in the soft areas is a terrible mistake, is basically unsound, poor scientific strategy, and one of the worst things that ever happened in the history of psychology. (Meehl, 1978, p. 817)

> My personal view is that *p*-values should be relegated to the scrap heap and not considered by those who wish to think and act coherently. (Lindley, 1999, p. 75)

7.1 Introduction

This chapter compares how the two schools of thought address four statistical issues. The first of these, discussed in Section 7.2, is about how the significant difference and significant sameness paradigms deal with model uncertainty concerns. Supporters of the former tend to consider this as mostly straightforward, something that can be handled by the relevant diagnostic tests in a single set of data. Adherents of the latter stress that model uncertainty issues call for the acquisition of new data. Section 7.3 looks at the nature of the predictions made in the two camps. Those faithful to the significant difference paradigm are content to settle for directional predictions, while champions of significant sameness argue for much greater predictive precision.

The importance of *p*-values in the two paradigms is considered in Section 7.4. Their reporting is mandatory according to the significant difference philosophy, whereas they play only an incidental role for those of a significant sameness bent. These roles are reversed when it comes to the reporting of effect sizes and confidence intervals (CIs). As Section 7.5 illustrates, in the significant difference model they are seldom provided, while their use is emphasized in the significant sameness approach. Concluding comments are made in Section 7.6.

7.2 Model Uncertainty

7.2.1 Significant Difference

Causal inferences are predicated on the assumption that a valid model has been postulated, meaning that specification errors addressed routinely in econometrics texts are not a worry. In other words, the feasibility of this assumption depends on such things as not omitting relevant and including irrelevant explanatory variables, capturing the appropriate functional form of the regression equation, correctly specifying the manner in which the stochastic disturbance term enters the equation, and so on (see, e.g., Kmenta, 1971, p. 392). The impasse is that the truthfulness of this assumption is beyond human recognition (Kingman, 1989). This is why it is impossible for statistical tests to prove anything (Freedman, Pisani, & Purves, 1978), why models can be used but not believed (Theil, 1971, p. vi), and why "all models are wrong" (Keuzenkamp & McAleer, 1995, p. 19).

Notwithstanding the above, Chatfield (1995) alleges that statisticians, never mind applied researchers, have not given the topic of model uncertainty the attention it deserves. Perhaps this inattention is due to the fact that model uncertainty issues often are taken to be unproblematic, in the sense that they can be dealt with by using the appropriate diagnostic statistical tests. Chief among these are model specification searches (see, e.g., Leamer, 1978), generally performed on the same data set that is used to both formulate and estimate the model, and which frequently can masquerade as data mining exercises. As Chatfield cautions, parameter values from such a model(s) are biased and overoptimistic. Model estimates arrived at by data mining lead to exaggerated claims of statistical significance (Lovell, 1983).

Null hypothesis significance testing exacerbates the model uncertainty dilemma. This is acknowledged by Freedman et al. (1978, p. 492):

> The procedure for carrying out a test [of statistical significance] doesn't force the investigator to check the model. This is a real flaw in the method, because usually the investigator sets things up so that what he wants to prove becomes the alternative hypothesis. He leaves its opposite—the null hypothesis—to take the heat and look ridiculous when a difference turns up. This is almost bound to happen if the model is wrong. . . . In a way, tests of significance reward investigators for getting the model wrong.

7.2.2 Significant Sameness

The model uncertainty problem can't be ignored and can be addressed only by *acquiring new data* (Chatfield, 1995, 2002; Kincaid, 1996, pp. 261–262; Ziliak & McCloskey, 2008, p. 120). Unfortunately, Chatfield (1995, p. 439) adds, statisticians seem mostly concerned with squeezing a single data set dry, which he calls a serious disease of statistical teaching. This would explain the concentration on developing statistical methods, for example, regression diagnostics in the model uncertainty case, for use on an isolated set of data. To underscore an earlier point, implicit in the statistician's focus on the analysis of single sets of data, rather than how to deal with many sets (Ehrenberg, 1990; Preece, 1987), is a commitment only to matters of internal validity.

Significant sameness, on the other hand, tackles the model uncertainty problem head-on by no longer assuming the very thing in doubt; instead, it explicitly puts any such concerns to the test. Lindsay and Ehrenberg (1993, p. 223) offer two yardsticks for dealing with model uncertainty: "When in doubt find out" and "Does the same result occur *despite* differences in conditions." Accordingly, researchers must nurture a replication tradition by seeking to establish significant sameness across many sets of data. They must also recognize that while statistical methods play a valuable role in determining significant sameness, they are nevertheless subordinate to matters of research design.

7.3 Nature of Predictions Made

7.3.1 Significant Difference

In what Haig (2014, p. 116) calls a "guess-and-test strategy," customary hypothetico-deductivism relies on predictive accuracy as the arbiter of theory success. Clearly, the more specific a theoretical relationship is described, the more useful and practical a generalization becomes. Given this, the statistical significance test is especially disappointing because it allows only a qualitative or *directional* prediction, for example, "as X increases, Y increases." Some proponents of statistical significance testing, such as Frick (1996) and Cortina and Dunlap (1997), maintain that this is all that social science theories are capable of. On this matter, Ziliak and McCloskey (2008, p. 114) write, censoriously, that the first American Nobel laureate in economics, Paul Samuelson, also was of this opinion. It is an opinion which continues to dominate to the present time. To see this it is enough to peruse the "comparative statics" section of any theoretical article in economics.

Pursuing this argument, McCloskey (2002, p. 55) is distraught over the harm inflicted on the discipline by the widespread commitment of what she labels the Two Sins of economics—being content with only qualitative predictions in *both* theory *and* applied work:

> The progress of economic science has been seriously damaged. You can't believe anything that comes out of the Two Sins. Not a word. It is all nonsense, which future generations of economists are going to have to do all over again. Most of what appears in the best journals of economics is unscientific rubbish. I find this unspeakably sad.

Sad indeed. She continued that, akin to difficult chess problems, the Two Sins of abstract, qualitative economic theorizing and (accept/reject) econometric analysis are mentally demanding activities. Yet without a focus on quantities, "they are worthless as science" (McCloskey, 2002, p. 55). I echo McCloskey's concerns. Samuelson's "qualitative calculus" is insufficient.

This acquiescence among researchers to settle for mere directional predictions is further borne out by Jeffrey Edwards and James Berry (2010, p. 670). They identified the 20 most cited articles dealing with theory development published in the *Academy of Management Review* for the period 1985–2009. These 20 theories yielded a total of 183 propositions, and not one of them predicted a point value or even a range of values.

Due to the limited aspirations concerning the nature of predictions shared by acolytes of the significant difference paradigm, Edwards and Berry's (2010) results are to be expected. Ortinau (2011, p. 153), in his "Writing and Publishing Important Scientific Articles: A Reviewer's Perspective," lends his approval to this practice: "A critical element to important scientific investigations is the generation of testable hypotheses that accurately identify the *directionality* of the relationships between constructs" (emphasis added). No mention of magnitudes here, only directionality. Ortinau goes on:

> Reviewers look for straightforward discussions that simply report the acceptance and rejection decisions of the hypothesized relationships along with support from summary tables and figures displaying the data and test results. Authors must . . . not waste space by repeating verbatim what tables and figures display but rather present the key results that support either the acceptance or rejection of each tested hypothesis. Reviewers are not expecting detailed discussions about insights, implications the results are suggesting, or explanations why the results came out as they did. (p. 155)

Such views are troublesome. In particular, the reader is *owed* precisely the information that is dismissed cavalierly in the last sentence of Ortinau's quotation.

Haig and McCloskey excepted, the opinions voiced thus far are injurious to progress in the social and management sciences. On the one hand, they consecrate the ideas that hypothetico-deductivism and statistical significance testing are of foremost importance in the research process. On the other hand, these views (conveniently) alleviate the investigator of the responsibility for subjectively *interpreting* the social or managerial relevance of a study's effect sizes. Ortinau (2011, p. 151) denies this latter necessity, claiming that "knowledge is verifiable objective facts . . . and subjectivity is not an element of a scientific process." For Ortinau and others, statistical and substantive significance are one and the same, and the "scientific method" is an objective cookbook in which it is sufficient to reject directional hypotheses and push on. Such a conception of science is naïve in the extreme.

7.3.2 Significant Sameness

Here, the goal is assessing the *predictive precision* or similarity of effect sizes and/or quantitative relationships across studies (Hubbard & Lindsay, 2013b, p. 1395). For instance, reporting effect sizes (e.g., *d, r*), while far more informative than making qualitative predictions, is less precise than a quantitative model statement such as $Y = 8X + 3 \pm 2$. This is because effect sizes do not point to how the variables are associated, and as a consequence there is no way of determining whether the relationships truly generalize across investigations (Lindsay & Ehrenberg, 1993). In particular, effect sizes are susceptible to any sampling methods which constrain the range or strength of the independent variable (what Abelson [1997, p. 125] dubs "cause size"). In short, studies with smaller cause sizes are prone to yielding smaller effect sizes. By way of comparison, regression slope coefficients (*bs*) tend to be relatively immune to changes in the variability of X (J. Cohen & Cohen, 1983; J. Cohen, 1994; Abelson, 1997). The relationship among the three terms means that the regression slope *b* is analogous to the ratio of effect size to cause size (Abelson, 1997).

Predictive precision is an axial concept because of its intimate connection with the "severity" of a test. A mainstay of Popper's (1959, 1963) philosophy is that a theory is better corroborated the less probable the theory's prediction is relative to our background knowledge. One disclosure of a severe test is making precise predictions or predictions that differ from what an alternative theory might make. The more precise the prediction, the easier it is to falsify our ideas or our previously established regularities, which provides the momentum for the discovery of new patterns, theory development, or the identification of methodological concerns. As Popper (1963, p. vii) explains:

> Criticism of our conjectures is of decisive importance; by bringing out our mistakes it makes us understand the difficulties of the problem which we are trying to solve. This is how we become better acquainted with our problem, and able to propose more mature solutions.[1]

The concept of predictive precision fundamentally distinguishes significant sameness from significant difference, a distinction that was raised in an article by Meehl (1967) titled "Theory-Testing in Psychology and Physics: A Methodological Paradox." Management and social scientists like to test point null hypotheses (of the form $H_0 : \theta = \theta_0$ versus $H_A : \theta \neq \theta_{0,}$ where θ_0 is a specific value of θ, almost always zero) because they call for only weak, directional tests—"like playing tennis with the net down," to borrow Blaug's (1992, p. 241) analogy. As the statistical power of a test approaches unity, it is a virtual certainty that the null will be rejected because, in practice, the point null hypothesis *always* is false (Marden, 2000, p. 1318; Meehl, 1967, pp. 108–110; Serlin, 2002, pp. 219–220), that is, misspecified, due to the so-called crud factor (Meehl, 1990, p. 204, 1997, pp. 402–403), whereby everything in the social sciences is correlated with everything else. This means that the researcher's alternative (research) hypothesis will be "supported" by the test—even if it is "totally without merit" (Meehl, 1967, p. 111).

But matters need not be this way. For example, in the physical sciences the researcher's prediction becomes the hypothesis under test (i.e., equivalent to the null in the management and social sciences) and typically is some quantitative estimate (Guttman, 1977, p. 86; Meehl, 1967). Under these circumstances, Meehl (1967, p. 112) proceeds, improvements in the accuracy of determining this quantitative magnitude via better measurement instruments and/or use of a larger sample size "has the effect of *narrowing* the band of tolerance about the theoretically predicted value." A successful prediction in this situation provides a great deal of corroboration for a theory or confidence in the generalization because so many other alternative explanations are ruled out by its exactitude. Contrast this with traditional null hypothesis testing, where many hypotheses, in addition to the researcher's own, can be consistent with the non-null state of affairs observed in a high-powered test.

While the fallacy of affirming the consequent, mentioned in Chapter 2, is a basic point in elementary logic, its critical importance in matters epistemological bears repeating. For any result confirming a particular hypothesis, an infinite number of mutually incompatible hypotheses exist which can explain or imply the data equally well. For practical purposes, however, this possibility is lowest when predictions are extremely precise or so counterintuitive that no other plausible theory can currently explain them (Cook, 1983).

In principle, there is no reason why theories in the management and social sciences cannot yield precise (or interval) predictions. As noted above, however, this line of thinking flies in the face of conventional wisdom that theories in these areas are unable to specify point predictions. Given this view, it is not surprising why the belief that a numerical prediction cannot be specified goes unchallenged—*we never try to establish one* (Lindsay & Ehrenberg, 1993, p. 219). But complete ignorance in an area can happen only once; after an initial result is found, the basis for making a numerical prediction, and therefore achieving greater predictive precision, exists (Bound & Ehrenberg, 1989; J. R. Edwards, 2008; J. R. Edwards & Berry, 2010; Ehrenberg, 1993a; Hubbard & Lindsay, 2013a, p. 1385; Lindsay & Ehrenberg, 1993; Van de Ven, 2007, p. 21). The significant difference approach wantonly leads to tests with low predictive precision because later studies disregard what we learn from earlier ones concerning the quantitative relationship among the variables of interest and/or the effect size obtained. The upshot of continuing down the accustomed path of positing only qualitative instead of quantitative predictions, argues Taagepera (2008, p. 237), a physicist and author of *Making Social Sciences More Scientific*, is that these disciplines will never attain scientific status.

7.4 The Role of *P*-Values

7.4.1 Significant Difference

As Chapter 2 showed beyond doubt, the *p*-value from a statistical significance test is exalted in this paradigm. Over the years a litany of works deeply critical of the trifling role of statistical significance testing and *p*-values in fostering cumulative knowledge development have been published (see, e.g., Carver, 1978; Hubbard & Meyer, 2013; Kline, 2004, ch. 3; Lindsay, 1995; Nickerson, 2000; Ziliak & McCloskey, 2008). The comments of some of these are summarized in Table 7-1. Deliberately or otherwise, subscribers to the significant difference paradigm largely are impervious to this chorus of criticism.

As was shown further in Chapter 3, even when taken on its own terms the *p*-value is a minor statistic totally undeserving of the center stage it occupies in empirical investigations. What is quite astonishing, however, is that owing to researcher misconceptions this inconsequential statistic has been invested with numerous powerful attributes it simply does not possess.

In fairness, it must be acknowledged that in many cases researchers pick up these distortions from the very textbooks charged with instructing them. Over three decades ago, for example, Kalbfleisch and Sprott (1976, p. 259) remonstrated that "tests of significance are . . . mishandled in practically all statistics textbooks, and careful consideration of the

Table 7-1 Overt Criticism of the Worth of Null Hypothesis Significance Testing (NHST)

Authors	*Quotations*
Bakan (1966, p. 436)	The test of significance in psychological research may be taken as an instance of a kind of essential mindlessness in the conduct of research.
Carver (1978, p. 378)	The emphasis on statistical significance over scientific significance in educational research represents a corrupt form of the scientific method.
J. Cohen (1990, p. 1310)	I believe . . . that hypothesis testing has . . . diverted our attention from crucial issues. Mesmerized by a single all-purpose, mechanized, "objective" ritual in which we convert numbers into other numbers and get a yes-no answer, we have come to neglect close scrutiny of where the numbers came from.
J. Cohen (1994, p. 997)	Null hypothesis significance testing (NHST; I resisted the temptation to call it statistical hypothesis inference testing).
J. Cohen (1994, p. 997)	I argue herein that NHST has not only failed to support the advance of psychology as a science but also has seriously impeded it.
Cox (1977, p. 60)	As noted . . . there are considerable dangers in overemphasizing the role of significance tests in the interpretation of data.
Cox (1982, pp. 327–328)	The criterion for publication should be the achievement of reasonable precision and not whether a significant effect has been found.
Cox (1986, p. 120)	It has been widely felt, probably for 30 years or more, that significance tests are overemphasized and often misused and that more emphasis should be put on estimation and predictions.
Falk & Greenbaum (1995, pp. 75–76)	Our position is that the prevalence of the significance-testing practice is due not only to mindlessness and the force of habit . . . there are profound psychological reasons leading scholars to believe that they cope with the question of chance and minimize their uncertainty via producing a significant result.

Authors	*Quotations*
Gigerenzer (2004, p. 587)	Mindless statistics.
Greenwald (1975, p. 19)	It is to be hoped that journal editors will base publication decisions on criteria of importance and methodological soundness, uninfluenced by whether a result supports or rejects a null hypothesis.
Guttman (1985, p. 4)	We shall marshal arguments against such [statistical significance] testing, leading to the conclusion that it be abandoned by *all* substantive science and not just by educational research and other social sciences which have begun to raise voices against the virtual tyranny of this branch of inference in the academic world.
Haig (2013b, p. 11)	Despite the plethora of critiques of statistical significance testing, most psychologists understand them poorly, frequently use them inappropriately, and pay little attention to the controversy they have generated.
Hubbard & Meyer (2013, p. 17)	That the *p*-value—a statistic of such limited consequence—lies at the heart of social "scientific method" is incredulous.
Hubbard, Parsa, & Luthy (1997, p. 552)	[Statistical significance] testing and reporting of *p* values is now virtually synonymous with empirical analysis.
Hubbard & Ryan (2000, p. 678)	It seems inconceivable to admit that a methodology as bereft of value as SST [statistical significance testing] has survived, as the centerpiece of inductive inference no less, more than four decades of criticism in the psychology literature.
Hunter (1997, p. 3)	Needed: A Ban on the Significance Test.
Ioannidis (2012, p. 647)	Young investigators are taught early on that the only thing that matters is making new discoveries and finding statistically significant results at all costs.

(Continued)

Table 7-1 (Continued)

Authors	*Quotations*
Kirk (2003, p. 100)	It is evident that the current practice of focusing exclusively on a dichotomous reject-nonreject decision strategy of null hypothesis testing can actually impede scientific progress. I suspect that the continuing appeal of null hypothesis significance testing is that it is considered to be an objective scientific procedure for advancing knowledge. In fact, focusing on *p* values and rejecting null hypotheses actually distracts us from our real goals: deciding whether data support our scientific hypotheses and are practically significant. The focus of research should be on our scientific hypotheses, what data tell about the magnitude of effects, the practical significance of effects, and the steady accumulation of knowledge.
Kline (2004. pp. 61–62)	I argue that the criticisms have sufficient merit to support the minimization or elimination of NHST in the behavioral sciences.
Kmetz (2011, p. 177)	The underlying problem is that NHST results are almost universally misinterpreted in the social-science literature. . . . Collectively, NHST and its emergent mythology have created a statistically and scientifically unsound basis for the evaluation and publication of [business] school research.
Lambdin (2012, p. 71)	Much of the blame lies with our journals, our statistics texts, and our graduate school statistics courses. The NHST orthodoxy is not only required for publication in most journals but its tenets continue to be proselytized from the pulpits of graduate school classrooms across the fruited plain. This trend has continued unabated since the 1960s.
Loftus (1996, p. 162)	I believe the reliance on NHST has channeled our field into a series of methodological cul-de-sacs.
Lykken (1968, p. 158)	The moral of the story is that the finding of statistical significance is perhaps the least important attribute of a good experiment: it is *never* a sufficient condition for concluding that a theory has been corroborated, that a useful empirical fact has been established with reasonable confidence—or that an experimental report ought to be published.
McCloskey & Ziliak (1996, p. 111)	We would not assert that every economist misunderstands statistical significance, only that most do, and these [are] some of the best economic scientists.

Authors	*Quotations*
Morrison & Henkel (1970, p. v)	Even their strongest proponents [of statistical significance testing] agree that there is much misuse, misinterpretation, and meaningless use of the tests.
Nelder (1999, p. 257)	The kernel of these non-scientific procedures is the obsession with significance tests as the end point of any analysis.
Nester (1996, p. 407)	Clearly, point hypothesis testing has no place in statistical practice. . . . This means that most paired and unpaired *t*-tests, analyses of variance . . . linear contrasts and multiple comparisons, and tests of significance for correlation and regression coefficients should be avoided by statisticians and discarded from the scientific literature.
Rosnow & Rosenthal (1989, p. 1277)	It may not be an exaggeration to say that for many PhD students, for whom the .05 alpha has acquired almost an ontological mystique, it can mean joy, a doctoral degree, and a tenure-track position at a major university if their dissertation *p* is less than .05. However, if the *p* is greater than .05, it can mean ruin, despair, and their advisor's suddenly thinking of a new control condition that should be run. . . . . [S]urely, God loves the .06 nearly as much as the .05.
Rozeboom (1960, p. 417)	The thesis to be advanced is that despite the awesome pre-eminence this method has attained in our experimental journals and textbooks of applied statistics, [NHST] is based upon a fundamental misunderstanding of the nature of rational inference, and is seldom if ever appropriate to the aims of scientific research.
Rozeboom (1997, p. 335)	Null-hypothesis significance testing is surely the most bone-headedly misguided procedure ever institutionalized in the rote training of science students.
Salsburg (1985, p. 220)	And it provides Salvation: Proper invocation of the religious dogmas of Statistics will result in publication in prestigious journals. This form of Salvation yields fruit in this world (increases in salary, prestige, invitations to speak at meetings) and beyond this life (continual references in the citation indexes).

(Continued)

Table 7-1 (Continued)

Authors	*Quotations*
F. L. Schmidt (1996, p. 116)	My conclusion is that we must abandon the statistical significance test.
F. L. Schmidt & Hunter (1997, p. 57)	Significance testing never makes a useful contribution to the development of cumulative knowledge.
F. L. Schmidt & Hunter (2002, p. 65)	Most researchers in the physical sciences regard reliance on significance testing as unscientific.
Shrout (1997, p. 1)	Significance testing of null hypotheses is the standard epistemological method for advancing scientific knowledge in psychology, even though it has drawbacks and it leads to common inferential mistakes.
Silva-Aycaguer et al. (2010, p. 2)	The statistical textbooks, with only some exceptions, do not even mention the NHST controversy. Instead, the myth is spread that NHST is the "natural" final action of scientific inference and the only procedure for testing hypotheses.
Stang, Poole, & Kuss (2010, p. 225)	The *P*-value is perhaps the most misunderstood statistical concept in clinical research. As in the social sciences, the tyranny of SST is still highly prevalent in the biomedical literature even after decades of warning against SST. The ubiquitous misuse and tyranny of SST threatens scientific discoveries and may even impede scientific progress. In the worst case, misuse of significance testing may even harm patients who eventually are incorrectly treated because of improper handling of *P*-values.
Walster & Cleary (1970, p. 16)	A virtual prerequisite for the publication of research in the social sciences is the attainment of statistical significance.
Young (2008, n.p.)	It's science's dirtiest secret: The "scientific method" of testing hypotheses by statistical analysis stands on a flimsy foundation.
Ziliak & McCloskey (2008, p. 2)	We . . . say that a finding of "statistical" significance . . . is on its own almost valueless, a meaningless parlor game.

logical principles involved is urgently needed." Some two decades ago, Freund and Perles (1993, p. 8) objected that differences in textbook definitions of *p*-values "abound." Today these criticisms remain equally cogent. Following my (Hubbard, 2011a, p. 1145) listing, several common misinterpretations of *p*-values are now addressed.

Some Common Misunderstandings About P-Values

A *P*-Value Is the Probability That the Null Hypothesis (H_0) Is True. Nickerson (2000, p. 246) argues that this misunderstanding of the *p*-value is among the most pervasive and widely criticized of all the false beliefs about null hypothesis significance testing (NHST). It is the idea that a *p*-value of .05 means that there is only a 5% probability that the results are due to chance given the data (x), or $Pr(H_0 \mid x)$. Statisticians Berger and Sellke (1987, p. 114) have no reservations about the almost universal acceptance of this fallacy: "Indeed, most nonspecialists interpret *p* precisely as $Pr(H_0 \mid x)$." If a *p*-value of .05 *did* mean that there is only a 5% probability that the results are attributable to chance, it would be a very helpful index. But it says no such thing. The *p*-value, recall, is defined as $Pr(x \mid H_0)$, the probability of the data (together with more extreme observations) conditional on a true null hypothesis, and not $Pr(H_0 \mid x)$. The latter is the posterior probability of the null from a Bayesian analysis. And Fisher (1973, p. 9), never shy of an opinion, disavowed the method of inverse probability (Bayesianism): "It will be sufficient in this general outline of the scope of Statistical Science to reaffirm my personal conviction, which I have sustained elsewhere, that the theory of inverse probability is founded upon an error, and must be wholly rejected."[2]

So prevalent is the belief that $p = Pr(H_0 \mid x)$, however, that Carver (1978, p. 383) calls it the "odds-against-chance" fantasy, Gigerenzer (1993, p. 330) refers to it as the "Bayesian Id's wishful thinking," and Falk and Greenbaum (1995, p. 78) label it the "illusion of probabilistic proof by contradiction." To appreciate that this is fantasy, wishful thinking, and illusion it is necessary to reassert the fact that the *p*-value is a *conditional* probability, one whose calculation is dependent on the truth of the null hypothesis. This seemingly innocuous caveat, which is almost always overlooked, has important repercussions. It says that the *p*-value for an outcome can only be computed *given* that H_0 is true, so it is *impossible* for a *p*-value to be the probability of H_0.

Recapping for maximum emphasis, confusion over the two conditional probabilities $Pr(x \mid H_0)$ and $Pr(H_0 \mid x)$ is endemic.[3] Nickerson (2000, p. 247) suggests that this mischief resembles that following commission of the *premise conversion error* in conditional logic, whereby *if P then*

Q mistakenly is deemed equivalent to *if Q then P.* In their denunciations of classical statistical significance testing, some authors label this error the *fallacy of the transposed conditional* (Wagenmakers, Wetzels, Borsboom, & van der Maas, 2011, pp. 428–429; Ziliak & McCloskey, 2008, p. 17) and the *inverse probability fallacy* (Cumming, 2012, p. 27).

Inevitably, misunderstandings such as the *p*-value connotes the probability of the truth of H_0 appear in the textbooks, an especially potent source of misguided statistical ideas. For once insinuated in this venue, errors assume the appearance of canonical truths for students, teachers, and researchers alike. This is seen in Ray Kent's (2007, p. 380) assertion that "the p-value is the probability that the null hypothesis is true." It is repeated by Taylor (2007, p. 125): "From the computer output the p-value is the probability of H_0 being correct." And again by Brace, Kemp, and Snelgar (2006, p. 10), who state that the *p*-value is "the probability that the null-hypothesis is correct." Unfortunately, the above textbook pronouncements are invalid.

A *P*-Value Is the Probability That the Alternative Hypothesis (H_A) Is True. Many investigators believe that a *p*-value of .05 means the probability of the alternative (research) hypothesis given the data, $Pr(H_A \mid x)$, being true is 1-*p*, or .95. In other words, there is a 95% probability that the observed results are *not* attributable to chance and that H_A is true. This argument is fallacious because if *p* is defined incorrectly, then by necessity so is its complement, 1-*p*.

Furthermore, as raised in Section 3.2.1, in Fisher's approach to statistical analysis there is no H_A. The alternative hypothesis was introduced by Jerzy Neyman and Egon Pearson (1928a, 1928b, 1933) in an effort to "improve" on Fisher's model. But Fisher never saw the need for H_A and opposed its introduction by Neyman–Pearson (J. Cohen, 1990, p. 1307; Hubbard & Bayarri, 2003, p. 172). Moreover, only Bayesians can give the probability of a hypothesis. Frequentists like Fisher and Neyman–Pearson, although of very different stripes, cannot.

Yet Gigerenzer (1993, p. 330) remarks that even foundational texts on statistical methods in psychology, such as those by Anastasi (1958, p. 11), George Ferguson (1959, p. 133), Guildford (1942, pp. 156–166)—probably the most highly read textbook in the 1940s and 1950s (Gigerenzer, 1993, p. 323)—and Lindquist (1940, p. 14), are guilty of the two misconceptions about *p*-values stated above. Additional examples of textbooks perpetuating such erroneous thinking are given by Carver (1978, p. 384) and Falk and Greenbaum (1995, pp. 82–84).

A *P*-Value Is the Probability That the Results Will Replicate. This illusion was exposed in Section 5.5.1. It is enough to mention here the central findings of Tversky and Kahneman's (1971) famous article "Belief in the Law of Small Numbers." They posed the following scenario to some members ($n = 84$) attending meetings of the Mathematical Psychology Group and of the American Psychological Association:

> Suppose you have run an experiment on 20 subjects, and have obtained a significant result which confirms your theory ($z = 2.23$, $p \leq .05$, two-tailed). You now have cause to run an additional group of 10 subjects. What do you think the probability is that the result will be significant, by a one-tailed test, separately for this group? (p. 105)

The median response was that this probability is .85!

Tversky and Kahneman (1971, p. 105) noted that, "apparently, most psychologists have an exaggerated belief in the likelihood of successfully replicating an obtained finding." They went on to advise that people have strong intuitions about random sampling, shared by novices and specialists alike, which can be fundamentally wrong and destructive to scientific advance. Specifically, Tversky and Kahneman

> submit that people view a sample randomly drawn from a population as highly representative, that is, similar to the population in all essential characteristics. Consequently, they expect any two samples drawn from a particular population to be more similar to one another and to the population than sampling theory predicts, at least for small samples. (p. 105)

Or as Tversky and Kahneman (1971, p. 106) jest, researchers believe that the law of large numbers applies to small numbers as well.

Interpretations like these underscore the misimpression that replicating results is a straightforward, almost guaranteed, matter and therefore not something to be terribly concerned about. We know, in fact, that this is not true. And, of course, Section 5.5.1 showed that using *p*-values to denote replication success is fraught with difficulties.

A *P*-Value Is a Measure of the Magnitude of an Effect. This myth is promulgated most transparently when researchers do not provide (all relevant) descriptive statistics in a study, opting instead to report only *p*-values. A common variation on this theme is seen in the language used by investigators such as $p < .05$ is a significant result, $p < .01$ is a very significant result, and $p < .001$ is an extremely significant result, generally accompanied by *, **, ***, respectively. Because science is concerned ultimately with magnitudes, it comes as no surprise to see how often *p*-values mistakenly are taken as surrogates for effect sizes (J. Cohen, 1990, p. 1309; Cumming, 2012, p. 39), something which Kruskal and Majors (1989, p. 3) find "depressing," and Bakker and Wicherts (2011, p. 666) error-prone.

Quantitative assessments of the extent of this misconception are available. In a piece cleverly titled "The Standard Error of Regressions," McCloskey and Ziliak (1996) inspected all full-length articles published in the *American Economic Review* (excluding the *Proceedings*) from January 1980 to December 1989 that used tests of statistical significance. Over this decade they found 70% of 182 articles made no distinction whatsoever between statistical significance versus economic or policy significance. Later results were worse. Repeating their study on the *American Economic Review* for the 1990s, they observed the failure to distinguish between statistical and tangible significance in 79% of 187 relevant articles (Ziliak & McCloskey, 2004).

Almost identical results to those in economics are seen in biomedicine. In what they termed the "significance fallacy," Silva-Aycaguer, Suarez-Gil, and Fernandez-Somoano (2010) discovered this misbelief of equating statistical with substantive significance to be standard. They reviewed articles from six journals in the English- and Spanish-speaking literatures in three medical areas. These were clinical specialties (*Obstetrics & Gynecology, Revista Espanola de Cardiologia*), public health and epidemiology (*International Journal of Epidemiology, Atencion Primaria*), and general and internal medicine (*British Medical Journal, Medicina Clinica*). Silva-Aycaguer et al. (2010, p. 7) recorded that across all three of these areas the "significance fallacy" was the norm in 70%–80% of 874 empirical papers for 1995–2006. Such results bolster Ioannidis's (2005b, p. 0101) contention that "there is a widespread notion that medical research articles should be interpreted based only on *p*-values."

But a *p*-value is not a measure of the size of an effect in the population; it is only what Ziliak and McCloskey (2008, ch. 2) refer to as "The Sizeless Stare of Statistical Significance." A trivial effect with a large enough sample will be statistically significant; a large effect with too small a sample will not. As might be imagined, this muddies the water to no end when trying to ascertain whether a result is of meaningful or practical, as opposed to statistical, significance in any given discipline.

A *P*-value Is a Measure of the Generalizability of a Result. This is another manifestation of wishful thinking and the capabilities of *p*-values.

A *p*-value contains no information about whether a finding observed under one set of conditions will generalize to others.

Despite this strict limitation, the attainment of statistically significant results in one study leads to over-optimism, of the 1-*p* kind, concerning the generalizability of findings. The outcomes of single studies with $p \leq .05$ results are credited with far greater applicability than they are due. This line of argument was made, in a broader connection, in Section 2.2.5.

A *P*-Value Is a Measure of the Causal Relationships Between Variables. Freedman (1999, p. 249) observes that the *p*-value is wrongly thought to contribute strong support for the causal relationships between variables. I agree wholeheartedly with Freedman's reservations on this matter, including his rebuke that scholars all too often ignore the complexities involved in establishing causal relations, relying instead on the passive recording of the size of the *p*-value to settle the issue. On this topic, Gigerenzer (1993, p. 318) reminds us that what is rejected in significance testing is the statistical null hypothesis, and not the presence or absence of a cause. Such cautionary advice aside, Quinlan's (2011, p. 402) textbook on business research methods nevertheless preaches this kind of wistfulness: "The probability of any outcome is called the p-value . . . with 0 equalling complete certainty that the prediction will not occur, and 1 equalling complete certainty that the prediction will occur."

Recall, moreover, that consistent with a critical realist philosophy, causal mechanisms must be imagined or invented. These mechanisms are hidden and have nothing to do with statistical significance tests and *p*-values. Returning to Freedman (1999, p. 253), establishing causality "rests not so much on the *P*-values, but more on the size of the effect . . . and on extensive replication both with the original research design and with many other designs." This, he notes, will require a huge investment in skill, intelligence, and hard work.

A *P*-Value Is a Type I Error Rate (α). At first blush misinterpreting the *p*-value as a Type I error rate might be thought to be nothing more than a minor, technical gaffe. Yet its implications are profound, and arguably the *greatest* cause of disarray among researchers when coming to grips with interpreting the outcomes of statistical significance tests. Because of this, and borrowing from Hubbard and Bayarri (2003, 2005) and Hubbard (2011b) in particular, this topic receives considerably more detailed examination in the pages that follow.

Confusion surrounding the reporting and interpretation of results of classical statistical tests is enormous among applied researchers, most of whom are of the mind that said tests are prescribed by a single coherent theory of statistical inference (Hubbard & Bayarri, 2003, p. 171). But as alluded to in Section 2.2.3, this is not so. Classical statistical testing is an anonymous hybrid—or more accurately a "mishmash" (Gigerenzer, 1993, p. 314)—of the rival and often contradictory formulations of Fisher on the one hand and Neyman and Pearson (N–P) on the other (see also in this regard Gigerenzer & Murray, 1987; Gigerenzer et al., 1989; Goodman, 1993; Hubbard, 2004; Hubbard & Armstrong, 2006; Lenhard, 2006). It is an amalgamation that neither Fisher nor N–P would have condoned. Of especial importance, this conflation of viewpoints has led to extensive chaos among investigators over the very meaning of "statistical significance" itself.

This is because the latter has two completely different interpretations. Fisher embraces an inductive philosophy of knowledge development, arguing from the particular to the general. For him, statistical significance is determined by using the *p*-value as a measure of *inductive evidence* against the null hypothesis, H_0. In direct opposition, N–P theory is infused with a deductive philosophy of knowledge procurement, arguing from the general to the particular. They see statistical significance as captured by α levels and these Type I error rates correspond to erroneous rejections of H_0.

The N–P model of hypothesis testing repudiates the very idea of inductive evidence, instead concentrating on statistical testing as a guide to making decisions and influencing *behavior*. In addition, while Fisher entertained only the null hypothesis, N–P offered two hypotheses, the null and the alternative, H_A, and their theory involves selecting between two courses of action, accepting H_0 or rejecting it in favor of H_A. Mistakes occur when deciding between accepting H_0 or H_A. From N–P's perspective, the significance level, or (upper bound on) the probability of a Type I error, α, is the false rejection of H_0, while a Type II error, β, is the false acceptance of H_0.

Neyman–Pearson theory focuses on error control and not on the collection of evidence. Pointedly, this error control is of a long-run nature. The N–P model, in contrast with Fisher's, does not apply to a specific investigation. Moreover, as revealed in Section 3.2.1, *p*-values have no valid long-run frequentist interpretation (see the applet www.stat.duke.edu/~berger). Lastly, Fisher's evidential *p*-value is a data-dependent random variable (Hubbard, 2004, p. 319; Hubbard, 2011b, p. 2619; Murdoch, Tsai, & Adcock, 2008), while N–P's α is part of the procedure itself and therefore is fixed (usually at the .05 level) *before* the study begins and does not change with the data. These stark differences between the two conceptions of statistical significance, p and α, are presented in Table 7-2.

When laid out as in Table 7-2, these two interpretations of "statistical significance" could hardly be more incongruous. The problem is that these differences seldom are explained in the literature. Fisher (1955, p. 74) complained, rightly, that his significance test was "assimilated" into the N–P hypothesis testing framework. This happened even though α levels play no part in Fisherian significance testing (Hubbard & Bayarri, 2003, p. 174) and p-values exercise no role in N–P tests (R. Christensen, 2005, p. 123; Hubbard & Bayarri, 2003, p. 175). Or as Royall (1997, p. 64) sees it:

> The distinction between Neyman–Pearson tests and [Fisher's] significance tests is not made consistently clear in modern statistical writing and teaching. Mathematical statistical textbooks tend to present Neyman–Pearson theory, while statistical methods textbooks tend to lean more towards significance tests. The terminology is not standard, and the same terms and symbols are often used in both contexts, blurring the differences between them.

It is this fusing of two distinct measures of statistical significance that has assured researchers that the p-value is an "observed" (Goodman, 1993, p. 489) or "data-dependent adjustable" (Hubbard & Bayarri, 2003, p. 175) Type I error rate, α. And because both ps and αs are tail area probabilities, this "blurs the division between concepts of evidence and error for the statistician, and obscures it completely for nearly everyone else" (Goodman, 1992, p. 879).

Confusion among researchers over ps and αs is close to total (Hubbard & Bayarri, 2003, p. 174). This is entirely understandable, though unfortunate, given the prevalence of the $p < \alpha$ yardstick in "most teaching texts" (Barnett, 1982, p. 489) as yet another definition of statistical significance. It is a yardstick *inviting* scholars to misconstrue p-values as "data-adjusted" Type I errors (αs).

A popular revelation of this misconstrual, necessitating elaboration at this juncture since it resurfaces in discussions hereafter, is seen in what Goodman (1993, p. 489) titles "roving alphas." These are a limited number of p-values, generally $p < .05$, $p < .01$, $p < .001$, and so on, that are misconceived as Type I error probabilities. We know these are roving alphas, as opposed to Fisherian p-values, because they are *not* the predetermined

Table 7-2 Contrasts Between *P*-Values and α Levels as Measures of "Statistical Significance"

P-Values	*α Levels*
Fisher's significance level	Neyman–Pearson's significance level
Significance testing	Hypothesis testing
Inductive philosophy—from particular to general	Deductive philosophy—from general to particular
Only the null hypothesis, H_0	Null hypothesis, H_0, and alternative hypothesis, H_A
Empirical evidence against H_0	Erroneous rejection of H_0—Type I error
Inductive inference—procedure for evaluating strength of evidence in data	Inductive behavior—prescriptions for making decisions between H_0 and H_A based on data
Data-dependent random variable with uniform distribution over the interval [0–1] under the null hypothesis	Predetermined fixed value
Characteristic of data	Characteristic of test procedure
Power of test only implicit	Power of test plays crucial role
Short-run orientation—applicable to each individual study	Long-run orientation—applicable only to ongoing, identical repetitions of original study, not to each individual study

Source: Adapted from: Hubbard, Raymond (2011b), "The Widespread Misinterpretation of *P*-Values as Error Probabilities," *Journal of Applied Statistics,* 38 (November), p. 2619 with permission of the publisher (Taylor & Francis Ltd, http://www.tandfonline.com).

error rates required by N-P orthodoxy. The N-P measure of statistical significance, α, must for any given study be *fixed* at a single value, say, .05, *before* that study is carried out so as to constrain the Type I error rate to that preselected level. Alpha, by definition, cannot be a variable rate—fluctuating between .05, .01, .001, or whatever—determined after the fact *solely by the data* within different sections of any investigation. This is perfectly legitimate for *p*-values, but not so for αs.[4]

Ironically, despite the overwhelming befuddlement concerning the meaning of statistical significance apparent among those in the significant difference camp, they nevertheless show little hesitation about how such testing should be carried out. Because of their blind allegiance to the view that the Fisher/N–P hybrid framework constitutes a single, *universal* means of statistical inference, researchers typically approach empirical investigations convinced that they are armed with the most rigorous scientific methods available. Susie Bayarri and I (Hubbard & Bayarri, 2003, p. 174) describe this armament as follows: The researcher specifies the null (H_0) and alternative (H_A) hypotheses, the Type I error rate/significance level, α, and purportedly—but in practice very rarely—computes the statistical power of the test (e.g., *z*). This sequence of steps is in line with N–P theory. Next, the test statistic is calculated for the sample data, and in an effort to have it both ways, a Fisherian *p*-value is determined. The *p*-value is then misinterpreted as a frequency-based observed Type I error rate, and at the same time as an incorrect (i.e., $p < \alpha$) measure of evidence against the null hypothesis.

Or put another way, for most researchers under the significant difference tent, the *p*-value is interpreted simultaneously according to N–P strictures as a deductive assessment of error in long-run repeated sampling situations, and from a Fisherian stance as a measure of inductive evidence in a single study. Clearly, this is asking the impossible (Goodman, 1999, p. 999), thus rendering the application of statistical significance tests meaningless (Hubbard, 2004, p. 318).

Should the reader feel that the extent of researcher mayhem over the interpretation of *p*s and αs has been overstated, information from several coteries demonstrating this is not the case is excerpted from Hubbard (2004, 2011b) below. These groups include statisticians themselves, articles critical of the worth of significance testing, textbooks, repositories of scholarly advice, surveys of researchers, and the statistical reporting behaviors of academicians.

Statisticians. Casella and Berger (1987, p. 133) admonish that "there are a great many statistically naïve users who are interpreting *p* values as probabilities of Type I errors." Unfortunately, however, sometimes even prominent statisticians can err in equating *p*-values with Type I error rates. Consider, for example, Gibbons and Pratt's (1975, p. 21) assertion, in their article titled "*P*-values: Interpretation and Methodology," that "reporting a *p*-value, whether exact or within an interval, in effect permits each individual to choose his own level of significance as the maximum tolerable probability of a Type I error." Barnard (1985, p. 7) similarly muddles the two entities: "For those who need to interpret probabilities as (of course, hypothetical) frequencies, a *P*-value 'measures' the possibility of an 'error of the first kind,' arising from rejection of H_0 when it is in fact true." As does Hinkley (1987, p. 128) when he misstates that "the interpretation of [the] *P* value as an error rate is unambiguously objective."

Again, Hung, O'Neill, Bauer, and Köhne (1997, p. 12) mix up *p*-values with Type I error probabilities: "The α level is a preexperiment Type I error rate used to control the probability that the observed *P* value in the experiment of making an error rejection of H_0 when in fact H_0 is true is α or less." Storey (2011, p. 504) likewise: "The upper threshold applied to the p-value . . . (often 5% in the scientific literature) determines the Type I error rate; i.e., the probability of making a Type I error when the null hypothesis is true." And the philosopher of statistics Mayo (2013, p. 12) follows suit: "Thus, whenever error probabilities, be they *p*-values." Obviously, if statistical authorities can occasionally misinterpret *p*s and αs, it stands to reason that others will do so as well.[5]

Articles Critical of NHST. In more than a twist of irony, a further source of specious statistical advice can be articles critiquing the abuse and limitations of statistical significance testing. These papers can be disproportionately influential in that they help (re)establish the standards for statistical reporting practices in their respective fields. Unfortunately, some of these articles perpetuate the same old mistakes, including the misrepresentation of *p*-values as error probabilities. Dyer's (1997, p. 260) instruction to healthcare professionals that "a small *P* value reduces the chance of making Type I errors" is just such a case.

The psychology discipline harbors some of the most outspoken reviewers of the merits of NHST (cf. Krantz, 1999). Yet here again critics in this area sometimes inadvertently pass along misinformation, including that *p*-values are Type I errors. Meehl's (1967, p. 107) inspired paper mentioned in the previous section is an early reminder of this: "[The researcher] gleefully records the tiny probability number '$p < 0.001$,' and there is a tendency to feel that the extreme smallness of this probability of a Type I error." And in what is without question the best and most detailed (60 pages) critique of the statistical significance testing debate thus far, Nickerson (2000, p. 243) does the same: "The value of *p* that is obtained as the result of NHST is the probability of a Type I error on the assumption that the null hypothesis is true." Nickerson commented later that "both *p* and α represent bounds on the probability of Type I error . . . *p* is the probability of a Type I error resulting from a particular test if the null hypothesis is true" (p. 259). Here, Nickerson sees the *p*-value as an "observed" or "data-adjusted" Type I error, an impossibility because error rates are confined to long-run hypothetical frequencies, not to actual individual studies. Neyman (1971, p. 13) was explicit on this score:

> It would be nice if something could be done to guard against errors in each particular case. However, as long as the postulate is maintained that the observations are subject to variation affected by chance (in the sense of frequentist theory of probability), all that appears possible to do is to control the frequencies of errors in a sequence of situations.

To repeat, the *p*-value is not a Type I error rate.

Textbooks—Major Sources of Statistical Misinformation. It is common to see in textbooks the mistake of defining the *p*-value as a Type I error probability. Table 7-3 provides quotations from a convenience sample of some of these.

Encompassing an array of disciplines—behavioral sciences, biology, biostatistics, econometrics, finance, health research, marketing research, psychology, quality control management, social research, sports studies, and statistics itself—many of these books (e.g., George, Rowlands, Price, & Maxey, 2005, p. 160; Gujarati, 2003, p. 137; Howell, 2008, p. 299; Kennedy, 2008, p. 48; Levine & Stephan, 2010, p. 147; Miles & Banyard, 2007, p. 88; Pindyck & Rubinfeld, 1998, p. 43) state unreservedly that the *p*-value is a Type I error rate. Health researcher Bowling's (2002, p. 169) determination that "if the *P* value for a test is less than 0.05 . . . [t]his means that there are less than five chances in 100 . . . that the result is a false positive (type I error)" is representative of this group.

Sometimes a *p*-value is described as an upper limit of committing a Type I error. This is witnessed in psychologist Fife-Shaw's (2006, p. 399) chapter in a text defining the *p*-value as "the maximum probability of making a Type I error."

Moreover, a number of textbook authors go amiss when calling the *p*-value the *exact* probability of committing a Type I error. This misunderstanding sees the *p*-value as a "data-adjusted" error rate, which, as referenced above, is impossible. Examples of this fallacy are contained in several textbooks (e.g., Gujarati, 2003, p. 137; Healey, 2007, p. 204; Howell, 2002, p. 191, 2008, p. 299; Parasuraman, Grewal, & Krishnan, 2004, p. 408; Salkind, 2008, p. 161). They are summarized by the econometrician Dougherty (2007, p. 105): "The *p* value approach is more informative than the 5 percentage/1 percentage approach, in that it gives the exact probability of a Type I error, if the null hypothesis is true." Such misinterpretation is preserved in the otherwise excellent book *The Cult of*

Table 7-3 Textbooks Misstating that *P*-Values are Type I Errors (αs)

Authors	*Quotations*
Bowling (2002, p. 169)	If the ***P* value** for a test is less than 0.05 . . . [t]his means that there are less than five chances in 100 . . . that the result is a false positive (type I error).
Brace, Kemp, & Snelgar (2006, p. 10)	Psychologists usually set the criterion (also known as the Type I error rate) at $p = 0.05$.
Brooks (2002, p. 81)	Informally, the *p*-value is also often referred to as the probability of being wrong when the null hypothesis is rejected.
Bryman (2004, p. 238)	Using a $p < 0.05$ level of significance means that we are more likely to make a Type I error than when using a $p < 0.01$ level of significance.
D. R. Cooper & Schindler (2006, p. 545)	The ***p* value** . . . represents the probability of a Type I error.
Dancey & Reidy (2004, p. 142)	Therefore, the *p*-value also represents the chance of your making a Type I error. If your *p*-value is 5% it means that you have a 5% probability of making a Type I error if you reject the null hypothesis.
S. F. Davis & Smith (2005, p. 259)	However, there is also a chance ($p = .021$ in this case) that we made a Type I error in rejecting the null hypothesis of no difference when it was actually true.
Dougherty (2007, p. 105)	The *p* value approach is more informative than the 5 percentage/1 percentage approach, in that it gives the exact probability of a Type I error, if the null hypothesis is true.
Dunn (2009, p. 292)	For most social psychological research, statistical significance is related to significance levels, which are based on probability or *p*-values (also referred to as alpha levels).

Authors	*Quotations*
Fife-Shaw (2006, p. 399)	This $p = 0.05$ figure, called an **alpha criterion**, is the maximum probability of making a Type I error.
Fowler, Cohen, & Jarvis (1998, p. 108)	In most statistical analyses the aim is usually to limit the probability of committing Type I errors . . . [so if] we reject H_0 at $P < 0.05$ . . . the risk of [Type I] error is accordingly reduced.
George, Rowlands, Price, & Maxey (2005, p. 160)	The *p*-value is the probability of making a Type I error.
Gratton & Jones (2004, p. 207)	For example, $p < 0.1$ indicates a likelihood of incorrectly rejecting a true null hypothesis one time in ten, whereas $p < 0.01$ suggests that this would be the case less than once in a hundred times.
Gray (2009, p. 303)	What are the chances of making a Type I error? . . . [I]f we set our significance level at 5 per cent (p = 0.05), we are willing to take the risk of rejecting the null hypothesis when in fact it is correct 5 times out of 100.
Gujarati (2003, p. 137)	The ***p*** **value** [is] **the exact probability of committing a Type I error.**
Healey (2007, p. 204)	*SPSS for Windows* reports . . . the "Sig. (2-tailed)" (.067) [i.e., *p*-value]. This last piece of information is an alpha level except that it is the *exact* probability of getting the observed difference in sample means if only chance is operating.
Howell (2002, p. 191)	Minitab computes the exact probability of a Type I error (the ***p*** **level**).
Howell (2008, p. 299)	Notice also that SPSS computes the exact probability of a Type I error (the ***p*** **value**).
Kennedy (2008, p. 58)	[The] *p* value [is] the smallest level of significance (type I error) for which the observed test statistic value results in a rejection of the null hypothesis.
Levine & Stephan (2010, p. 147)	You can consider the *p*-value the actual risk of having a type I error for a given set of data.

(Continued)

Table 7-3 (Continued)

Authors	*Quotations*
Miles & Banyard (2007, p. 88)	This mistake is called a **type I error** and we estimate the probability of making this error because it is given in our *p*-value.
Nolan & Heinzen (2008, p. 556)	**Alpha is another name for the p level [value].**
Parasuraman, Grewal, & Krishnan (2004, p. 408)	The actual significance level, or *p*-value . . . implies that there is about a 16% chance that Karen would be making a mistake . . . [i]f Karen finds this . . . probability of a Type I error of .16 acceptable.
Pindyck & Rubinfeld (1998, p. 43)	The *p*-value measures the likelihood of a Type I error.
Rosner (2006, p. 232)	The ***p*-value** for any hypothesis test is the α level at which we would be indifferent between accepting or rejecting H_0 given the sample data at hand.
Salkind (2008, p. 161)	Software such as SPSS gives you the exact probability, such as $p = .013$ or .158, of the risk you are willing to take that you will commit a Type I error.
Ziliak & McCloskey (2008, p. 99)	[This] underscores the arbitrariness of Fisher's 5 percentage ideology—the Type I error was about 12 percent ($p \leq .12$).

Source: Adapted from: Hubbard, Raymond (2011b), "The Widespread Misinterpretation of *P*-Values as Error Probabilities," *Journal of Applied Statistics*, 38 (November), P. 2621 with permission of the publisher (Taylor & Francis Ltd, http://www.tandfonline.com).

Note: All bolded and italicized parts of the quotations are from the original works.

Statistical Significance (Ziliak & McCloskey, 2008, p. 99): "[This] underscores the arbitrariness of Fisher's 5 percentage ideology—the Type I error was about 12 percent ($p \leq .12$)."

The volume on quality control management by George et al. (2005) also misdefines the *p*-value as a Type I error rate. This allows me the opportunity to further differentiate between the two measures of statistical significance. Even Fisher (1955, p. 69) acknowledged that N-P's α level can be extremely helpful in quality control situations, where the aim is to restrict flaws in manufactured goods to some fixed, preassigned amount. What is not generally realized, however, is that while N-P's α *can* be fixed at some prespecified (e.g., .05) level, this same capability does not extend to *p*-values (see Berger & Delampady, 1987, p. 329, along with Section 3.2.1, as to why this is so). Therefore, George et al.'s belief that the *p*-value is a Type I error rate is disconcerting in quality control contexts.

Repositories of Scholarly Advice. The most authoritative sources of possibly unsound information are various oracles, such as academic publication manuals and the "Suggestions to Contributors" sections of scholarly journals, dispensing advice to their readerships. These can include the literal dictation of how to report statistical findings.

Take, as an example, the psychology discipline. Here the American Psychological Association (APA) *Publication Manual* sets the standards for editorial practice for 27 APA journals and at least a thousand others in psychology and related fields (Fidler, 2002). Inspection of the three most recent of these manuals portrays ongoing disorder and equivocation about the interpretation of *p*s and αs. It has been emphasized that a preselected, fixed Type I error rate (α) cannot be viewed simultaneously as a measure of evidence (*p*) once the result is observed. As I have written elsewhere (Hubbard, 2004, p. 312), however, this mistake is officially sanctioned in the first of these three guidebooks:

> For example, given a true null hypothesis, the probability of obtaining the particular value of the statistic you computed might be [$p =$] .008. Many statistical packages now provide these exact [*p*] values. You can report this distinct piece of information in addition to specifying whether you rejected or failed to reject the null hypothesis using the specified alpha level.
>
> With an alpha level of .05, the effect of age was statistically significant, $F(1, 123) = 7.27$, $p = .008$. (APA, 1994, p. 17)

Further complicating the situation, psychologists are presented with mixed messages when they are told: "Before you begin to report specific results, you should routinely state the particular alpha level you selected for the statistical tests you conducted: [for example,] An alpha level of .05 was used for all statistical tests" (APA, 1994, p. 17). This recommendation is sound enough as far as it goes in that it is consistent with N-P principles. The problem is that it does not go far enough, being followed immediately by poor advice: "If you do not make a general statement about the alpha level, specify the alpha level when reporting each result" (APA, 1994, p. 17). The latter virtually orders the habitual use of unacceptable roving alphas. But α, recall, must be decided upon before, not after, the study commences.

A later APA (2001) *Publication Manual* trod in the footsteps of its predecessor by acknowledging, albeit without explanation, that there are two kinds of probabilities, *p*s and αs, connected with statistical significance testing. This edition similarly offers incorrect counsel:

> The APA is neutral on which interpretation [*p* or α] is to be preferred in psychological research. . . . Because most statistical packages now report the *p* value . . . *and because this probability can be interpreted according to either mode of thinking,* in general it is the exact probability (*p* value) that should be reported. (p. 24; emphasis added)

This passage assumes that *p*s and αs are interchangeable concepts. But this is not so. The *p*-value is not a Type I error rate, and an α level is not evidential (Hubbard, 2004, p. 312).

The APA (2001, p. 25) *Publication Manual* goes on to announce:

> There will be cases—for example, large tables of correlations or complex tables of path coefficients—where the reporting of exact probabilities could be awkward. In these cases, you may prefer to identify or highlight a subset of values in the table that reach some prespecified level of statistical significance. To do so, follow those values with a single asterisk (*) or double asterisk (**) to indicate $p < .05$ or $p < .01$, respectively. When using prespecified significance levels, you should routinely state the particular alpha level you selected for the statistical tests you conducted:
>
> An alpha level of .05 was used for all statistical tests.

Suggestions like these are highly confusing to authors (Hubbard, 2004, p. 313). After just being advised to use exact, data-dependent, *p*-values wherever feasible, the reader is next told to employ *both* *p*-values and α levels to indicate fixed (yet variable?) degrees of statistical significance. This practically forces the analyst to see, wrongly, a *p*-value as an "observed" Type I error rate.

Strangely, the latest edition of the APA (2010) *Publication Manual* is mute on the topic of the existence of two probabilities, p and α, and their roles in the reporting of statistical significance tests. The potential author simply is asked to provide exact p-values in the text of his or her paper, and to revert to "the '$p <$' style if using exact probabilities would make it difficult to comprehend the graphic [or tables]" (p. 139). Alpha appears only once, and without remark, in a stylized example in this most recent APA *Publication Manual* (p. 140).

Because they often impose how the outcomes of statistical tests are to be presented, the "Guidelines to Contributors" sections of scholarly journals can be extremely powerful sources of misleading advice. For example, the manuscript submission guidelines for the *American Sociological Review* (www.sagepub.com/journals/Journal201969/manuscriptSubmission) tell prospective authors: "Use asterisks *, **, and *** to indicate significance at the $p < .05$, $p < .01$, and $p < .001$ levels, respectively." Comparable instructions are given to contributors to the *Academy of Management Journal* (2011, p. 1083). This use of "fixed" p-values strongly suggests to researchers that they are, in fact, roving alphas.

Conversely, when it comes to reporting statistical findings in the *Journal of Advertising Research* (*JAR*; "Note to Contributors," 2008, p. 165), authors are informed that

> the Neyman–Pearson approach to statistical deduction will be followed. Null hypotheses should be established and statistical tests used to reject them and accept or implicitly support the alternative hypotheses. Unless the author offers a reasonable argument for a different level of significance, the standard alpha = .05 will be used for all tests.

In principle at least, then, *JAR* policy traditionally has registered a commitment to an N-P, fixed α, methodology in analyzing and presenting results. Yet owing to massive confusion among researchers over the meanings of ps and αs, especially the belief that the p-value is an observed Type I error rate, this is contradicted in practice.

For instance, I examined the contents of two randomly selected issues of the *JAR* for every year since its inception in 1960 through 2002 to determine how the results of statistical tests are conveyed (Hubbard & Armstrong, 2006, pp. 116–118). This procedure uncovered 279 data-based papers employing such testing. Of these, a trivial 2 (0.7%) actually carried out a statistical test in the manner prescribed by N-P theory. In contrast, though use of the p-value violates the underpinnings of the N-P model, its presence in some form was apparent in 218 (78.1%) published articles.

Surveys of Researchers. Anecdotal support for the widespread interpretation of *p*-values as error probabilities is offered by Kline (2004, p. 64), who says that many researchers mistakenly believe that "if $p < .05$, there is less than a 5% chance that the decision to reject the null hypothesis is a Type I error." Or as Pollard and Richardson (1987, p. 160) comment, "Our informal inquiries within a wide and varied cross section of our professional colleagues" reveal a dominant belief that *p*-values are Type I errors.

More formal survey research confirms this view. As an example, Oakes (1986, p. 80) reported that 86% (60) of 70 British academic psychologists thought that the *p*-value is the probability of making a Type I error. In a further study with dismaying implications for statistical literacy, Gigerenzer, Krauss, and Vitouch (2004, p. 395) discovered that 67% of 39 German psychology faculty members not teaching statistics courses and 73% of 30 faculty *teaching* them made the same mistake as their British counterparts!

Statistical Reporting Practices of Academicians. In addressing this issue I have compiled data on how researchers in the fields of psychology and marketing report the outcomes of tests of statistical significance (Hubbard, 2011b). As might be anticipated from the above, the misinterpretation of *p*-values as error probabilities is rampant. Consider, first, the data from psychology. These were obtained by content-analyzing a randomly chosen issue of each of 12 American Psychological Association journals (*American Psychologist, Developmental Psychology, Journal of Abnormal Psychology, Journal of Applied Psychology, Journal of Comparative Psychology, Journal of Consulting and Clinical Psychology, Journal of Counseling Psychology, Journal of Educational Psychology, Journal of Experimental Psychology—General, Journal of Personality and Social Psychology, Psychological Bulletin,* and *Psychological Review*) for every year from 1990 through 2005. This resulted in a total of 2,168 empirical articles, of which 2,043 (94.2%) employed significance testing.

As depicted in Table 7-4, fully 1,357 (66.4%) of the 2,043 articles used untenable roving alphas. A further 75 (3.7%) articles reported specific *p*-values (e.g., $p = .083$), that is, data-adjusted "exact probabilities of Type I errors" (recall Table 7-3). Still another 491 (24.0%) investigations presented a combination of exact *p*-values and roving alphas. Additionally, 89 (4.4%) articles couched their outcomes as "fixed" *p*s, rather than αs, and 8 (0.4%) simply reported findings as being "significant," with no details provided beyond that. A mere 16 (0.8%) papers conveyed their results in terms of the preassigned, fixed αs demanded by N-P theory.

Table 7-4 Reporting the Outcomes of Statistical Significance Tests in Psychology and Marketing

			"Fixed" Level Values				
Roving Alphas	*Exact P-values*	*Combination of Exact P-values and Roving Alphas*	*ps*	*"Significant"*	*αs*	*Unspecified*	*Total*
Psychology: 1990–2005							
1,357 (66.4)	75 (3.7)	491 (24.0)	89 (4.4)	8 (0.4)	16 (0.8)	7 (0.3)	2,043 (100)
Marketing: 1990–2000							
312 (71.7)	19 (4.4)	61 (14.0)	21 (4.8)	4 (0.9)	5 (1.1)	13 (3.0)	435 (100)

Source: Adapted from: Hubbard, Raymond (2011b), "The Widespread Misinterpretation of *P*-Values as Error Probabilities," *Journal of Applied Statistics*, 38 (November), p. 2623 with permission of the publisher (Taylor & Francis Ltd, http://www.tandfonline.com).

Note: Values in parentheses are percentages.

The marketing data were collected from an annual random sample of two issues of three leading periodicals (*Journal of Consumer Research, Journal of Marketing,* and *Journal of Marketing Research*) for 1990 through 2000. It consisted of 478 empirical articles, of which 435 (91.0%) involved significance testing (see Table 7-4). Marketing's results parallel those in psychology. For example, 312 (71.7%) employed roving alphas, 19 (4.4%) used exact *ps*, and 61 (14.0%) presented findings as a mixture of roving alphas and exact *p*-values. A further 21 (4.8%) papers reported "fixed" *ps* (not αs), and 4 (0.9%) mentioned that results were "significant" without going beyond this. As with psychology, studies using the fixed α method, an obligatory component of N-P theory, are conspicuous by their absence: only 5 (1.1%) papers heeded this approach.

The above outcomes reflect an almost total failure to comply with the stipulation of, ostensibly, N-P statistical convention to use preselected, fixed α levels in empirical investigations. On the one hand, this translates to a stunning, not to say incomprehensible, finding. But as I explain, it is a finding easy enough to reconcile when the research community believes, as has been shown abundantly throughout the preceding pages, that the *p*-value is an "observed" Type I error rate (Hubbard, 2011b, p. 2623).

Addendum. It was claimed by several researchers in Section 2.4 that empirical work in the business and social sciences likely is infected with Type I errors, that is, erroneous rejections of H_0. This claim now requires clarification. The above accounts have shown that *p*-values, and not fixed αs, are the almost universal means of determining the statistical significance or otherwise of results. But the *p*-value is not a Type I error rate, α. Therefore, while questionable research practices aimed at achieving statistically significant outcomes means that empirical literatures remain littered with errors, they are not, strictly speaking, of the Type I variety.

Summing Up

Unresponsive to a swelling literature pointing to the manner in which researchers in the social and management sciences, as a matter of course, incorrectly imbue the *p*-value with a raft of capabilities it does not possess, such abuses nonetheless continue apace. Thus, we have seen that the *p*-value is misinterpreted as the probability that both the null (H_0) and alternative (H_A) hypotheses are true and that the results of a study will replicate. The *p*-value also is wrongly seen to be an indicator of effect sizes, the generalizability of results, the causal relationships between variables of interest, and observed Type I errors. These misconceptions are pure flights of fancy and raise serious questions about the teaching of methods of statistical inference.[6]

It is saying something to admit that many investigators are unaware of just exactly what that most cherished research desideratum of all—statistically significant results—means. Tryon's (1998, p. 796) unease over the general misunderstandings of statistical tests by psychologists (and others) is palpable:

> The fact that statistical experts and investigators publishing in the best journals cannot consistently interpret the results of these [NHST] analyses is extremely disturbing. Seventy-two years of education have resulted in miniscule, if any, progress toward correcting this situation. It is difficult to estimate the handicap that widespread, incorrect, and intractable use of a primary data analytic method has on a scientific discipline, but the deleterious effects are undoubtedly substantial and may be the strongest reason for adopting other data analytic measures.

It is an unease felt by Kmetz (2011, p. 172), who is also fearful that wholesale bungled interpretations of the results of statistical significance tests have produced "an unsound, unscientific, [and] typically incorrect body of findings." He is despondent that such a fallout has meant the effective loss of 50 years of research efforts. I share Tryon's and Kmetz's sentiments. As, no doubt, does McCloskey (2002).

7.4.2 Significant Sameness

P-values play a minor part in the significant sameness model. In these circumstances their role is of a more heuristic kind and not the end point of a study (Hubbard & Lindsay, 2013a, p. 1386). Their functions include determining the goodness-of-fit of substantive (non-null) models to data as well as detecting observations (data sets) that are outliers to the general pattern.

7.5 The Role of Effect Sizes and Confidence Intervals

7.5.1 Significant Difference

In psychology, an earlier APA (1994, p. 18) *Publication Manual* advised that *p*-values were not acceptable indices of effect sizes and "encouraged" contributors to supply information on the latter. What has been the impact of this encouragement? A number of empirical studies on this topic conclude not much.

Kirk (1996), for instance, inspected the 1995 volumes of four APA journals to calculate the percentage of articles using inferential statistics that also reported effect size information. He grumbled that these percentages, in parentheses, were too low—*Journal of Applied Psychology* (*JAP*; 77%), *Journal of Educational Psychology* (55%), *Journal of Experimental Psychology, Learning and Memory* (12%), and *Journal of Personality and Social Psychology* (47%)—and warned that the higher figure for *JAP* is anomalous because of the greater use of regression analysis in that journal, with software routinely providing a measure of R^2. Bruce Thompson and Patricia Snyder (1997) noted that 64% of articles in the 1994–1995 and 1995–1996 issues of the *Journal of Experimental Education* reported effect sizes together with significance tests. They voiced disappointment with this figure, however, given that an entire 1993 theme issue of this same journal had dealt with the shortcomings of statistical significance testing and of the need to report effect sizes. Vacha-Haase, Nilsson, Reetz, Lance, and Thompson (2000) indicated concern over the fact that, on average for the period 1990–1997, only 48% of empirical papers in *Psychology and Aging* supplied measures of effect magnitude, and 43% did so in the *Journal of Counseling Psychology.*

The 2001 edition of the APA *Publication Manual* also called for a greater focus on the provision of effect sizes in scholarly papers. Did this call have the desired effect? According to Hoekstra, Finch, Kiers, and Johnson (2006) it did not. While their investigation of 266 empirical articles published in the 2002–2004 issues of *Psychonomic Bulletin & Review* found that 259 (97.4%) used statistical significance testing, only 3 (1.2%) of these presented *standardized* (e.g., Cohen's *d*) effect sizes (Hoekstra et al., 2006, p. 1035).

Table 7-5 displays information gathered on the publication frequency of effect size reporting in marketing (Hubbard & Lindsay, 2013a). This table is derived from a simple random sample of one issue each of the *Journal of Consumer Research* (*JCR*), *Journal of Marketing* (*JM*), and *Journal of Marketing Research* (*JMR*) for every year from 2000 through 2007. It shows that the documentation of one or more of some 40 measures of effect sizes (excluding simple correlation coefficients) found in Kirk (1996) receive short shrift.

For all three journals, 28.0% of empirical articles supplied an effect magnitude. The breakdown of these results by journal is as follows: *JCR* (17.5%), *JM* (37.7%), and *JMR* (33.0%). Much like Kirk's (1996) concerns over the results for *JAP*, these numbers are misleading; they are affected disproportionately by R^2 and $\bar{R}^2$ measures. Removing these indexes yields a far spottier portrayal of effect size reporting behavior: *JCR* (8.2%),

Table 7-5 Reporting of Effect Magnitudes and Confidence Intervals in Marketing: 2000–2007

Journals[a]	*Number of Empirical Studies*	*Number Using Effect Magnitudes*	*%*	*Number Using Confidence Intervals*	*%*
JCR	97	17	17.5	1	1.0
JM	61	23	37.7	0	0
JMR	88	29	33.0	3	3.4
Total	246	69	28.0	4	1.6

Source: Adapted from Hubbard, Raymond and R. Murray Lindsay (2013a), "From Significant Difference to Significant Sameness: Proposing a Paradigm Shift in Business Research," *Journal of Business Research*, 66 (September), p. 1385 with permission from Elsevier.

[a]*JCR = Journal of Consumer Research, JM = Journal of Marketing, JMR = Journal of Marketing Research.*

JM (3.3%), and *JMR* (2.3%). It is obvious that greater attention to the inclusion and *interpretation* of effect sizes in marketing's published empirical literature is required.[7]

While reservations have been aired over the insufficient reporting of effect sizes in the literature, there is abject failure to provide CIs around point estimates. In cataloging this, Finch, Cumming, and Thomason (2001) analyzed the first 30 articles from the *Journal of Applied Psychology* for the years 1940, 1955, 1970, 1985, and 1999. They also included 30 articles from the 1999 volume of the *British Journal of Psychology* (*BJP*). Of the 150 *JAP* articles, only 4 (2.7%) reported CIs, and only 1 of 30 (3.3%) *BJP* papers did so. Kieffer, Reese, and Thompson (2001) found that none of the 251 empirical articles published in the *American Educational Research Journal* over the period 1988–1997 used CIs, while only 1 of 506 (0.2%) such papers did so in the *Journal of Counseling Psychology.* Bonett and Wright's (2007) findings are remarkably similar. They divulged that of more than 130 empirical articles featured in the *Academy of Management Journal* and *Administrative Science Quarterly* during 2003–2004, just 1 (0.8%) reported a CI. Finally, Hubbard and Lindsay (2013a) tell of close to a total void of presenting CIs in their random sample of work published in leading marketing journals. As Table 7-5 reveals, only 4 of 246 (1.6%) empirical articles listed CIs. Understandably, there was great consistency among the journals: *JCR* (1.0%), *JM* (0%), and *JMR* (3.4%).

The penultimate APA (2001) *Publication Manual* went beyond its predecessor in recommending, rather than merely encouraging, the inclusion of effect sizes in submitted manuscripts. It was also the first one to strongly suggest the reporting of CIs. Despite the suggestion to report CIs, Hoekstra et al. (2006, p. 1035) were disillusioned to find that only 13 (5.0%) of 259 empirical articles published in the 2002–2004 issues of *Psychonomic Bulletin & Review* recorded CIs. They concluded that "changing the practice of reporting statistics seems doomed to be a slow process" (p. 1033).

Cumming et al. (2007) likewise were disappointed that this advice to employ CIs was being less than enthusiastically adopted by authors of articles published in the following 10 leading international psychology journals: *Acta Psychologica, Child Development, Cognition, Journal of Abnormal Child Psychology, Journal of Abnormal Psychology, Journal of Consulting and Clinical Psychology, Journal of Experimental Psychology: General, Journal of Personality and Social Psychology, Psychological Science*, and *Quarterly Journal of Experimental Psychology*. For these journals they coded 40 empirical papers from each appearing in 1998, 2003–2004, and 2005–2006. On average, the use of null hypothesis significance testing for these three time periods is almost universal at 97.8%, 97.7%, and 96.9%, respectively. The corresponding figures for the reporting of CIs are 3.7%, 9.2%, and 10.6%; only 24.1% of articles with CIs attempted to interpret them.

The latest APA (2010, p. 34) *Publication Manual* employs stronger language still than previous editions when advocating use of CIs:

> The inclusion of confidence intervals . . . can be an extremely effective way of reporting results. . . . [T]hey are, in general, the best reporting strategy. The use of confidence intervals is therefore strongly recommended. . . . Wherever possible, base discussion and interpretation of results on point and interval estimates.

Whether this will be enough to persuade researchers in psychology to change the manner in which they analyze data and present findings remains to be seen. But for the sake of a more relevant psychology, and by extension the social and management sciences in general, I hope that such recommendations finally are taken on board.

7.5.2 Significant Sameness

As told in Section 3.2.3, this paradigm insists on the reporting and interpretation of sample statistics, effect sizes, and the CIs around them.

7.6 Conclusions

The significant difference paradigm commissions statistical practices which thwart cumulative knowledge development. It does so by assuming that diagnostic statistical procedures applied to single data sets are all that is required to satisfactorily deal with the validity of inferences made in the face of the model uncertainty problem. Moreover, members of this paradigm have been all too willing to back the blinkered view that social and management theories are capable only of making qualitative predictions. The elevation of mindless NHST and its minor statistic, the *p*-value, to utter dominance in this paradigm, even though poorly understood, is both dispiriting and embarrassing. More useful statistical information such as providing effect sizes, and particularly CIs around point estimates, too often is ignored.

The significant sameness paradigm, on the other hand, fosters cumulative knowledge growth. Model uncertainty issues can be resolved only via the collection of more data sets (replications) to see whether the same or similar results hold. Of critical interest, significant sameness calls for quantitative predictions—or what Ziliak and McCloskey (2008) call "oomph." That management and social science theories are not sophisticated enough to make point or interval predictions is something that has been accepted largely without challenge. And this acceptance, in turn, has ossified the role of statistical significance testing. Significant sameness pushes for the reporting of sample statistics, effect sizes, and their CIs. Indeed, as presented in Chapter 3, overlapping CIs between two similar studies is how to judge (but not in a machinelike fashion) significant sameness.

The significant sameness model is a far truer account of the untidy manner in which science works in the real world than the pristine, detached, textbook version characteristic of the significant difference approach. It is a model in need of adoption by those in the social science and business disciplines.

Notes

1. For a more recent exposition in the Popperian tradition, see Mayo (1996).

2. Here, Fisher is incorrect. As Johnstone (1986, p. 491) underlines, Bayes' theorem is beyond reproach. In fact, a recent book by McGrayne (2011) attests to the resurgence of Bayes' theorem in the scientific community, something I welcome despite being a frequentist by training.

3. See Krämer and Gigerenzer (2005) for documentation of how conditional probabilities *in general* confuse consumers of statistical methods.

4. An aside is needed to explain why α must be fixed before the study is carried out. Hence the following extensive quotation from an article of mine (Hubbard, 2004, pp. 310–311):

> Royall (1997, pp. 119–121) addresses the rationale for prespecifying α, and I paraphrase him here. Suppose a researcher decides not to supply α ahead of time, but waits instead until the data are in. It is decided that if the result is statistically significant at the .01 level, it will be reported as such, and if it falls between the .01 and .05 levels, it will be recorded as being statistically significant at the .05 level. In the Neyman–Pearson model, this test procedure is improper and yields misleading results, namely, that while H_0 will occasionally be rejected at the .01 level, it will *always be rejected* any time the result is statistically significant at the .05 level. Thus, the research report does not have a Type I error probability of .01, but only .05. If the prespecified α is fixed at the .05 level, no extra observations can legitimize a claim of statistical significance at the .01 level. This, Royall (1997) points out, makes perfect sense within Neyman–Pearson theory, although no sense at all from a scientific perspective, where additional observations would be seen as preferable. And this deficiency in the Neyman–Pearson paradigm has encouraged researchers, albeit unwittingly, to *incorrectly* supplement their testing procedures with "roving alpha" *p* values.
>
> The Neyman–Pearson hypothesis testing program strictly disallows any "peeking" at the data and (repeated) sequential testing (Cornfield, 1966; Royall, 1997). Cornfield (1966, p. 19) tells of a situation he calls common among statistics consultants: suppose, after gathering *n* observations, a researcher does not quite reject H_0 at the prespecified $\alpha = .05$ level. The researcher still believes the null hypothesis is false, and that if s/he had obtained a statistically significant result, the findings would be submitted for publication. The investigator then asks the statistician how many more data points would be necessary to reject the null. And, in Neyman–Pearson theory, the answer is *no amount of (even extreme) additional data points* would allow rejection at the .05 level. As incredible as this answer seems, it is correct. If the null hypothesis is true, there is a .05 chance of its being rejected after the first study. As Cornfield (1966) remarks, however, to this chance we must add the probability of rejecting H_0 in the second study, conditional on our failure to do so after the first one. This, in turn, raises the total probability of the erroneous rejection of H_0 to over .05. Cornfield shows that as the *n* size in the second study is continuously increased, the significance level approaches .0975 (= .05 + .95 x .05). This demonstrates that no amount of collateral data points can be gathered to reject H_0 at the .05 level. Royall (1997) puts it this way: "Choosing to operate at the 5% level means allowing only a 5% chance of erroneously rejecting the hypothesis, and the experimenter has already taken that chance. He spent his 5% when he tested after the first *n* observations" (p. 111). But once again, despite being in complete accord with Neyman–Pearson theory, this explanation runs counter to common sense. And once again, researchers *mistakenly* augment the Neyman–Pearson theory with Fisherian *p* values, thereby introducing a thicket of "roving" or pseudo-alphas in their reports.

5. More generally, a small scale ($n = 25$) empirical assessment by Lecoutre, Poitevineau, and Lecoutre (2003, p. 42) was sufficient for them to conclude that "it is not actually an easy task, even for professional statisticians, to interpret *p* values [in NHST situations]."

6. Unhappily, thoughtlessness about what statistical significance and *p*-values are can be expected to worsen. As an illustration of this, in psychology a national survey of undergraduate programs in the United States (Friedrich, Buday, & Kerr, 2000), and of PhD programs in both the United States and Canada (Aiken, West, Sechrest, & Reno, 1990), found that less emphasis is being placed on quantitative methods in the curriculum. These surveys are consistent with Kirk's (2001, p. 216) condemnation, as a writer of five statistics textbooks in the education/psychology areas, that publishers pressure authors to "dumb down" the texts and put greater reliance on software packages. The consequence is that the student, and future researcher, is predestined to view the outcomes of statistical analyses as a "black box" to be accepted unquestionably on faith (Searle, 1989, p. 189). Researchers end up treating statistics as if it were a religion, and "like any good religion, it involves vague mysteries capable of contradictory and irrational interpretation" (Salsburg, 1985, p. 220). This mysticism attains full bloom in Saunders, Lewis, and Thornhill's (2012, p. 512) business research textbook: "Once you have entered data into the analysis software, chosen the statistic and clicked on the appropriate icon, an answer [e.g., *p*-value] will appear as if by magic."

7. On matters related to effect magnitudes, Combs (2010, p. 9) tells, disapprovingly, that these have been decreasing in size over time in the management area. Drawing on information culled from the size of Pearson correlation coefficients (*r*s) published in the *Academy of Management Journal* for 2007–2008 versus 1987–1988, he concludes that, on average, they fell from .22 to .17, a 23% decline. Combs is troubled by the fact that these shrinking effect sizes, because of the larger samples involved ($\bar{x}$ = 300 to $\bar{x}$ = over 3,000 between 1987–1988 and 2007–2008), continue to achieve the gold standard of statistical significance, but perhaps at the cost of masking their theoretical and managerial relevance.

WHITHER THE ACADEMY?

> It is probably fair to say that the current [academic publish or perish] system has led to conditions under which a significant contribution to knowledge is not at the forefront of most participants' thoughts as they engage in the research and publication process. (AMA Task Force, 1988, p. 6)
>
> KEEP DOING WHAT YOU'RE DOING. (Iacobucci, 2005)

8.1 Introduction

The legitimacy of the social and management disciplines as bona fide sciences has been, and continues to be, a topic of debate. Of prime concern is the question as to whether these fields contribute useful knowledge to their various constituents. For the most part I believe that they do not, nor do I see this happening in the future. This is because the vast majority of people in these disciplines share a view of the research endeavor which promotes bad science.

What is the likelihood of the significant sameness paradigm supplanting, or at least paralleling, that of significant difference? This is a tall challenge, one made worse by two formidable and interrelated barriers. The first is that members of the significant difference paradigm cherish their conception of science and tend to look at alternative approaches to knowledge procurement with an air of suspicion, or even dismissiveness. They like the way their methodological world is set up and wish to keep doing things the same way they have been done for decades. As a consequence, the sheer weight of academic inertia, fortified by researcher unawareness of how science makes headway, acts as a powerful antidote against the need for change of any kind. The second barrier is that all too often academicians are preoccupied with enriching their careers as scholars and are unconcerned with real-world knowledge development. These themes are amplified in Section 8.2.

Section 8.3 suggests how one might help cultivate a significant sameness mentality among management and social science researchers. This consists of some recommendations aimed at a realignment of academic priorities that are conducive to knowledge production as well as some nuts-and-bolts specifics for institutionalizing replication research. Unbeknown to many readers, Section 8.4 sketches a time—the 19th century—when the concept of significant sameness was invaluable in the nurturing of the embryonic statistics and social studies disciplines. Conclusions follow in Section 8.5, where I call upon the research community to readopt significant sameness as a way to progress in the management and social sciences.

8.2 Obstacles to the Implementation of the Significant Sameness Paradigm

8.2.1 Academic Inertia and Unawareness

Most individuals in the social and business sciences appear to be sincere in their unalloyed commitment to the significant difference paradigm. It has been mentioned repeatedly that they see it as the application of *the* scientific method, a method drummed into them in their graduate school educations. As Locke (2007, p. 867) parodies:

> Everyone who publishes in professional journals in the social sciences knows that you are supposed to start your article with a theory, then make deductions from it, then test it, and then revise the theory. At least that is the policy that journal editors and textbooks routinely support.

The most rigorous tests in the arsenal of the dominant hypothetico-deductive (H–D) model are, of course, those of the null hypothesis variety, with $p \leq .05$ outcomes being especially desirous.

Thus, attacks on the integrity of the significant difference paradigm and its methods of testing (see Table 7-1 for a taste of these) are met with, at best, some irritation, but mostly with indifference. It is as if, through ignorance and/or evasion—Meehl's (1967, p. 107) conspiracy of silence?—those in the management and social sciences share an unquestioning obedience to doing things the way they have been done for many years, criticisms be damned (cf. Macdonald & Kam, 2007, pp. 650–651).

To get some appreciation of the depth of the apathy, denial, and resistance among members of the significant difference camp on the need for educational reform, I provide several examples of the outright failure to implement previous specific recommendations to improve research practices. These examples do not bode well for the future of significant sameness, which demands the acceptance of *all* of them.

Failure to Improve Statistical Power. In a seminal contribution on this topic, Jacob Cohen (1962) assessed the power of some 2,088 statistical tests found in 70 articles of the 1960 issues of the *Journal of Abnormal and Social Psychology.* Cohen found that the mean power levels in these articles were "far too low" (p. 153). The probabilities of detecting small, medium, and large effect sizes were only .18, .48, and .83, respectively. If one assumes, and some (e.g., Bakker, van Dijk, & Wicherts, 2012, p. 544) may dispute this as overly optimistic, that medium effect sizes characterize the social and management science disciplines, then a researcher has only about a 50-50 chance of rejecting a false null hypothesis.

Sadly, Cohen's (1962) article illuminating the problem of low statistical power in the empirical psychology literature, together with his 1969 book and its 1988 second edition aimed at redressing this state of affairs, appear to have had no remedial impact whatsoever on researchers' shoddy ways. This, despite the fact that the books have attracted impressive citation counts of 778 and 508, respectively.[1]

For example, 27 years after Cohen's (1962) original study, Sedlmeier and Gigerenzer (1989) analyzed the power exhibited in 54 articles of the 1984 volume of the (by now renamed) *Journal of Abnormal Psychology* and found the average probabilities of uncovering small, medium, and large effect sizes in the population—.21, .50, and .84, in turn—to be virtually identical to Cohen's. The same is true of Rossi's (1990) conclusions arising from a power analysis of 6,155 statistical tests in 221 articles appearing in the 1982 volumes of the *Journal of Abnormal Psychology, Journal of Consulting and Clinical Psychology,* and *Journal of Personality and Social Psychology.* Here, the power to discern small, medium, and large effects was .17, .57, and .83. Ostensibly, both of these works helped further reveal the plight of underpowered research designs in the field, the former gaining 505, and the latter 179, citations.

Subsequent to the above investigations, Jacob Cohen's (1992) article "A Power Primer" has enjoyed a staggering 11,944 citations! Moreover, the ease of performing power analyses has been greatly enhanced by the publication of various software packages, including, but certainly not limited to, those of Erdfelder, Faul, and Buchner (1996), Borenstein, Rothstein, and Cohen (1997), and SPSS. To no avail, it seems. Thus, Bezeau and Graves (2001) examined the statistical power in a representative sample of 66 articles in the 1998 and 1999 issues of the *Journal of Clinical and Experimental Neuropsychology, Journal of the International Neuropsychology Society,* and *Neuropsychology.* They found that the median (chosen instead of the mean) power to detect a medium effect was only .50. Bezeau and Graves further reported that barely 3% of the reviewed studies conducted a priori power calculations.

So in the face of numerous appeals to raise power levels, and the means at hand to readily achieve this, the passage of 50 years has seen no improvement in psychology research. The same generally is true in other disciplines (recall Table 2A-1). This is why Maxwell (2004, p. 161) warns that until the issue of low statistical power is dealt with, "the published literature is likely to contain a mixture of apparent results buzzing with confusion." It is a view seconded by Schimmack (2012, p. 564).

Failure to Report/Interpret Confidence Intervals. Frank Schmidt (e.g., 1992, 1996) and Schmidt and Hunter (e.g., 1997, 2002) have done more than most to show how feckless devotion to statistical significance testing is ruinous to cumulative knowledge development. In common with the orientation put forward here, Schmidt (1996, p. 115) advocates replacing significance testing with point estimates and confidence intervals (CIs) in individual studies and using meta-analysis to integrate the findings of multiple studies. Schmidt (1996, p. 128) explains how at Michigan State University, John Hunter and Ralph Levine changed the graduate statistics course sequence in psychology to reflect the above views. The result, Schmidt goes on, was protests among faculty from the significant difference school. They did not claim that these new methods were erroneous, but were concerned that their graduate students would not be able to publish their work unless it used statistical significance testing. And those of a significant difference orientation don't need to be reminded that publication means everything. Following Hunter's death, however, this attempt at statistical reform at Michigan State has petered out.

A second example, also from psychology, further illustrates the problems encountered in trying to amend statistical reporting practices, such as using CIs rather than *p*-values. Fidler, Thomason, Cumming, Finch, and Leeman (2004) write of the difficulties Loftus (1993) grappled with in his attempts to decrease the obsession with null hypothesis significance testing (NHST) as editor of *Memory & Cognition.* During his tenure in this role, Fidler et al. remark, the percentage of articles employing error bars (both CIs and standard errors) increased to 41%, compared with 7% under his forerunner. Yet when Loftus was no longer editor of *Memory & Cognition,* this proportion dropped to 24%. Without constant editorial prompting, people tend to revert to old and familiar customs.

The frequency of articles employing CIs in the literature at large hovers at dismal levels. Based on an admittedly small number of studies, Chapter 7 showed that in social science journals these estimates range between 0% and 10.6%, while for management sciences it is 0.8%–1.6%. Additionally, few of these attempt to interpret CIs.

Failure to Conduct/Publish Replications. Still another example indicates how academic inertia prevents scholars from changing their epistemological ways. This one involves a follow-up of Hubbard and Armstrong's (1994) article by Evanschitzky, Baumgarth, Hubbard, and Armstrong (2007). Our initial study analyzed the publication frequency of replications and extensions in 835 empirical articles obtained from an annual random sample of issues of the *Journal of Consumer Research* (*JCR*), *Journal of*

Marketing (*JM*), and *Journal of Marketing Research* (*JMR*) for the period 1974–1989. It revealed that a mere 20 of these 835 papers, or 2.4%, qualified as replications and extensions. The breakdown by journal was *JCR* (2.3%), *JM* (3.4%), and *JMR* (1.9%).

The paper by Evanschitzky et al. (2007, p. 411) set out to answer the following question: "What has happened to the publication rate for replications and extensions in marketing in the years following [the Hubbard and Armstrong] study?" It turns out that the article by Scott Armstrong and me has been quite influential, at least as measured by the counting of citations, attracting some 275 of these. Given this, it's not unreasonable to anticipate that our study, particularly when allied with others on this topic, such as Lindsay and Ehrenberg (1993), Hubbard and Vetter (1996), and Hubbard, Vetter, and Little (1998)—with citation scores of 280, 137, and 169—might have a positive effect on the occurrence of replication research in the field. As it happens, this has not been the case; the numbers actually went down. Evanschitzky et al.'s census of 1,389 empirical studies published in these same three journals for the period 1990–2004 shows that only 16, or 1.2%, were replications and extensions. In every one of these journals—*JCR* (1.7%), *JM* (1.2%), and *JMR* (0.6%)—the publication incidence of such work for 1990–2004 was lower than it was for 1974–1989.

Failure to Curb NHST. Since the 1960s there has been a loudening drumbeat of criticism over the limitations and chronic misuse of NHST in the behavioral and business sciences. Yet marked censure of NHST has failed to stem its usage. On the contrary, its presence has increased steadily. This is highlighted in Table 8-1, showing on a decade-by-decade basis how even high-impact articles and books (in terms of citation rates) severely critical of NHST have had no effect at all on aggregate researcher behavior.

In the 1960s, when roughly 50%–55% of data-based research in the social and management sciences featured significance testing, Rozeboom (1960), Bakan (1966), and Lykken (1968)—with 1,981 citations between them—wrote some of the most scathing reviews of this practice. The beginning of the next decade saw Morrison and Henkel's (1970) anthology, *The Significance Test Controversy,* "popularize" the issue, while Carver (1978) weighed in with, in my opinion, one of the finest rebukes of significance testing. Yet in comparison with the 1960s, a pronounced increase in employment of the latter in both the management (80%) and social (72%) sciences is apparent from Table 8-1. And so it went in the 1980s, with rates climbing to 89% and 84%, respectively.

Table 8-1 Citation Counts of Selected Articles and Books Critical of NHST and Its Nonetheless Growing Prevalence (%) in Empirical Work: 1960–2007

Years	*Article/Book*	*Citation Count*[a]	*Management Sciences*[b]	*Social Sciences*[b]
1960–1969	Bakan (1966)	741	52%	56%
	Lykken (1968)	742		
	Rozeboom (1960)	498		
1970–1979	Carver (1978)	746	80%	72%
	Greenwald (1975)	631		
	Morrison & Henkel (1970)	564		
1980–1989	Guttman (1985)	88	89%	84%
	Rosnow & Rosenthal (1989)	609		
	Salsburg (1985)	157		
1990–1999	J. Cohen (1990)	1,396	92%	92%
	J. Cohen (1994)	2,460		
	Harlow, Mulaik, & Steiger (1997)	461		
	Kirk (1996)	871		
	Schmidt (1996)	827		
	Wilkinson & APA Task Force (1999)	1,935		
2000–2007	Gigerenzer (2004)	194	93%	92%
	Kline (2004)	584		
	Nickerson (2000)	486		
	Ziliak & McCloskey (2008)	369		

[a]Google Scholar citation counts are from date of publication.

[b]These figures show the percentage of empirical work using significance testing. They are approximations based on data presented in Chapter 2.

Table 8-1 reveals how the 1990s witnessed the publication of several (overwhelmingly) anti-NHST works with astonishing citation counts. These range from a "low" of 461 for Harlow, Mulaik, and Steiger's (1997) edited book, *What If There Were No Significance Tests?*, to a high of 2,460 for Jacob Cohen's (1994) "The Earth is Round ($p < .05$)." The average number of citations earned by the six contributions from the 1990s listed in the table is a remarkable 1,325. And still the publication incidence of statistical significance testing in empirical research forged ahead in the social and management sciences, both now at 92%. It continued to inch ahead in the 2000s.

The 19 articles and books referred to in Table 8-1 have accumulated a total of 14,359 citations, or an average of 756 apiece. This would make them seemingly impossible to ignore in academic circles. Yet ignored they are when it comes to changing statistical analysis and reporting habits. Jeff Gill (1999, p. 670) comments that many social scientists have the impression that a testing procedure which is so pervasive and long-lived cannot be seriously flawed. I agree with Gill and add that this is a profoundly destructive example of circular reasoning whereby collective researcher misunderstandings of what testing is about serve to fuel its popularity.

Summing Up. It is apparent from the above examples that effecting changes in the manner in which things are done by those in the significant difference school, including how numerical evidence is obtained, analyzed, and presented in the literature, will be a daunting task. It is a task exacerbated by the likes of Iacobucci (2005), then editor of the *Journal of Consumer Research,* who on the topic of reporting quantitative results simply informed her readership, "KEEP DOING WHAT YOU'RE DOING." Hence one of the quotations introducing this chapter, and an authoritative rubber-stamping of business as usual. Iacobucci's reaction is consistent with Macdonald and Simpson's (2001, p. 130) more general observation that "in management research, criticism is not welcome." In the face of entrenched opinions like these, it is easy to understand Freese's (2007, p. 166) take that many members of his discipline (sociology) are surprisingly fatalistic about their ability to change communal research practices.

Complicating matters is the idea that adherents of the significant difference paradigm may not even be motivated by a desire to create knowledge applicable by those in their ambit. Put another way, the creation of usable knowledge succumbs to the allure of career advancement, as examined next.

8.2.2 Knowledge Development Versus Career Advancement

At first glance the heading seems to pose an absurd trade-off. Of course, one would reply, people in these disciplines are concerned with providing usable knowledge for their fields. The management disciplines, in particular, share elements in common with areas like medicine and engineering, where a major part of their raison d'être, presumably, is the formulation of knowledge to guide everyday practice. Likewise, one would think that the behavioral sciences are firmly in the vanguard of yielding robust information to aid in the understanding of society and the construction of effective social policies.

Yet on closer scrutiny the comparison suggested in the heading of this section is not an outrageous one, but rather a meaningful inquiry deserving of serious attention. The emergence of the business fields affords an opportunity to explore this issue. At the end of the 1950s two influential reports appeared, funded by the Carnegie Corporation (Pierson, 1959) and the Ford Foundation (Gordon & Howell, 1959), respectively, which were critical of the low educational standards exhibited by business schools and their faculties. They called for greater intellectual rigor in course content and from those teaching the courses. It was suggested that business faculty be given reduced teaching loads in order to engage in "scientific" research, with cognate disciplines like economics, psychology, and sociology acting as role models to emulate. In what surely counts as an exemplar of the "law of unintended consequences," the recommendations of the Ford Foundation and Carnegie Corporation helped set in motion a process leading inexorably to the dysfunctional situation found today. First, business researchers were initiated into a highly simplistic representation of *the* scientific method—the significant difference paradigm—which, by its very nature, obstructs the development of apposite, usable knowledge. Second, the growing fixation with scholarly research has driven a seemingly irreversible wedge between business faculty, on the one hand, and the very people they were charged with enlightening, practitioners, on the other. It is no stretch to say that academics and practitioners live in two different worlds.

The Academic World: Publish or Perish

For a couple of generations now, business (and social) science faculty have been all too keenly aware that career enhancement in the academy comes almost entirely from the publication of their work in prestigious, refereed journals (see, e.g., Giner-Sorolla, 2012, p. 569; Hubbard & Armstrong, 1994, pp. 243–244; Hubbard & Lindsay, 2002, p. 396; N. L. Kerr, 1998, p. 213; Macdonald & Kam, 2007, p. 640; Mahoney, 1985, p. 31; McCloskey, 2002, p. 57; McGrath, 2007, p. 1366; Nosek, Spies, & Motyl, 2012, pp. 616–617; Pettigrew, 2005, p. 973; Pfeffer, 1993, p. 610, 2007, p. 1334; Starbuck, 2006, p. 93). This publish-or-perish zeitgeist is true in general, but is especially pronounced in top-tier business schools whose members populate, disproportionately, the editorial boards of these same journals.[2, 3]

Some manifestations of the Darwinian publish-or-perish mentality immanent in the significant difference model are (1) slavish dedication to authoring a large number of "original" research articles which (2) are judged to have a major impact on the field but which, for the most part, (3) are not truly original at all. These issues are taken up below.

Quantity of Published Original Research. In the significant difference paradigm, publishing novel research in high-profile refereed journals epitomizes academic accomplishment. The more the merrier. While nothing like the situation seen today, particularly in the social and management sciences, even back in the 1950s there were telltale signs in some areas that volume of research output was becoming increasingly decisive to academic advance. In an article in *Science* titled "The Competitive World of the Pure Scientist," Reif (1961, p. 1960) put it this way:

> [There is the need] to publish as many papers as possible. . . . No longer does a scientist study a topic at some length before publishing his findings in a paper or monograph. Instead, he tries to publish a note on a subject as soon as he obtains any result worth mentioning—and occasionally even before. The threat of someone else's getting there first is too great.

Because "the road to academic heaven is paved with publications" (Reif, 1961, p. 1959), anything and everything, regardless of its value, must end up in print. Over time, this has resulted in the proliferation of what Broad (1981, p. 1137) calls the Least Publishable Unit (LPU), as when more and more papers are squeezed from a given research project. These LPUs are all in the service of padding one's résumé and typically make negligible contributions to knowledge (Hamilton, 1990, p. 1331). Published articles are no more than piecework, mere widgets rolling down the assembly line (Waters, 2004, p. 35).

Otherwise expressed, subscribers to the significant difference paradigm are rewarded heavily on the basis of publication counts (Hubbard, 1995b, p. 671), while little attention is focused on the development of useful knowledge. The irony is that the all-consuming insistence on boosting the volume of research and publication has backfired; rather than expediting the growth of sound knowledge, it has led to an "eclipse of scholarship" (Waters, 2004) and the production of empirical literatures whose integrity and utility must be called into question (see, e.g., Andreski, 1972, p. 30; Asendorpf et al., 2013, p. 115; Bakker et al., 2012, p. 543; Bedeian, Taylor, & Miller, 2010, p. 716; Fanelli, 2009, p. 6, 2010, p. 4; C. J. Ferguson & Heene, 2012, p. 558; Francis, 2012b, p. 151; Gerber & Malhotra, 2008, p. 3; Greenwald, 1975, p. 15; Guttman, 1985, p. 5; J. K. Hartshorne & Schachner, 2012, p. 1; Holcombe & Pashler, 2012, p. 355; Howard et al., 2009, p. 148; Hubbard, 1994, pp. 257–258; Hubbard & Armstrong, 1992, pp. 128–129; Hubbard & Vetter, 1996, p. 154; Hubbard, Vetter, & Little, 1998, pp. 243–244; Ioannidis, 2005b, p. 0101, 2011, p. 16, 2012, p. 645; Ioannidis & Doucouliagos, 2013, p. 997; Ioannidis & Trikalinos, 2007, p. 245; John, Loewenstein, & Prelec, 2012, p. 527; N. L. Kerr, 1998, p. 200; Lambdin, 2012, p. 68; Leamer, 1983, pp. 36–37; Lipsey & Wilson, 1993, pp. 1194–1195; Masicampo & Lalande, 2012, pp. 2272–2273; McCloskey, 2002, p. 55; McLeod & Weisz, 2004, p. 235; Meehl, 1967, p. 114; Pashler & Harris, 2012, p. 535; Pashler & Wagenmakers, 2012, p. 528; Schimmack, 2012, p. 553; Simmons, Nelson, & Simonsohn, 2011, p. 1359; Sovacool, 2008, p. 272; Stroebe, Postmes, & Spears, 2012, p. 681; Van de Ven, 2007, p. 17; Waters, 2004, pp. 18–21; Ziliak & McCloskey, 2008, p. 2). In the significant difference world the noise increasingly dwarfs the signal.

Worse yet, the crucial necessity of publishing one's work can all too easily result in an openly *corrosive* empirical literature. Under this kind of pressure, science can mutate into a winner-take-all game tempting scholars to falsify and fabricate data. This was intimated for the social and business sciences in the work reported by several investigators (e.g., Bedeian et al., 2010, p. 719; Carpenter, 2012, p. 558; Enders & Hoover, 2004, p. 489, 2006, p. 93; Fanelli, 2009, p. 6; Honig & Bedi, 2012, p. 101; Honig, Lampel, Siegel, & Drnevich, 2013, p. 119; John et al., 2012, p. 527; Stroebe et al., 2012, p. 683) in Chapters 2 and 4. It is observed also in bold relief in medical fields, where a piece in the *New York Times* (Zimmer, 2012) tells of the sharp upturn (a tenfold increase over the previous decade) in the number of articles being retracted by journals.[4] Such actions are symptomatic of a broken research climate. To quote a key contributor to the *New York Times* article, Dr. Ferric C. Fang, professor at the University of Washington School of Medicine: "What people do is they count papers, and they look at the prestige of the journal in which the research is published. . . . It's not about the quality of the research" (p. 3). But suppose, for the moment, that it *was* the quality and not the quantity of published papers which mattered. How would the quality of this research be assessed? This issue is explored next.

Quality of Published Original Research. Evaluation of the quality of novel research in the significant difference paradigm is conceived almost exclusively in terms of yet another count, this time of citations of one's work by other members of the academy.[5] This metric, which suffers from a number of problems even as an *academic* barometer of impact (see, e.g., Geuens, 2011, p. 1106), nevertheless is regarded as the gold standard of scholarly performance.

When employing this yardstick it quickly becomes apparent that most original published research sinks into oblivion (Hubbard, 1992, p. 34; Waters, 2004, p. 18). As proof of this, Laband's (1986) examination of the citation frequency of some 5,880 articles and notes published in 40 economics journals from 1974 to 1976 demonstrated that most of them are cited rarely, if ever. For the following period of 1977–1982, 84% of these works were cited fewer than 10 times, while only 0.3% were cited over 100 times. Laband (1990) states that every economics journal publishes articles that never get cited, and Phelan (1999) remarks that most academicians garner fewer than three citations during a lifetime. Putting things further into perspective, a mere 2% of over 32 million papers that were cited at least once between 1945 and 1988 were cited more than 50 times (Campanario, 1995, p. 306). In a similar vein, Earl Hunt (2013, p. 127) refers to Bensman (2008) as saying that 88% of the 1,745 articles in the 2005 Social Science Citation Index earned fewer than 2 citations, while only 4 had more than 10. The fact is, 91% of history, 49% of sociology (van Dalen & Henkens, 2001, p. 462), and 90% of political science (King, 1995, p. 445) articles go uncited.

Critically, it must be reasserted that using citation counts as a measure of the importance of published research is flawed. It is simply one index reflecting how academicians judge (or at least mention) the efforts of their peers. It says nothing about how academic scholarship influences practice, something crucial in purportedly *applied* disciplines and therefore revisited later in the chapter.

In closing, that most original work is ignored—even by other academicians—is disquieting. It raises the suspicion that perhaps most published research is not really original, or informative, at all.

Is Published Academic Research Truly Original? Evidence suggests that peer review is an inefficient mechanism for identifying original work and is in fact inclined to treat genuinely innovative, risk-taking, research harshly (Hubbard, Norman, & Parsa, 2010, pp. 670–671; Mahoney, 1985, p. 34; Nicholson & Ioannidis, 2012, pp. 34–36; Nosek et al., 2012, pp. 622–623; Stephan, 2012, p. 14; Waters, 2004, p. 53). Bearing in mind the caveat that citation counts approximate only *academic* recognition of published studies, consider the results presented next on the initial rejection of important scholarship.

Sometimes highly original manuscripts are dismissed because reviewers are unable to discern their true novelty and how they contribute to a field's body of knowledge. For example, Gans and Shepherd (1994) asked more than 140 prominent economists (including all living Nobel Prize and John Bates Clark Medal winners) whether they could provide examples, if they existed, of journals rejecting their papers. Following a 60% response rate, Gans and Shepherd reported that "many papers that have become classics were rejected initially by at least one journal—and often, more than one" (p. 166).

Some colleagues and I (Hubbard, Norman, & Parsa, 2010, p. 671) accessed via the Institute of Scientific Information Web of Science the total number of citations accumulated by some of these classics from their inauguration through 2007. We also computed their average citations per annum (ACPA) scores by dividing this total by the number of years since publication. Among these classics, with the above data contained in brackets, are Akerlof's (1970) offering on the economics of information [1,725; 45.4], Becker's (1965) theory of the allocation of time [1,553; 36.1], Black and Scholes's (1973) option-pricing formula [2,740; 78.3], Lucas's (1972) outlining of rational expectations [789; 21.9], and Sharpe's (1964) exposition of the capital asset pricing model [1,676; 38.1]. In spite of such novel and provocative works, Gans and Shepherd (1994) convey that some of the authors in their sample complained of enduring off-hand comments from referees: the results are "well-known and not interesting" (p. 172); "the ideas were 'already known' somehow" (p. 173); and "referees tell me that it's obvious, it's wrong, and anyway they said it years ago" (p. 178).

Other inventive economics articles have had a difficult time getting into print. One said to have suffered a "humiliatingly negative referee report" from the first journal to which it was submitted is Barten's (1969) piece (presently at 968 citations) on the estimation of a system of demand equations (Campanario, 1995, p. 311). Another is Nelson and Plosser's (1982) study on trends and random walks in macroeconomic time series. Rejected for publication by the *Journal of Political Economy,* this contribution currently has 4,572 citations.

Landmark books also faced publishing hardships. For example, Maddala's (1983) *Limited Dependent and Qualitative Variables in Econometrics,* with 13,672 citations, initially was declined by five publishers, including MIT Press, Academic Press, and McGraw-Hill, before being accepted by Cambridge University Press (Campanario, 1995, p. 319). Similarly, the Brookings Institution turned down first rights to publish Williamson's (1975) nonmainstream account *Markets and Hierarchies.* It was published eventually by the Free Press, whose staff, in anticipation of low sales, had planned to discontinue production after the first run (Campanario, 1995, p. 319). This book has gained a phenomenal 27,350 citations.

Straub (2008, p. vi) maintains that top journals reject original papers because editors are risk-averse; they are reluctant to overturn reviewer recommendations about publication decisions. The problem, Straub continues, is that the latter tend to concentrate on *methods* rather than on *ideas.* As an example of this bias, had the editor of *Information Systems Research* followed reviewer advice, he would not have published the most cited article, by far, in the information systems literature between 1990 and 2004: DeLone and McClean's (1992) piece on information systems success [432; 28.8]. Straub's (2008, p. vi) adamance that "good ideas should always prevail over good methods, other things being equal" is not the reaction of some disaffected scholar. He is a past—singularly courageous in my eyes—editor of *MIS Quarterly,* and the above insights are from one of his editorials.

Straub's (2008) perspective is shared by members of the AMA Task Force (1988, p. 5), as well as by a commentator on their efforts:

> Consequently, we often seem to apply increasingly greater methodological sophistication to increasingly less important problems. . . . Designing a "tight study" to investigate some relatively minor variation in an accepted conceptual framework is usually easier and less risky than designing a study that cannot be methodologically as tight, but which addresses a more important issue. (Churchill, 1988, p. 31)

The focus on methodological rigor over fresh ideas and relevance in academic work is one that has continued to the present time (see, e.g., D. R. Lehmann, McAlister, & Staelin, 2011). Carson (1995, p. 663) describes this aversion of researchers to tackle really original research as follows:

> As a Journal Editor I do indeed often reflect on the "poverty" of new ideas coming forward. Where are the new ideas? Very few authors have genuinely new ideas. Most authors are comfortable and safe with the mainstream uncontroversial thought patterns. Authors do not want to be unconventional, in fact I would contend that most are afraid of being unconventional.

Or in a playfully blunt way, Rousseeuw (1991, p. 41) expresses it thus: "One is sometimes under the uneasy impression that editors have as their policy distinguishing high quality manuscripts from junk, and then publishing the junk." Invoking an old cliché, he objects that there is too much garbage *in*, resulting in too much garbage *out*, something he lays squarely at the door of the publish-or-perish culture.

The fact of the matter, revealed in an examination of 68 empirical papers on the topic, is that peer review suppresses the publication of truly innovative research (Armstrong, 1997, p. 63).[6] This sad state of affairs is recognized by Macdonald and Kam (2007, p. 649) and says much about the efficacy of the peer review process. The trick to publishing one's work is to make the submission appear to be different, but certainly not too different, from the pack.[7]

Neglect of the Practitioner World

On the other side of the coin, incontrovertible evidence supports the view that academic business editors, reviewers, and researchers disregard the needs of practitioners because for tenure, promotion, and other rewards, all that really matters is publishing research within the scholarly community. Here, I offer only a few recent instances of this long, ongoing phenomenon. Rynes (2007, p. 1380), for example, as the 2005–2007 editor of the *Academy of Management Journal,* candidly admits that some business school faculty do not see it as their responsibility to provide work that would be useful for practitioners. McGrath (2007, p. 1366) minces no words on this score: "I can note that there is a clear consensus among observers that having an impact on real-world practice is largely irrelevant to the career prospects of business school faculty." Or as Pettigrew (2005, p. 973) describes the situation: "Indeed, [real-world] impact has become the dog that doesn't bark in the social and management sciences. The loudest-barking dog is just output, and often one form of output, that which is made available in 'A'-rated scholarly journals." In the academic world, priorities are crystal clear.

Neil Anderson, Peter Herriot, and Gerald Hodgkinson (2001) make a convincing case that the practitioner-academic divide in industrial, work, and organizational psychology—easily generalized to the management and social sciences as a whole, as they suggest—leads to the production by academics of knowledge that is irrelevant for practitioners. More specifically, give thought to their fourfold typology of research shown in Figure 8-1 based on the two dimensions of practical relevance and methodological rigor. They allege, and I agree, that academic and consulting

Figure 8-1 A Typology of Managerial and Social Science Research

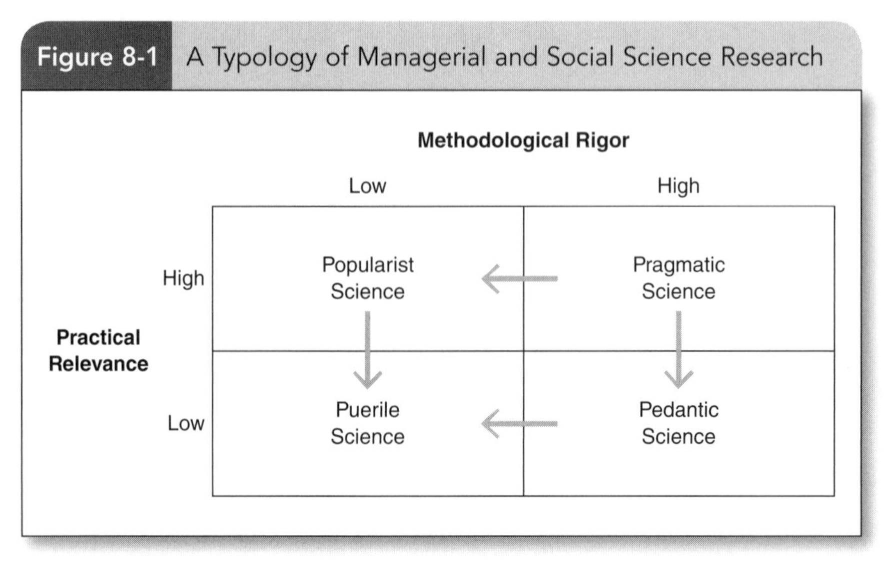

Source: Adapted from N. Anderson, Herriot, & Hodgkinson (2001, p. 394).

Note: The arrows represent pressures toward different quadrants faced by academics and practitioners.

research has largely given up on *pragmatic* science (high in terms of both relevance and rigor) and descended into *popularist* (high relevance–low rigor), *pedantic* (low relevance–high rigor), and, ultimately, *puerile* (low relevance–low rigor) science.

In the business (and social) sciences, knowledge acquisition takes a back seat to the pursuit of personal ends (Starbuck, 2006, p. 74; Ziliak & McCloskey, 2008, p. 32). So academicians write for other academicians, the "incestuous, closed loop" as Hambrick (1994, p. 13) put it in his presidential address to the Academy of Management. Academicians talk only to themselves and not to their supposed real-world audiences (as confirmed by, e.g., Cahill, 1994, p. 11; D. J. Cohen, 2007, p. 1017; Crosier, 2004, pp. 541, 552; Forster, 2007, p. 24; Geuens, 2011, p. 1104; Khurana, 2007, p. 311; Kmetz, 2011, p. 175; Mayer, 1993, p. 10; McKenzie, Wright, Ball, & Baron, 2002, p. 1206; Reibstein, Day, & Wind, 2009, p. 3).[8] This is why the high-impact articles that, for example, Huber (2007) alludes to are not measured by their influence on applied work but by their citation counts from other academicians. It also helps to explain why practitioners don't read academic journals; they don't see them as being relevant. For instance, Rynes, Colbert, and Brown's (2002, p. 161) survey of 5,000 human resource professionals (959 replies) found that less than 1% of them read the three most research-oriented journals in their field—*Journal of Applied Psychology, Personnel Psychology,* and *Academy of Management Journal.*[9] Speaking from a position of authority as chief knowledge officer for the Society for Human Resource Management, Debra Cohen (2007, p. 1015) understands why this is the case: There's not a great deal of overlap between issues near and dear to practitioners (e.g., leadership development, succession planning, talent recruitment and retention, employee engagement, health care costs) and topics that academicians (e.g., goal setting, intelligence, personality) like to write about. Furthermore, Rynes, Bartunek, and Daft (2001, p. 346) report there are signs that even practitioners with doctoral degrees cease reading academic journals once immersed in the business world.

As an additional demonstration of this lack of concern for producing useful knowledge aimed at practitioners (or academics for that matter), Bartunek (2007, pp. 1324–1326) mentions that to be published in the *Academy of Management Journal* authors must include a section in their manuscripts highlighting the significance of their findings for the management field. She comments, however, that many published articles fail to include such a section, while for those which do, this section often is desultory in nature.

Yet another indicator of the polarization between academic and practitioner groups when it comes to communicating knowledge is offered by the marketing discipline. In the late 1970s the *Journal of Marketing* was repositioned from an academic/practitioner periodical to one focused chiefly on academicians (Kerin, 1996). Much later, in 1992, the American Marketing Association created a new journal, *Marketing Management,* targeted at practitioners. Yet segregating the literature in this fashion between "them" and "us" simply reinforces for the practitioner the idea that academicians write only for themselves. In support of this argument, few kudos are earned by academicians publishing in *Marketing Management,* whereas career prospects blossom for having an "acceptance" in the *JM.*

And still a further measure of the disconnect between business academics and practitioners is seen in Murray Lindsay and my figures on the dwindling number of journal articles published by the latter (Hubbard & Lindsay, 2002, p. 389). Our data are based on an annual content analysis of two randomly chosen issues of nine marketing journals—*European Journal of Marketing* (1971), *International Journal of Market Research* (1966), *Journal of the Academy of Marketing Science* (1973), *Journal of Advertising Research* (1960), *Journal of Consumer Research* (1974), *Journal of Marketing* (1936), *Journal of Marketing Research* (1964),

Journal of Retailing (1945), and *Marketing Science* (1982)—for every year indicated in parentheses through 2000. Whereas 43% of the articles published in these journals prior to 1960 were contributions from practitioners, and 44.2% were written by practitioner-academic alliances, these numbers were only 8.2% and 13.7%, respectively, for 1990–2000.

I updated these figures for three leading marketing journals—*Journal of Consumer Research, Journal of Marketing, Journal of Marketing Research*—for 2001–2010 (see Table 8-2). They are based on a simple random sample of one issue each year from all three journals and show an almost total separation between the academic and practitioner camps. Not a single one of the 345 articles published in these journals was authored solely by practitioners. Only 11 (3.2%), in total, involved joint efforts between practitioners and academics. And even here, as occurred in previous years, the "practitioners" often were newly minted PhDs publishing with their dissertation advisors. For *JCR* there was no input at all from practitioners. Data supporting the disappearance of practitioner and practitioner–academic collaborative work in journals also is given by Neil Anderson, et al. (2001, p. 398).

In addition, published practitioner research is cited less than academic offerings. Signs of this come from Hubbard and Norman's (2007) inspection of the contents of two randomly selected issues of each of five marketing journals—*Journal of Advertising Research, Journal of Consumer Research, Journal of Marketing, Journal of Marketing Research, Journal of Retailing*—for each year of the period 1970–2000. We showed that the average citations per annum for papers authored by academics ($n = 186$) is 1.39, while that for practitioners ($n = 99$) is only 0.56.

Table 8-2 Proportion of Papers Authored by Practitioners in Marketing: 2001–2010

Journals	*Number of Papers*	*Number of Papers by Practitioners*	*%*	*Number of Papers by Practitioner–Academic Alliances*	*%*
JCR	139	0	—	0	—
JM	92	0	—	7	7.6
JMR	114	0	—	4	3.5
Total	345	0	—	11	3.2

Note: JCR = Journal of Consumer Research, JM = Journal of Marketing, JMR = Journal of Marketing Research.

Moreover, think about the shameful statistic, recorded by Pfeffer (2007, p. 1336) and based on the (at the time) unpublished work of Mol and Birkinshaw, that *none* of the world's 50 most important management innovations arose from academic research.[10] Comparable assessments are seen in accounting, with Kasanen, Lukka, and Siitonen (1993, p. 249) concluding that most management accounting and control breakthroughs have originated from companies and consulting firms rather than from the halls of academe. Views critical of the pertinence of scholarly research are raised also in marketing by the likes of Garda (1988) and Cornelissen (2002), and resonate in the AMA Task Force quotation published a quarter of a century ago that introduces this chapter.

The latest empirical study with which I am familiar points to an ever-widening gap between the perceived suitability of academic research for both student and practitioner needs. In a longitudinal investigation, Pearce and Huang (2012, p. 249) traced the decline in the pages of scholarly journals of what they call "actionable" research, that is, research from which any person(s) could possibly take organizational action by knowing its outcomes. Specifically, these authors examined the content of all 420 data-based articles appearing in the entire volume (year) of *Administrative Science Quarterly* (ASQ) and *Academy of Management Journal* (AMJ) for every 10 years from 1960 through 2010. They discovered that whereas 65% of the 1960 empirical papers in *ASQ* were judged to be actionable, this figure dropped to 35% by 1990 and was only 19% by 2010. Corresponding results for *AMJ* are 43%, 36%, and 24%. Pearce and Huang are saddened that

> despite the decades of laments about the lack of relevance of our research, the many thoughtful practical suggestions for addressing the problem, editors' requirements to include statements about the implications of the research for practice, and an increasing proportion of experienced students in our classrooms, we [find] that the practical relevance of our research actually has gotten worse. (p. 248)

As indeed it has.[11]

Interestingly, based on Pearce and Huang's (2012) findings, Roger Martin (2012, p. 294) estimates the cost to the business academy of producing a single "A-journal" actionable article to be $1.5 million. He further calculates that the cost of producing not-actionable A-journal business articles to be approximately $600 million per year.

Taken as a whole, the material presented in this section on the value of academic business research for commercial practice amounts to nothing less than a damning indictment of a broken system. All too predictably, an analogous situation regarding the dubiousness of academic research for real-world application exists in the social sciences. The psychologist Meehl (1967, p. 114) recognized this earlier than most:

> Meanwhile our eager beaver researcher . . . relying blissfully on . . . statistical hypothesis testing, has produced a long publication list and been promoted to a full professorship. In terms of his contribution to the enduring body of psychological knowledge, he has done hardly anything. His true position is that of a potent but sterile intellectual rake, who leaves in his merry path a long train of ravished maidens but no viable scientific offspring.

This colorful account says it nicely.

In a book he called a "sociology of non-knowledge," Andreski (1972, p. 12) was every bit as devastated as Meehl about the worth of academic social science research:

> If we look at the beliefs widely shared among social scientists, we see that they contain little if anything that could be attributed to a superior professional understanding [such that] . . . the social sciences appear as an activity . . . where anybody can get away with anything. (pp. 30, 16)

So much for the creation of scientific knowledge. Later reservations by a number of authorities (e.g., Bauer, 1994, p. 128; Elms, 1975, p. 967; Loftus, 1996, p. 161; Pashler & Wagenmakers, 2012, p. 528; Shweder & Fiske, 1986, p. 1; Taagepera, 2008, p. 236) indicate that little has changed since Andreski's day.

8.3 Cultivating a Significant Sameness Tradition

I have acknowledged the enormity of the challenge of dislodging, let alone replacing, the significant difference school of thought with that of a significant sameness perspective. This is because to do so involves nothing less than instilling in researchers a complete rethinking about how science works. I believe these changes to be vital if the management and social sciences are to make progress in contributing knowledge that is more than common sense or intuition and that is capable of practical application.

Clearly, this will be a difficult and time-consuming process. In the interim, therefore, I offer two sets of suggestions which, added up, will assist in bringing about a better way of conducting research. The first set of recommendations is aimed at facilitating knowledge accumulation by changing the research habits of those beholden to the tenets of the significant difference model. The second set of directives applies more specifically to helping institutionalize replication research in the academy. It is obvious that in neither case are these proposals meant to be exhaustive, but they are useful starting points.

8.3.1 Suggestions for Encouraging Knowledge Accumulation

- Inform scholars, and especially graduate students, that there is an intellectual world beyond the confines of the H–D model and its accompanying dependence on statistical significance tests in search of "original" results. Editorial-reviewer urgings for novelty have done as much as anything to stifle cumulative knowledge growth. Not every empirical journal article has to be strewn with hypotheses. Statisticians rarely are in the business of testing hypotheses (Chatfield, 2002, p. 7); only management and social scientists are relentless in this aim.

  Adding insult to injury, this compulsion to test statistical hypotheses is a methodology which flatters to deceive by seemingly conferring scientific respectability on its users. In truth it is little more than a charade—referred to in the 1970s as "playometrics" by the pioneering Norwegian econometrician Ragnar Frisch (cited in Mayer, 1980, pp. 168–169), in the 1990s as "game playing" by Blaug (1992, p. 244), and more recently termed a mere parlor game by Ziliak and McCloskey (2008, p. 2)—engaged in by, and for the benefit of, academicians looking to boost their list of publications and, in reward, career options. Seldom do these publications contribute to the establishment of knowledge, being instead of the "pretend" variety so opposed by Wells (2001). Or as Lambdin (2012, p. 67) has it, "the continuing popularity of significance tests in our peer-reviewed journals is at best embarrassing and at worst intellectually dishonest." That almost everyone plays this game in no way sanctions its legitimacy. In Macdonald and Kam's (2007, p. 640) words, it is a game no longer worth playing. And it is a game that will require "some pounds of courage" (Gigerenzer, 2004, p. 604) among editors and reviewers to bring it to a halt. Nelder's (1999, p. 261) candid appraisal that "the most important task before us in developing statistical science is to demolish the *P*-value culture, which has taken root to a frightening extent in many areas of both pure and applied science, and technology" accentuates the above positions.
- Some might argue that a shift away from an H–D, significant difference, framework to one of significant sameness will lead to a severe retrenchment in the number of empirically testable hypotheses found in the literature. Yes it will. And this is good. There are hundreds of thousands of hypotheses tested each year in the social and management sciences. Can anyone be found who honestly believes that these vast numbers of "original" ideas—see Section 8.2.2—make solid contributions to knowledge and are deserving

of publication? Whatever happened to Ockham's razor? Drastically reducing the volume of testable hypotheses, not their endless profusion, is what is required. We are drowning in an exploding literature, most of which goes unread and, if read, unacknowledged. We need less testing of hypotheses, not more. Retain and nurture those ideas that are worthy of further attention, a theme which describes significant sameness.

Does this mean that occasionally some useful hypotheses may be overlooked? No doubt. But this does not imply that they will be overlooked forever. Science is reflective; it does not proceed in a lockstep manner.

- Teach students exactly what a *p*-value is and what it is not. *Explicit* and *repeated* instruction in statistics and methods courses about the *p*-value's emphatic limitations in the acquisition of knowledge should go a long way to baring its inconsequential nature. That many (most?) researchers would be hard pressed to define algebraically (as shown, e.g., in Section 2.2.1) an index they use with abandon is deeply troubling. That investigators continue to bestow on this index numerous "magical" properties it does not have, despite a decades-long literature attesting to this, is inexcusable.

  Preach to students about the need to focus on the magnitude of effects. Science deals with the size of phenomena. Accordingly, students must be taught about the importance of CIs around sample statistics and effect sizes, and why they are superior to *p*-values. It speaks poorly about the vision of graduate education programs that they fail to prepare social science and management students about how to engage in useful research.

  The publication bias against $p > .05$ results has seriously undermined confidence in the probity of the management and social science empirical literatures. Furthermore, determining whether scientific progress has or has not occurred in terms of imperceptible changes in the size of a *p*-value—$p = .05$ is "good," $p = .06$ is "bad"—is silly. This is what Rosnow and Rosenthal (1989, p. 1277) meant when they said, "Surely, God loves the .06 nearly as much as the .05." So the publication bias problem must be addressed.

  This could be achieved readily, in principle, by editors' willingness to publish work with adequately powered "negative" outcomes. An added benefit of such a policy would be requiring authors to include power calculations as integral parts of their manuscript submissions, something which we have seen happens rarely in the significant difference paradigm. If authors insist on using statistical tests, at least have them incorporate acceptable power levels. More importantly, negative results can be valuable.

  Another way of dealing with the publication bias against null outcomes is to consider *no results* reviewing of manuscripts. By this I mean that authors submit to journals comprehensive research designs prior to data collection and analysis, in a manner akin to the evaluation of research grant proposals. A practical advantage here is shorter manuscripts to review. More consequentially, this approach would direct attention to the substance (or otherwise) of the topic the researcher is studying. All too often the welter of significance testing demanded in empirical reports conceals the fact that the research problem itself is unimportant (Hubbard, 1995c).

- Convey to students the benefits of empirical, rather than statistical, generalization. Doing so will remind them of the need to focus on quantitative rather than directional predictions. Underline for students the pivotal role that overlapping CIs—never employed in a mechanical fashion, but always in tandem with sound professional judgment—play in this process.
- Downplay the demands for random samples. Sometimes the impression is given that only results based on probability samples are to be trusted. Of course, if use of such samples is feasible then do so, because they permit the legitimate employment of formal methods of statistical inference. However, true random samples are difficult to come by. In any case, the use of nonrandom samples has secured the bulk of knowledge development for centuries. On the other hand, restrict the publication of articles based solely on the opinions of undergraduate or graduate students. These are fine for some *genuine* exploratory research. But they need to be followed up with "real people," something which hardly ever is done (Cahill, 1994) in the significant difference world.
- Change the requirements for tenure and promotion. Granted, this is a topic deserving of a separate book (see Waters, 2004, in this regard) and will only be touched on here. Increasingly, scholars have been herded into a single, utterly congested, avenue for job security—publishing copiously in top-tier journals. The typical novice researcher has a maximum of 6 years on his or her tenure clock in which to do so.[12] Relying mostly on publication (and citation) counts has led to a range of researcher attitudes and behaviors, highlighted throughout this book, which are inimical to good scientific practice.

  In the unseemly rush to print whatever we can and by whatever means, taken for granted today, it is difficult to visualize an earlier period when this was not the case. For example, A. William Phillips authored only 8 papers in his many years at the London School of Economics (Blaug, 1985, pp. 199–201). However, one of these, published in 1958, resulted in the eponymous "Phillips Curve" showing the inverse relationship between unemployment levels and changes in money wage rates (later prices). By coincidence, 1958 also was the year that James D. Watson gained promotion to the rank of associate professor of biochemistry at Harvard. At that time, Broad (1981, p. 1137) tells us, his curriculum vitae listed a mere 18 papers. But one of them, appearing in 1953, described the structure of DNA. In contrast, Broad alleges that by the early 1980s someone seeking promotion in this area might have needed to produce 50 or perhaps even 100 papers. These days, punishment for the failure to publish an abundance of journal articles has reached absurd proportions. Peter Higgs, discoverer of the Higgs boson and 2013 Nobel laureate in physics, confessed: "Today, I wouldn't get an academic job. It's as simple as that. I don't think I would be regarded as productive enough" and had become "an embarrassment to [Edinburgh University] when they did research assessment exercises" (as cited in G. Gigerenzer &

J. Marewski, 2014, p. 436). Disappointingly, a lesson which appears to have been forgotten is that for hundreds of years university researchers have made enormous contributions to knowledge without the whip of journal publication counts (Hubbard, 1995b, p. 672).[13]

Evaluation procedures for tenure and promotion should move away from the glorification of rapidly publishing for its own sake, something so injurious to the life of the mind (Waters, 2004, p. 19). Instead, sufficient time should be allotted for faculty to work on important, as opposed to publishable, topics (Hubbard, 1995b, p. 671). Universities should be places for the serious *contemplation* of ideas, some of which, allowed to reach fruition, may further our understanding of the world. Over time, the scientific community advances knowledge by weeding out poor conjectures and continuing with their more fecund rivals. Science does not make headway by charging the presses at every opportunity.

Manifest recognition of the part that time plays in the careful incubation of new ideas would argue persuasively for elevating the status of the scholarly *monograph*. Being less encumbered by the adversarial, even "masochistic" (AMA Task Force, 1988, p. 14), nature of the journal peer review system, the monograph affords a potentially ideal vehicle for transmitting innovative, important, and controversial ideas.

Sadly, the pressure to publish lots of refereed articles usually means that researchers do not have the time to write such monographs. If they elect to do so as untenured assistant professors, they will likely end up looking for another position because "books are no longer valued in . . . tenure decisions" (MacInnis, 2011, p. 151). After being tenured it is still the lure of journal publication counts which points the way to a full professorship and attendant perquisites of academic life. Even for those safely ensconced at the top of the academic ladder and sympathetic to the role of the monograph, it is an all-too-human reaction that given their own grueling ordeals to attain present status, those behind them can expect to be judged similarly. So journal publication counts remain everything, and monographs underappreciated. Or as I once cracked: "Under the present academic research-publication system, it is all too easy to imagine Keynes being admonished at Cambridge for 'working on that damned book,' when his journal output was lacklustre. Too bad" (Hubbard, 1995b, p. 672).[14]

Faced with the demands exerted by a publish-or-perish world, together with high rejection rates of 70%–90% or more for manuscripts in management and social science journals (see American Psychological Association, 2010; Honig et al., 2013; "How Science Goes Wrong," 2013; Nosek et al., 2012; Zuckerman & Merton, 1971), publishing in these areas is inordinately difficult. A strategy to help take the sting out of such a world is to make publishing easy. Using digital journals, this approach has been advocated by Nosek et al. (2012, p. 623) as well as others.

Of course, publications in print journals are accorded much greater recognition than those in online periodicals. For now. The way in which we disseminate information is in the midst of a revolution. And digital journal *PLoS ONE* (http://plosone.org), which publishes work from all scientific fields, is an exciting challenge to the established order. As Nosek et al. (2012, p. 622) tell us, this journal, introduced in 2006, focusing on methodological soundness rather than the perceived "importance" of the topic, and with a 70% acceptance rate, published 13,798 articles in 2011. This makes it the largest journal in the world. So things can change; give credit to faculty for publishing in online journals.[15]

Reward, don't marginalize, scholars who are trying to supply knowledge that is useful to practitioners. Numerous articles and presidential addresses (e.g., to the Academy of Management) have beseeched their colleagues to take this task seriously. They have been met repeatedly with nonchalance and worse. It is abundantly clear from Chapter 2 that such advice has made no observable dent in business (and social science) academicians' enthusiasm for theory and its empirical assessment via statistical significance testing. On the contrary, this fervor has intensified over time.

In promulgating the view that management and social scientists should encourage research that is applicable to the real world—we already have a surfeit of theory which practitioners find unhelpful—I can do no better than recommend Andrew Van de Ven's (2007) important book on engaged scholarship. Paraphrasing Van de Ven (p. ix), engaged scholarship is a participative genre of research which depends heavily on the combined inputs of various stakeholders such as researchers, practitioners, clients, users, and the like. Its success revolves around including these stakeholders in each of four major aspects of the research design: (1) grounding the research problem in reality; (2) developing alternative possible explanations concerning this problem; (3) designing sound research plans to appraise, empirically, the worth of these explanations; and (4) applying the research findings to address the problem. Fittingly, Van de Ven announces: "My argument assumes, of course, that the primary motivation of engaged scholars for undertaking research is to understand this complex world, rather than to get published and promoted" (p. 29). Ah, there's the rub, unfortunately. But I admire and support wholeheartedly Van de Ven's stance.

8.3.2 Institutionalizing Replication Research

- As Helfat (2007) advises, impress on members of the academic community that *replication* indeed is not a dirty word. That this needs to be pointed out says it all, really. It has been shown throughout this book that replication research is *the* means for establishing significant sameness among phenomena. This is accomplished by first detecting empirical regularities, which takes a great deal of time and many studies. The discovery of empirical regularities is a major scientific achievement in its own right. As noted, these regularities will not be of the universal kind envisioned by the positivists in their search for veridical truth. They will, however, be of sufficient durability to impel serious attempts at their explanation, thus stimulating theory building. Such a strategy also will teach students about the virtues of deductive, inductive, and especially abductive reasoning. That editors and reviewers schooled in the significant difference paradigm see little of value in this work, and resist publishing it, is mystifying.

- Far more attention needs to be paid to the issue of replication in the methodological literatures of the management and social sciences than presently is the case. Too often this topic is treated superficially or ignored altogether (S. Schmidt, 2009, p. 90; Tsang & Kwan, 1999, p. 761). For instance, a convenience sample of 14 marketing research textbooks showed that only 1 included a reference to the activity in the index. Similarly, only 1 of 15 consumer behavior texts indexed "replication" (Hubbard & Armstrong, 1994, p. 245).
- Modify journal editorial policies to facilitate the publishing of replication research. This bullet point and several following it show increasing engagement with the replication cause. Revise journal editorial procedures to help authors access the data which may be needed to conduct replications. Greenwald (1976), when editor of the *Journal of Personality and Social Psychology,* was an early voice for this practice. He introduced a policy requiring that authors retain their raw data and copies of the details of the procedures employed in their work for 5 years following its publication (Hubbard & Armstrong, 1994, p. 244). At first controversial, this compact has been endorsed for many years now by all American Psychological Association journals.

  The American Marketing Association (see Kinnear, 1992) supports a similar arrangement for increasing replication efforts, as does the *American Economic Review,* flagship journal of the American Economic Association. Cited by Freese (2007, p. 155), this advisory begins:

  > It is the policy of the *American Economic Review* to publish papers only if the data used in the analysis are clearly and precisely documented and are readily available to any researcher for purposes of replication. Authors of accepted papers that contain empirical work, simulations, or experimental work must provide to the *Review,* prior to publication, the data, programs, and other details of the computations sufficient to permit replication. These will be posted on the *AER* Web site. The Editor should be notified at the time of submission if the data used in a paper are proprietary or if, for some other reason, the requirements above cannot be met.

  And hold those at the journals accountable for enforcing these policies. As was seen in Sections 4.4.1 and 5.5.2, however, all too often these requirements are brazenly disregarded by authors. When this is allowed, such efforts to promote replications may be worse than having no standards in the first place, coming across as toothless editorial gestures rather than a dedication to safeguarding the literature.

- Editorial policies favoring special issues of entire journals, or portions thereof, devoted to replications would be helpful. Recent examples of this are a 2012 issue of *Perspectives on Psychological Science* and a 2013 issue of the *Journal of Business Research*. Likewise, the *American Journal of Business* is planning two special issues of research replicating influential studies for 2014 and 2015.

  Hugely welcome as they are, a possible downside to such contributions is precisely the fact that they are "special" issues. They are not the norm; they are seen as sporadic, isolated occurrences (e.g., the only other special issue of the *Journal of Business Research* dealing with replication research appeared in 2000, 13 years earlier). As such, they are easily overlooked by pragmatic researchers inoculated against the need to bother with replications.
- To combat the weakness noted above, journal editorial practices must be amended to incorporate replication research on an ongoing basis. In this context, Section 4.4.1 told of a number of journals whose platforms are sympathetic to the calls for more replications. When serious editorial commitment to the publication of replications is made well known, impressive results can be realized. Take the case of the *Quarterly Journal of Business and Economics* (*QJBE*), whose editors in the Winter 1984 issue expressed a desire to publish replications carried out independently. Hubbard and Armstrong (1994, p. 244) content-analyzed a 50% probability sample of all issues of the *QJBE* appearing over the time period 1978–1989. None of the articles published before the 1984 editorial policy intervention were replications, whereas such work constituted 17.3% of empirical studies featured in the journal between 1984 and 1989. For 1984–1995, this figure was 22.5% (Fuess, 1996). When editorial intent to publish replications is unambiguous and enforced, researchers are more apt to perform them.
- Appoint a replications editor (Hubbard & Armstrong, 1994, p. 244), as has been done, for example, at the *Journal of Applied Econometrics* (Evanschitzky et al., 2007, p. 413). In the past, some editors of prominent business journals (e.g., Mick, 2001; Monroe, 1992a, 1992b; Mouncey, 2011; Winer, 1998) said they would consider publishing replications. The drawback is that such initiatives are dependent on the intellectual dispositions of the particular editors themselves; their successors in no way are obligated to follow suit. Ironically, an excellent example of this is provided by none other than the *Journal of Personality and Social Psychology* (*JPSP*). Anthony Greenwald's (1976) commendable efforts as editor of *JPSP* to facilitate replication research, just noted, were not honored by a recent editor of this same journal, Eliot Smith, who as explained in Section 5.5.8 was adamant that *JPSP* does not publish such work. But the appointment of a replications editor would help further institutionalize the centrality and ongoing nature of replication research in the generation of scientific knowledge.

  This editor would be expected to actively solicit replications, especially of works thought to be relevant (Hubbard & Armstrong, 1994, pp. 224–225).

While recognizing the subjective nature of the criterion of "relevance," in the marketing area, for example, substantial assistance is offered by the U.S. Marketing Science Institute's priority listing of topics for research. Additional suggestions as to which empirical works in the management and social sciences are judged to be important could be determined by surveys of both academicians and practitioners (as was done by Armstrong & Hubbard, 1992). Crucially, Nosek et al. (2012, p. 622) report that members of the Open Science Collaboration (see later in this chapter) are working to develop metrics to identify replication value (RV). An RV index, involving citation impact and the precision of the effect, would have a fundamental role to play in determining which articles to replicate. Other thoughts on this issue received coverage in Section 5.5.6.

The replications editor also would be concerned with replications of previous studies that yielded interesting, surprising, and/or controversial findings. Furthermore, she or he would be charged with proposing guidelines for reporting such research. As an example, replications confirming earlier results could probably be shorter pieces than those contradictory to the original paper (Hubbard & Armstrong, 1994, p. 245). And the font size for this research could be smaller—say, like the Methods section in American Psychological Association journals—so as to maximize information per page.

As well as modifying editorial outlooks, efforts could be undertaken to change reviewer behaviors too. For instance, Uncles (2011, p. 581) nominates the delisting of reviewers who disparage replication research.

- Another editorial initiative would be to reserve some portion of journal space for the publication of replications. Since Section 5.5.8 demonstrated unilaterally that the publication of novel research is hardly threatened by replications, the question is, what should this appropriation be? While not written in stone, based on a survey of 1,292 psychologists, Fuchs, Jenny, and Fiedler's (2012, p. 641) recommendations of about 20% for print and 27% for online journals seem reasonable. An earlier estimate by Neuliep and Crandall (1993, p. 28) was 15%. According to *The Economist* ("How Science Goes Wrong," 2013, p. 30), the journal *Perspectives on Psychological Science* (*PoPS*) will soon have a section committed to replications.

  Indeed, Spellman (2013a, 2013b) explains that *PoPS* has begun a Registered Replication Reports project. The gist of this project has researchers proposing a study to be replicated and why it justifies the time and effort involved. Then, together with original author(s), the researchers design a robust protocol to enable an "exact" replication. This protocol is then posted and other labs are encouraged to conduct the studies. Importantly, the replication findings will be published in *PoPS* no matter the outcome.
- Teach graduate students about the absolute primacy of replication research in scientific advance. In an article debating whether economics is a science,

Mayer (1980, p. 175)—who believes it is not (yet)—recommended that some foundation should finance a program enabling graduate students to rerun each year about 10% of the empirical work published the previous year or so in the journals. Kmetz (2011, p. 186) holds similar views.

Dewald, Thursby, and Anderson (1986, p. 600) explain how graduate students in advanced econometrics classes at Ohio State University were asked to replicate and extend a published study as part of their training. In the advertising field, Reid, Soley, and Wimmer (1981, p. 11) have argued for the granting of more master's theses and doctoral dissertations for replication research. Meanwhile, focusing on the replication of prior studies is a staple in the master's degree program in marketing at the University of Auckland, New Zealand (Hubbard, Brodie, & Armstrong, 1992, pp. 1–2). Likewise, Frank and Saxe (2012, p. 600), at Stanford University and Massachusetts Institute of Technology, in order, speak excitedly of the benefits of having their students replicate published research in experimental methods courses in psychology (see also Asendorpf et al., 2013, p. 114). They are convinced that an apprenticeship of this kind provides students (future researchers) with object lessons about the scientific process and acquaints them with the value of academic openness. Even some undergraduate psychology students at Tilburg University in the Netherlands are being taught the virtues of conducting replication research (Ijzerman, Brandt, & van Wolferen, 2013, p. 129).

In addition, requiring students to check on the accuracy of published research allows them to get their hands dirty as they seek to reconstruct the databases, and employ the same methods, as the original authors. This is good training in real-world research, far more edifying than working through end-of-chapter textbook exercises with their usually impeccable data sets.

While admittedly rare, sometimes these reanalyses of existing articles make headlines. This was the case recently when the *Wall Street Journal* (Cronin, 2013) reported that, as part of a University of Massachusetts graduate econometrics class, Herndon, Ash, and Pollin (2013) failed to replicate Reinhart and Rogoff's (2010) "seminal" paper "Growth in a Time of Debt." Detected by Herndon et al., a coding error by Reinhart and Rogoff, which omitted several countries from the investigation, changed their results noticeably. As first portrayed by Reinhart and Rogoff, economies with ratios of public debt to gross domestic product exceeding 90% were found to *contract* by 0.1% annually, an outcome potentially supportive of the need for austerity in both Europe and the United States. The reexamination by Herndon et al. showed instead that these economies *grew* by 2.2% annually. It pays to check the validity of published empirical results.

- Asendorpf et al. (2013, p. 115) recommend that granting agencies could signal the importance of replication by insisting that some portion of any given grant's budget be set aside to replicate key aspects of the research. Alternatively, they suggest that these agencies fund their own consortium of investigators to carry out such work.

- Make greater use of the Internet as a tool for increasing replication research (Uncles, 2011, p. 581). This could be accomplished in two broad ways. First, use the Internet as a means of archiving the data and code needed to permit replication attempts. Second, create online journals to publish replications.

  Using the Internet as a vehicle for storing and accessing data and related materials was mentioned earlier in connection with the *American Economic Review.* It is a strategy followed also by the *Journal of Applied Econometrics* (Koenker & Zeileis, 2009, p. 845). Other journals have adopted similar policies. For example, a condition for publication in the *Journal of Conflict Resolution* since 2002 is that authors must sign a statement that they have made available on a dedicated website the necessary information to allow others to replicate their work (Evanschitzky et al., 2007, pp. 412–413). A variation on this theme is in place at the *American Journal of Political Science,* where a requirement of publication is that authors provide footnotes indicating where the data and methods may be retrieved on the Internet (Evanschitzky et al., 2007, p. 413). Worried about the accuracy of sociology's empirical findings, Freese (2007, p. 154) also champions use of the Internet as an opportunity for increasing the transparency and believability of results. More specifically, Freese (p. 159) notes the further advantage that authors can sidestep dealings with individual journal editors. This can be achieved by depositing data and/or code directly to, for example, the Interuniversity Consortium for Political and Social Research Publications-Related Archive. Following correspondence by Freese, the latter has confirmed its ability to do so. As usual, the success of these editorial actions lies in their enforcement.

  Aware of the fact that recent accounts show many data-based results to be unreliable, Joshua Hartshorne and Adena Schachner (2012, p. 1) propose the establishment of an open-access journal specializing in the publication of replication attempts. Thus, they argue, the measure of an academician's reputation would no longer be restricted to a tallying of the number of publications she or he has, and their accompanying citation rates, but much more importantly would be judged by their *credibility* (replicability). Absent such credibility, Hartshorne and Schachner (p. 3) spell out, the current reward system in academe encourages the publication of spurious results.

  Another idea paralleling the above is the website PsychFileDrawer.org. Instigated by Hal Pashler at the University of California, San Diego, this site allows psychologists to easily post, in brief form, successful and unsuccessful replications (Carpenter, 2012, p. 558; Stroebe et al., 2012, p. 682). Efforts like this deserve great applause.

- Needless to say, replication research *must* be rewarded (Ioannidis & Doucouliagos, 2013, p. 1001). One way to ensure this is to see that it is cited. With this in mind, Koole and Lakens (2012, p. 611) suggest that journals could create a "Replication File" composed of the original paper and the replication attempts following it. Each time the original study gets cited, there would be a co-citation of the Replication File. In this way replications earn formal recognition. Another, more pragmatic, way to underline the value of replication research is to make it count in faculty tenure and promotion decisions (Asendorpf et al., 2013, pp. 115–116). We need more confirmatory and less original research from the academy.

- Because in the significant difference paradigm it is not in an individual researcher's best career interests to undertake replications, some kind of *community* effort toward this end would be ideal. Improbably, an initiative of this sort is at present underway. The Reproducibility Project is a collaborative endeavor involving 72 volunteers from 41 universities around the world, whose aim is to determine both the frequency and predictors of direct/exact replication success in psychology (Open Science Collaboration, 2012, p. 657). Coordinated by Brian Nosek at the University of Virginia, these scholars have organized to replicate studies from among the first 30 articles published in the 2008 issues of three prominent journals: *Journal of Experimental Psychology: Learning, Memory, and Cognition, Journal of Personality and Social Psychology,* and *Psychological Science.* Each of these 72 contributors plays an important, but limited, role by offering his or her expertise as part of a team performing one replication. This may include team members soliciting feedback on their research designs from the original authors before data collection (Open Science Collaboration, 2012, p. 658).

  In the interests of clarity, Open Science Collaboration (2012, p. 659) encourages readers to visit its website (http://openscienceframework.org/project/EZcUj) to examine any facet of the project. It also wishes to recruit additional scholars so as to increase the number of journals whose articles are made available for replication purposes.

  A precedent like this truly is inspiring. If sustained, and especially if adopted by those in other fields, it offers serious promise of nurturing grassroots replication traditions in the social and management sciences. Those participating in the Reproducibility Project are to be roundly commended for their bold actions. I hope their idea proves to be highly contagious.
- Finally, a recent development in the medical area deserves acclaim. In 2014 Stanford University opened its Meta-Research Innovation Center (METRICS), cofounded by John Ioannidis and Steven Goodman. Included in its charges, the METRICS laboratory and its collaborators will tackle various publication bias concerns by trying "to design new publishing practices that discourage bad behaviour among scientists" ("Metaphysicians: Combating Bad Science," 2014, p. 74). This would involve a greater focus on the need for replication. Given that a number of articles in the prestigious journal *Lancet* claimed that a breathtaking 85% of world spending on medical research in 2010, or some $200 billion, was "squandered" on flawed, redundant, never published, and badly reported investigations ("Metaphysicians," 2014, p. 74), METRICS laboratory has its work cut out.

Sections 8.3.1 and 8.3.2 offered several directions for the grooming of a significant sameness outlook among members of today's academy. What is not generally appreciated is that the significant sameness approach predates the significant difference model. This antecedent from the 19th century is sketched in Section 8.4.

8.4 Retrospective: Empirical Regularities and the Emergence of 19th Century Social Statistics and Social Science

The way forward for both the management and social sciences, as I see it, is to build explanatory theory around the discovery of empirical regularities or demi-regularities. Clearly, such a strategy presupposes that it is the location of these regularities that drives theory creation. Whereas the notion of first uncovering empirical regularities is largely eschewed in the modern academy, it was in vogue for much of the 19th century. Indeed, it was critical to the evolution of contemporary statistics: "The idea of statistical regularity was thus of signal importance for the mathematical development of statistics" (T. M. Porter, 1986, p. 17). It was correspondingly instrumental in the rise of the social sciences themselves.

Most early statisticians and social scientists were influenced heavily by the French positivism of Henri Saint-Simon and August Compte. This philosophy maintained that all phenomena, natural and social, are governed by universal laws. Compte underscored that it was the researcher's job to discover the laws shaping the behavior of societies.

As devotees of this philosophy, 19th century social statisticians detected many laws, revealed as empirical regularities, in social data. Foremost among these was Adolphe Quetelet, a Belgian astronomer, credited also as the founder of both sociology (together with Compte) and statistics (Tankard, 1984, p. 32; see also Raison, 1963).[16] Unlike Compte, however, Quetelet was keen to apply probability theory to discern these empirical regularities. In doing so, Quetelet used the astronomer's law of error, now known as the Gaussian or normal curve. He fit this distribution to human characteristics such as height, weight, shoe size, birth and death rates, and compared them by age, sex, profession, and place of residence (T. M. Porter, 1986, p. 46). This normal curve regularity also was observed for nonbiological data, including rates of drunkenness, insanity, suicide, crime (Stigler, 1986, p. 169), illnesses (Tankard, 1994, p. 32), marriage (T. M. Porter, 1986, p. 46), and even musical (Howie, 2002, p. 35) and poetic (Hacking, 1990, p. 110) abilities. These regularities were, of course, of an approximate sort (Stigler, 1986, p. 203).

Thus, Quetelet was able to show that while the behavior of individuals may not be predictable, in the aggregate it is lawlike, obeying a normal distribution. In this manner he saw the statistical regularity as the key to understanding social science. He saw, further, Poisson's "law of large numbers" as the fundamental axiom of his "social physics" (T. M. Porter, 1986, p. 52).

Quetelet heralded the mean of the normal distribution of errors, or *L'homme moyen*. His "average man" represented the perfect form of the species (Howie, 2002, p. 36). Because, Quetelet reasoned, "all our qualities, in their greatest deviations from the mean, produce only vices" it follows that "one must consider him [average man] all that is beautiful and all that is good" (cited in T. M. Porter, 1986, p. 103).

Quetelet's ideas were popularized by Henry T. Buckle, a "prodigiously erudite" man (Hacking, 1990, p. 125), in the first volume of his *History of Civilization in England,* published in 1858. In this magisterial work Buckle reinforced the view that statistical regularities could portray the universal laws of society, just as regularity in the physical world unmasked the laws of nature (Stigler, 1986, p. 227). For much of the 19th century, then, statistics was the numerical science of society (Gigerenzer et al., 1989, p. 38). Even history could be transformed into a science, a belief held by John Stuart Mill (cited in S. Gordon, 1991, p. 403) and later by Hempel (1942).

As the following brief examples show, the impact of Quetelet and Buckle's ardor for the universality of the error law on subsequent 19th century science and statistics is hard to exaggerate. For instance, James C. Maxwell was an admirer of Quetelet. His work on statistical physics in the 1860s was an attempt to see whether Quetelet's normal law could be transferred from its social setting of large numbers of people to molecules in a gas (Howie, 2002, p. 42). This analogy of physicists adopting methods applied in social statistics is fascinating; usually the borrowing is the other way around. Maxwell's kinetic gas theory helped to establish the probabilistic nature of the physical world.

There is also the case of the polymath Francis Galton, who first introduced the correlation coefficient.[17] Stigler (1986, p. 267) points out that, as a statistician, Galton was a direct descendant of Quetelet. He thought of the normal distribution of the most interesting traits of people as an autonomous statistical law (Hacking, 1990, p. 186). He parted company with Quetelet, however, in important respects. While Quetelet saw the dispersion of the normal distribution as error, Galton saw it as genuine variation in traits. It was this variability that interested Galton in his studies on heredity, whereas Quetelet trumpeted the mean.

Consider, in addition, two other late 19th century disciples of Quetelet. Gustav Fechner was the founder of psychophysics, acknowledged in Section 3.4. He devised what came to be known as the Weber–Fechner law, which shows that people's ability to discriminate using their sense modalities generally follows a Gaussian distribution. Similarly, it was common for Francis Y. Edgeworth, the mathematical economist and statistician, to cite Quetelet approvingly.

Finally, Theodore Porter (1986, pp. 304–305) notes, Karl Pearson was a true follower of Quetelet in that both believed that the goal of research was to study the laws of social development and that statistics was above all else a practical science. Among other things, Pearson coined the term *normal curve* and proposed the chi-square distribution to detect whether large samples deviated from it and other curves.

By any measure, Galton, Edgeworth, and Pearson were extraordinarily gifted scientists who made striking advances in mathematical statistics. In doing so they admitted a profound intellectual debt to Quetelet and the centrality of empirical regularities.[18]

Like Galton and Pearson, most of the late 19th and early 20th century social scientists remained wholly committed to the positivist doctrine that universal laws, manifested as stubborn facts or empirical regularities, governed societal phenomena. In doing so they embraced increasingly large-sample descriptive statistical methods as the means for conveying these regularities.[19] They took solace from Pearson that science was a matter of collecting and classifying facts, and formulating laws from their critical scrutiny (Ross, 1991, p. 157). Proponents of this view included, for example, the economists Arthur Bowley, Richard T. Ely, William S. Jevons, Wesley C. Mitchell, and Henry L. Moore (Morgan, 1990, pp. 45, 49, 140; Ross, 1991, p. 111); the political scientists Henry J. Ford and Abbott L. Lowell (Ross, 1991, pp. 290, 293); and the sociologists Emile Durkheim, Franklin Giddings, and William G. Sumner (S. Gordon, 1991, p. 443; Ross, 1991, p. 85). Still others among their number, like Edgeworth, Ernst Engel, Wilhelm Lexis, and Adolph Wagner (Hacking, 1990, p. 129; T. M. Porter, 1986, pp. 168, 244, 261), were not content to establish empirical regularities, but wished to understand their *causes*. Importantly, what Pearson and the early statisticians and social scientists had in common was their endorsement of what has been called here the principle of significant sameness.[20]

Simplifying things, the significant sameness orientation lost ground, from which it has never recovered, with the arrival of Fisher. Rather than the emphasis on large-scale descriptive studies favored by his predecessors, Fisher drew attention to small-sample theory, methods of inference, and tests of statistical significance. This smoothed the way for the growth of the significant difference paradigm which has dominated how we statistically analyze data for several decades, right up to the present time.

8.5 Conclusions

This chapter has unveiled some exacting barriers to the adoption of a significant sameness vision when conducting research. Most academicians tend to be highly indisposed to changes in the way things are done, especially when they see little or nothing wrong with their long-held epistemological leanings. That such inclinations fuel career prospects simply gilds the appeal of playing by the dictates of the significant difference model.

At the same time, however, researchers understandably are concerned about the unyielding publish-or-perish ethos and its enormous impact on their livelihoods. As a consequence there are keen incentives to publish as much of one's original research as possible in a rather short period of time. Regrettably, evidence suggests that much of the boom in the quantity of published original work with statistically significant results in fact lacks genuine quality and is far from being truly original. Further, the absolute necessity of surviving in the dog-eat-dog publishing world characteristic of the significant difference paradigm makes it incumbent on investigators to pay homage only to academe and to ignore the needs of practitioners. It also encourages them to employ dubious practices which, collectively, drain confidence in the soundness of the literature.

But if much of what appears in print isn't really original, and is likely tainted, why the resistance to publishing replication research? Given that replications take up a negligible amount of journal space, then *even by default* an argument can be made supporting a much greater emphasis on their visibility. When the far more important collateral argument is laid out that replication research is the only path to the discovery of empirical regularities and the theories to explain them, its justification is no longer of an apologetic, derived kind; it is pressing in its own right. Therefore, several recommendations, straightforward enough to enact if the will is there, were offered to help pave the way for transitioning from a significant difference to a significant sameness philosophy. In doing so we would be well served by going back to the future and learning from our illustrious 19th century predecessors.

Notes

1. Unless stated otherwise, these citations and all others in this chapter were obtained from Google Scholar in August 2013. Moreover, to put these counts into perspective it is vital to remind, as is shown in Section 8.2.2, that most published work goes uncited.

2. Interestingly, Stremersch, Verniers, and Verhoef (2007, p. 179) estimated that, on average, 63% of all articles published in the *International Journal of Research in Marketing, Journal of Consumer Research, Journal of Marketing, Journal of Marketing Research,* and *Marketing Science* between 1990 and 2002 involved an editorial board member of one of these journals.

3. Indeed, Honig et al. (2013, p. 124) claim that for the past two decades this same pressure to show a list of publications has extended to PhD students looking to be hired. A major downside of this urgent need even for students to publish is that many of them don't have the time "to acquire the deep knowledge of their field that was common for previous generations" (Honig et al., 2013, p. 124).

4. Alarmingly, for 2000–2010 about 80,000 patients were involved in clinical trials based on research that was later retracted ("How Science Goes Wrong," 2013, p. 13).

5. And by citing liberally one's own research. An estimate has it that self-citations account for 5%–20% of all citations (Hamilton, 1990, p. 1331).

6. In the interests of equity it must be admitted that because of referees' negative evaluations, path-breaking works also are rejected initially for publication in the physical sciences (cf. Campanario, 1995, 1996). Similar reservations apply in the biomedical areas. For example, a *Newsweek* obituary for Dr. Jonas Salk, creator of the first polio vaccine, observed: "The scientific establishment never embraced Salk. He was never elected to the National Academy of Sciences; academicians sniffed that his work was not 'original'" (S. Begley, 1995, p. 63).

7. Another indicator of the limitations of peer review to detect, prospectively, exceptional work is seen when examining the citation records of prize-winning articles. In marketing, for example, the editorial review boards of some leading journals select annually the "best" contributions published that year in terms of their anticipated far-higher-than-average impact on marketing theory and/or practice. Three of these journals, with the six prestigious awards in parentheses, are the *Journal of Consumer Research* (*JCR*/Best, Ferber), *Journal of Marketing* (Maynard, MSI/Root), and *Journal of Marketing Research* (Green, O'Dell). Using citation count data obtained from the ISI Web of Science, Hubbard, Norman, and Parsa (2010, p. 676) found the ACPA score of 159 award-winning articles to be 6.6 versus 1.8 for 92 "typical" articles (randomly selected from the same journal issues as the award winners) over the approximate time period 1970–2004. Thus, marketing's "Oscars," as we named them, on average outperform regular research published in these journals.

But this does not tell the full story. First, many of the Oscars have only modest citation frequencies. For example, 31% (50/159) have ACPA scores below 3. Second, there are many non-award-winners with outstanding scores. For example, only the top 4 of the 159 award-winning papers would merit inclusion in a list of the 20 most cited non-award-winning articles published in the *JCR, JM,* and *JMR.* In addition, *every* comparison of ACPA counts between the 159 award-winning and top 159 non-award-winning works, by 20-article cohorts, reveals the latter outscoring the former by an average ratio of almost 3:1 (Hubbard, Norman, & Parsa, 2010, p. 680). So much high-impact research is overlooked routinely by marketing's Oscars search committees. No doubt similar results would be observed in other management and social science fields.

8. To underline the imperativeness for career advancement of publishing in scholarly journals, I offer the following anecdote. When hiring new business (or social science) faculty, it is standard procedure in evaluating candidates to say that she or he has two _____, one _____, and a _____ (insert the journals of your choice), but rarely is the *substance* of these articles raised, let alone discussed.

9. Roger Martin (2012, p. 295) facetiously seconds such findings: "Some business people will regularly read academic journals in their specific field. I have met two in my life, so I know they exist." *The Economist* ("Those Who Can't, Teach," 2014, p. 66), on the other hand, criticizes the usefulness of business school research more directly: "Oceans of papers with little genuine insight are published in obscure periodicals that no manager would ever dream of reading." Addendum: in leading journals as well.

10. Now published as Mol and Birkinshaw (2008).

11. In addition Pearce and Huang are dismayed to find that research from top scholarly management journals fails perceptibly to have an impact in high-quality non-academic publications. For example, after inspecting all 2006–2010 issues of *The Economist*—a magazine Pearce and Huang (2012, p. 256) acknowledge is targeted at an educated practitioner audience, or "exactly the audience we seek to reach as scholar-teachers"—they found only one occasion of work being cited from *AMJ* and none from *ASQ*. During this same time period a comparison journal they used, *Psychological Science,* was referenced 18 times.

12. Paradoxically, for business school faculty at least, this emphasis on (company) short-run objectives is something decried in the classroom.

13. In what now seems impossibly quaint, there was even a time when people like Newton and Darwin had the temerity to *delay* for years the publication of their brilliant works. Such restraint in our day would be most welcome.

14. Walsh's (2011, pp. 217–218) anecdote reinforces convincingly this lopsided academic demand for articles over books:

> I once asked a very distinguished colleague in a well-known business school why he never wrote a book that pulled together his many years of research into a single powerful statement. I will never forget his quick answer: "If I don't write for our top journals, I might as well be writing a letter to my mother!"

15. As further evidence of this change, consider Nobel laureate Randy Schekman's accusation that elite print periodicals in the field of medicine, like *Cell, Nature,* and *Science,* are "luxury journals" concerned more with citation impact factors than publishing the best research ("What's Wrong with Science," 2013). From his base at the University of California, Berkeley, Schekman edits the open-access medical journal *eLife* aimed at competing with the above-listed giants.

16. In fact, Quetelet was one of the founding members of the Royal Statistical Society of London in 1834 (Tankard, 1984, p. 32).

17. Whose form in modern statistics textbooks is that suggested by Karl Pearson.

18. It must be added that in their push for greater mathematical rigor in statistics, Galton, Edgeworth, and Pearson were very much in the minority. In the early 19th century (and well into the 20th), a mere handful or so of statists, as they were called then, were mathematically inclined. Therefore, only a few papers appearing in the first 100 years of the *Journal of the Royal Statistical Society* were mathematical (Hill, 1984). Members of the Royal Statistical Society predominantly were economists and saw their job as the accumulation and sorting of *facts* which were burgeoning with the publication of

population censuses and other government reports. Of interest, because this process was to be an integral part of social reform, it had to be seen to be an objective one free of conjecture and opinion. Accordingly, the Royal Statistical Society made this philosophy explicit in its motto *aliis exterendum*—"to be threshed by others." Speculation of any kind was to be avoided.

19. This they shared with the statists who likewise favored description based on complete counts of data. Indeed, early statisticians were suspicious of sampling as a means of inference about populations because it seemed to them a form of speculation or opinion which contravened the mandate of the Royal Statistical Society.

20. A word of clarification is needed. While the views of 19th century social scientists/statisticians and those put forward in this book both uphold the notion of significant sameness, they do so from different philosophical vantages. These 19th century adherents touted what is now widely accepted as an outmoded, naïve positivism, whereas Chapter 3 made clear that the present rationale follows a critical realist perspective.

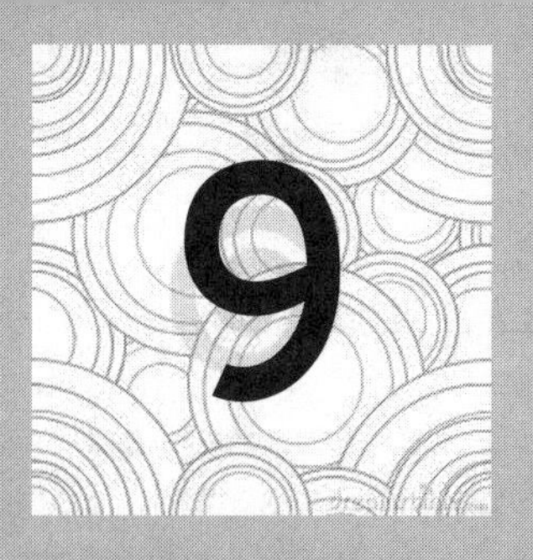

EPILOGUE

> Although I applaud the motivation of attempts to eliminate *P*-values, they have failed in the past and I predict that they will continue to fail. This is because they treat the symptoms and not the underlying mindset, which must be our target. We must change how we think about science itself. (Goodman, 2001, p. 295)

> So, we can continue along the same comfortable, well-worn, and increasingly marginalized path [of producing insufficient relevant knowledge], or we can play the game differently. (Jaworski, 2011, p. 223)

In this book I have compared and contrasted over a number of important dimensions—philosophical, methodological, and statistical—the efficacy of two paradigms with respect to knowledge production. In doing so, it has been made transparent that members of the significant difference paradigm, with their virtual monopoly on research carried out in the management and social sciences, hold a vulgarized conception of how knowledge is attained. Steeped in hypothetico-deductivism, those in this essentially positivistic paradigm deify a model of knowledge acquisition that fails on its own terms as a satisfactory explanation of scientific advance. For this reason it is a framework condoned by neither philosophers nor practitioners of science. This seems immaterial to researchers in the significant difference school, who view H–D prescriptions as a "cookbook" formalizing the workings of *the* scientific method.

In brief, the lone investigator (or with a coauthor or two) in this paradigm first comes up with a theory, usually over a rather short duration of time, to address a problem thought to be of current interest to the field. Crucially, research on this theory/problem must be positioned as original if it is to stand a chance of being published, which is the major driving force behind its undertaking. From this theory a number of hypotheses believed to account for the problem at hand are "deduced" and enumerated in scrupulous, neat and tidy detail.

Next, the soundness of these merely directional hypotheses must be evaluated. The most rigorous way of accomplishing this is via application of the wildly misunderstood tests of statistical significance. So the emphasis is on statistical generalization—from sample *to* population (and the concomitant necessity of random sampling)—a methodology easy to implement in textbooks but exceedingly difficult to do in practice. It is obligatory that the outcomes of (enough of) these tests show statistically significant ($p \leq .05$) results because, in addition to preferences for original research, editors and reviewers likewise are partial to such findings and unimpressed by "negative" ($p > .05$) ones. The researcher therefore is anxious to obtain $p \leq .05$ outcomes supporting his or her novel contribution(s). Once accepted by a journal, the investigator in the publish-or-perish significant difference paradigm skips to another original research topic if career options are to flourish.

An approach to conducting research that is intrinsically flawed is bad enough. In tandem with publication biases in favor of one-off studies promoting original work full of $p \leq .05$ results, however, it is a recipe for the self-destruction of embryonic sciences. It produces literatures consisting in the main of uncorroborated, brittle, overgeneralized, and error-prone results as academicians strive to present to editors the attractive packages they desire. The further editorial-reviewer bias against replications means that these errors are unlikely to be detected—recall that on the few occasions when replications *are* published, about one-half are said to contradict initial research—thereby robbing true science of its ultimate authority, its self-correcting nature. When this is the situation there is ample room for suspicion about the integrity of the management and social science empirical literatures. In the final analysis, the model of research venerated in the significant difference world consists largely of a misguided game or naïve ritual—a kind of methodological posturing, if you will—geared more toward career than knowledge enhancement. This also helps explain why practitioners see negligible relevance in the collective output of academe.

The philosophical underpinnings of the significant sameness paradigm rest on abductive, as opposed to hypothetico-deductive, reasoning. Abduction is the principal means of theorizing in the critical realist philosophy of science which informs the significant sameness approach. Critical realism for the most part is based on how scientists *actually* behave, rather than the H–D account of how they *supposedly* go about their activities.

The importance of replication research permeates every aspect of the significant sameness paradigm. For example, by detecting errors in studies, replications help boost confidence in the credibility of a discipline's empirical record. As noted above, they are the mechanism responsible for the self-correcting nature of good science.

Replication is the only means of deriving empirical regularities or stubborn facts, which are major intellectual achievements in their own right. And it cannot be avoided that their discovery requires a time-consuming, onerous process involving multiple studies (replications) by research teams. Moreover, because magnitudes are vital to science, in the significant sameness paradigm the focus isn't on *p*-values, but instead on estimating effect sizes and the confidence intervals (CIs) around them in single investigations and observing whether the same or similar results occur in the various replications of the initial study as gauged by the criterion of overlapping CIs. So the focus is on quantitative as opposed to directional predictions and on empirical rather than statistical generalization, strategies less beholden to the frequently unworkable demand for random samples.

In due course, the discernment of empirical regularities provides the inspiration for theory development because their relative stubbornness is worthy of concerted efforts at interpretation. This is not the case in the significant difference paradigm, where the results of predominantly isolated analyses are written up based on both premature theorizing and inadequate data.

Again unlike the significant difference school, rather than being viewed with disappointment, negative results, that is, those running counter to expectations, can play a valuable role in the significant sameness paradigm. They do this by refining theory because explanations must be sought for their anomalous character.

Replications are the means for establishing the validity of the entire research process. Thus, such research helps appraise the internal (determination of the effect of X on Y) and external (generalizability) validity of results. Replications facilitate the development of causal theory by analyzing the validity of hypothetical constructs and their interrelationships. To wit, replications examine a construct's convergent, discriminant, and nomological validities. To paraphrase an earlier remark, it doesn't get more important than this. When seen in this light, that replication research is denigrated by adherents of the significant difference school is extraordinary.

In closing, this book argues for nothing less than a complete overhaul in ideas by those in the social sciences and management disciplines about how science works. To this end, I have elaborated on an alternative paradigm—significant sameness—for guiding research in these areas. Unlike the highly caricatured significant difference approach, with its hopes of lending a patina of scientific credence invested in the rote application of null hypothesis significance testing badly misplaced, I have endeavored to show the idea of significant sameness to be an accurate description of the messy way in which real science progresses.

So will the significant sameness paradigm gain purchase in these disciplines? Clearly, this will be difficult to do. There are entrenched interests held by those in the significant difference camp to keep doing things the way they have always been done. If these researchers succeed it would be depressing in the utmost, for, as demonstrated throughout this book, the significant difference model institutionalizes corrupt research.

References

Aaker, D. A., & Keller, K. L. (1990). Consumer evaluations of brand extensions. *Journal of Marketing, 54*(1), 27–41.

Abelson, R. P. (1997). A retrospective on the significance test ban of 1999: (If there were no significance tests, they would be invented). In L. L. Harlow, S. A. Mulaik, & J. H. Steiger (Eds.), *What if there were no significance tests?* (pp. 117–141). Mahwah, NJ: Lawrence Erlbaum.

Abramowitz, S. I., Abramowitz, C. V., Jackson, C., & Gomes, B. (1973). The politics of clinical judgment: What nonliberal examiners infer about women who do not stifle themselves. *Journal of Consulting and Clinical Psychology, 41*, 385–391.

Academy of Management Journal. (2011). Style guide for authors. *Academy of Management Journal, 54*, 1081–1084.

Aiken, L. S., West, S. G., Sechrest, L., & Reno, R. R. (1990). Graduate training in statistics, methodology, and measurement in psychology. *American Psychologist, 45*, 721–734.

Akerlof, G. A. (1970). The market for "lemons": Quality, uncertainty and the market mechanism. *Quarterly Journal of Economics, 84*, 488–500.

Akerlof, G. A., & Shiller, R. J. (2009). *Animal spirits: How human psychology drives the economy, and why it matters for global capitalism.* Princeton, NJ: Princeton University Press.

Alba, J. (1999, December). President's column: Loose ends. *Association of Consumer Research Newsletter,* p. 2.

Aldag, R. J., & Stearns, T. M. (1988). Issues in research methodology. *Journal of Management, 14*, 253–276.

Aldhous, P. (2011, May 5). Journal rejects studies contradicting precognition. *New Scientist.* Retrieved from http://www.newscientist.com

Allen, M., & Preiss, R. (1993). Replication and meta-analysis: A necessary connection. *Journal of Social Behavior and Personality, 8*(6), 9–20.

Allison, D. B., & Faith, M. S. (1996). Hypnosis as an adjunct to cognitive-behavioral psychotherapy for obesity: A meta-analytic reappraisal. *Journal of Consulting and Clinical Psychology, 64*, 513–516.

AMA Task Force on the Development of Marketing Thought. (1988). Developing, disseminating, and utilizing marketing knowledge. *Journal of Marketing, 52*(4), 1–25.

American Psychological Association. (1994). *Publication manual of the American Psychological Association* (4th ed.). Washington, DC: Author.

American Psychological Association. (2001). *Publication manual of the American Psychological Association* (5th ed.). Washington, DC: Author.

American Psychological Association. (2010). *Publication manual of the American Psychological Association* (6th ed.). Washington, DC: Author.

Amir, Y., & Sharon, I. (1990). Replication research: A "must" for the scientific advancement of psychology. *Journal of Social Behavior and Personality, 5*(4), 51–69.

Anastasi, A. (1958). *Differential psychology* (3rd ed.). New York, NY: Macmillan.

Anastasi, A. (1982). *Psychological testing* (5th ed.). New York, NY: Macmillan.

Anderson, C. A., Lindsay, J. J., & Bushman, B. J. (1999). Research in the psychological laboratory: Truth or triviality? *Current Directions in Psychological Science, 8*(1), 3–9.

Anderson, J. C. (1978). The validity of Haire's shopping list projective technique. *Journal of Marketing Research, 15*, 644–649.

Anderson, N., Herriot, P., & Hodgkinson, G. P. (2001). The practitioner–research divide in industrial, work and organizational (IWO) psychology: Where are we now, and where do we go from here? *Journal of Occupational and Organizational Psychology, 74*, 391–411.

Anderson, R. G., & Dewald, W. G. (1994). Replication and scientific standards in applied economics a decade after the *Journal of Money, Credit and Banking* project. *Federal Reserve Bank of St. Louis Review, 76*(November/December), 79–83.

Andreski, S. (1972). *Social sciences as sorcery*. New York, NY: St. Martin's Press.

Angell, M. (1989). Negative studies. *New England Journal of Medicine, 321*, 464–466.

APA Presidential Task Force on Evidence-Based Practice. (2006). Evidence-based practice in psychology. *American Psychologist, 61*, 271–283.

Aram, J. D., & Salipante, P. F. (2003). Bridging scholarship in management: Epistemological reflections. *British Journal of Management, 14*, 189–205.

Ariely, D. (2008). *Predictably irrational*. New York, NY: HarperCollins.

Ariely, D. (2009, July). The end of rational economics. *Harvard Business Review*, 78–84.

Armstrong, J. S. (1983). Cheating in management science (with discussion). *Interfaces, 13*(4), 20–29.

Armstrong, J. S. (1996). Management folklore and management science—On portfolio planning, escalation bias, and such. *Interfaces, 26*(July–August), 25–55.

Armstrong, J. S. (1997). Peer review for journals: Evidence on quality control, fairness, and innovation. *Science and Engineering Ethics, 3*(1), 63–83.

Armstrong, J. S., & Collopy, F. (1996). Competitor orientation: Effects of objectives and information on managerial decisions and profitability. *Journal of Marketing Research, 33*, 188–199.

Armstrong, J. S., & Hubbard, R. (1992). *Importance of empirical research on consumer behavior: Expert opinions* (Working paper). Philadelphia: University of Pennsylvania, Wharton School.

Armstrong, J. S., & Schultz, R. L. (1993). Principles involving marketing policies: An empirical assessment. *Marketing Letters, 4*(3), 253–265.

Arndt, J. (1973). Haire's shopping list revisited. *Journal of Advertising Research, 13,* 57–61.

Asendorpf, J. B., Conner, M., De Fruyt, F., De Houwer, J., Denissen, J. J. A., Fiedler, K., . . . Wicherts, J. M. (2013). Recommendations for increasing replicability in psychology. *European Journal of Personality, 27*, 108–119.

Atkinson, D. R., Furlong, M. J., & Wampold, B. E. (1982). Statistical significance, reviewer evaluations, and the scientific process: Is there a (statistically) significant relationship? *Journal of Counseling Psychology, 29*(2), 189–194.

Babbie, E. (1992). *The practice of social research* (6th ed.). Belmont, CA: Wadsworth.

Bagozzi, R. P., & Yi, Y. (1991). Multitrait-multimethod matrices in consumer research. *Journal of Consumer Research, 17*, 426–439.

Bailey, C. D., Hasselback, J. R., & Karcher, J. N. (2001). Research misconduct in accounting literature: A survey of the most prolific researchers' actions and beliefs. *ABACUS, 37*(1), 26–54.

Baird, D. (1988). Significance tests, history and logic. In S. Kotz & N. L. Johnson (Eds.), *Encyclopedia of statistical sciences* (pp. 466–471). New York, NY: Wiley.

Bakan, D. (1966). The test of significance in psychological research. *Psychological Bulletin, 77*, 423–437.

Bakker, M., van Dijk, A., & Wicherts, J. M. (2012). The rules of the game called psychological science. *Perspectives on Psychological Science, 7*, 543–554.

Bakker, M., & Wicherts, J. M. (2011). The (mis)reporting of statistical results in psychology journals. *Behavior Research Methods, 43*, 666–678.

Balakrishnan, R., Linsmeier, T. J., & Venkatachalam, M. (1996). Financial benefits from JIT adoption: Effects of customer concentration and cost structure. *Accounting Review, 71*, 183–205.

Ball, A. D., & Sawyer, A. G. (2013). Issues involving the use of significant sameness in testing replications and generating knowledge. *Journal of Business Research, 66*, 1389–1392.

Bamber, L., Christensen, T. E., & Gaver, K. M. (2000). Do we really "know" what we think we know? A case study of seminal research and its subsequent overgeneralization. *Accounting, Organizations and Society, 25*, 103–130.

Bamberger, P. A., & Pratt, M. G. (2010). Moving forward by looking back: Reclaiming unconventional research contexts and samples in organizational scholarship. *Academy of Management Journal, 53*, 665–671.

Bandura, A. (1986). *Social foundations of thought and action: A social cognitive theory.* Englewood Cliffs, NJ: Prentice Hall.

Bandura, A. (2005). The evolution of social cognitive theory. In K. G. Smith & M. A. Hitt (Eds.), *Great minds in management: The process of theory development* (pp. 9–35). New York, NY: Oxford University Press.

Bangert-Drowns, R. L. (1986). Review of developments in meta-analytic method. *Psychological Bulletin, 99*, 388–399.

Bargh, J. A., Gollwitzer, P. M., Lee-Chai, A., Barndollar, K., & Trötschel, R. (2001). The automated will: Nonconscious activation and pursuit of behavioral goals. *Journal of Personality and Social Psychology, 81*, 1014–1027.

Barlow, D. H., & Hersen, M. (1984). *Single case experimental designs: Strategies for studying behavior change* (2nd ed.). New York, NY: Pergamon Press.

Barnard, G. A. (1985). *A coherent view of statistical inference.* Ontario, Canada: University of Waterloo, Department of Statistics and Actuarial Science.

Barnett, V. (1982). *Comparative statistical inference* (2nd ed.). New York, NY: Wiley.

Baroudi, J. J., & Orlikowski, W. J. (1989). The problem of statistical power in MIS research. *MIS Quarterly, 13*(1), 87–106.

Barten, A. P. (1969). Maximum likelihood estimation of a complete system of demand equations. *European Economic Review, 1*(1) 7–73.

Bartunek, J. M. (2007). Academic–practitioner collaboration need not require joint or relevant research: Toward a relational scholarship of integration. *Academy of Management Journal, 50*, 1323–1333.

Barwise, P. (1995). Good empirical generalizations. *Marketing Science, 14*(3), G29–G35.

Bass, F. M. (1993). The future of research in marketing: Marketing science. *Journal of Marketing Research, 30*, 1–6.

Bass, F. M. (1995). Empirical generalizations and marketing science: A personal view. *Marketing Science, 14*(3), G6–G19.

Bass, F. M., & Wind, J. (1995). Introduction to the special issue: Empirical generalizations in marketing. *Marketing Science, 14*(3), G1–G5.

Bauer, H. H. (1994), *Scientific literacy and the myth of the scientific method.* Urbana: University of Illinois Press.

Bayarri, M. J., & Mayoral, A. M. (2002). Bayesian design of "successful" replications. *The American Statistician, 56*, 207–214.

Bearden, W. O., Netemeyer, R. G., & Mobley, M. F. (1993). *Handbook of marketing scales: Multi-item measures for marketing and consumer behavior research.* Newbury Park, CA: Sage.

Beck, A. T. (1993). *Cognitive therapy of depression: A personal reflection.* Aberdeen, Scotland: Scottish Cultural Press.

Becker, G. S. (1965). A theory of the allocation of time. *Economic Journal, 75*, 493–517.

Bedeian, A. G., Taylor, S. G., & Miller, A. L. (2010). Management science on the credibility bubble: Cardinal sins and various misdemeanors. *Academy of Management Learning & Education, 9*, 715–725.

Begg, C. B., & Berlin, J. A. (1988). Publication bias: A problem in interpreting medical data. *Journal of the Royal Statistical Society, Ser. B, 151*, 419–463.

Begley, C. G., & Ellis, L. M. (2012). Raise standards for preclinical cancer research. *Nature, 438*, 531–533.

Begley, S. (1995, July 3). The savior of summer. *Newsweek*, p. 63.

Belia, S., Fidler, F., Williams, J., & Cumming, G. (2005). Researchers misunderstand confidence intervals and standard error bars. *Psychological Methods, 10*, 389–396.

Bem, D. J. (2011). Feeling the future: Experimental evidence for anomalous retroactive influences on cognition and affect. *Journal of Personality and Social Psychology, 100*, 407–425.

Bennis, W. G., & O'Toole, J. (2005, May). How business schools lost their way. *Harvard Business Review*, 96–104.

Bensman, S. J. (2008). Distributional differences of the impact factor in the sciences versus the social sciences: An analysis of the probabilistic structure of the 2005 journal citation reports. *Journal of the American Society for Information Science and Technology, 59*, 1366–1382.

Berelson, B., & Steiner, G. A. (1964). *Human behavior: An inventory of scientific findings*. New York, NY: Harcourt, Brace & World.

Berger, J. O. (2003). Could Fisher, Jeffreys and Neyman have agreed on testing? (with discussion). *Statistical Science, 18*, 1–32.

Berger, J. O., & Berry, D. A. (1988). Statistical analysis and the illusion of objectivity. *American Scientist, 76*(2), 159–165.

Berger, J. O., Boukai, B., & Wang, Y. (1997). Unified frequentist and Bayesian testing of a precise hypothesis (with discussion). *Statistical Science, 12*, 133–160.

Berger, J. O., & Delampady, M. (1987). Testing precise hypotheses (with discussion). *Statistical Science, 2*, 317–352.

Berger, J. O., & Sellke, T. (1987). Testing a point null hypothesis: The irreconcilability of *P* values and evidence (with discussion). *Journal of the American Statistical Association, 82*, 112–139.

Berkson, J. (1942). Tests of significance considered as evidence. *Journal of the American Statistical Association, 37*, 325–335.

Berthon, P., Ewing, M., Pitt, L., & Berthon, J. P. (2003). Reframing replicative research in advertising. *International Journal of Advertising, 22*, 511–530.

Beveridge, W. I. B. (1951). *The art of scientific investigation*. London, UK: William Heinemann.

Bezeau, S., & Graves, R. (2001). Statistical power and effect sizes of clinical neuropsychology research. *Journal of Clinical and Experimental Neuropsychology, 23*, 399–406.

Bhaskar, R. (1978). *A realist theory of science*. Hassocks, UK: Harvester Press.

Bhaskar, R. (1979). *The possibility of naturalism*. Hassocks, UK: Harvester Press.

Bird, C. P., & Fisher, T. D. (1986). 30 years later: Attitudes toward the employment of older workers. *Journal of Applied Psychology, 71*, 515–517.

Birnbaum, A. (1962). On the foundations of statistical inference (with discussion). *Journal of the American Statistical Association, 57*, 269–326.

Bishop, R. C. (2007). *The philosophy of the social sciences*. New York, NY: Continuum.

Blachowicz, J. (2009). How science textbooks treat scientific method: A philosopher's perspective. *British Journal for the Philosophy of Science, 60*, 303–344.

Black, F., & Scholes, M. (1973). The pricing of options and corporate liabilities. *Journal of Political Economy, 81*, 637–654.

Blair, E., & Zinkhan, G. M. (2006). Nonresponse and generalizability in academic research. *Journal of the Academy of Marketing Science, 34,* 4–7.

Blankenship, A. B., & Breen, G. E. (1993). *State of the art marketing research.* Lincolnwood, IL: NTC Business Books.

Blaug, M. (1985). *Great economists since Keynes.* Cambridge, UK: Cambridge University Press.

Blaug, M. (1992). *The methodology of economics: Or how economists explain* (2nd ed.). Cambridge, UK: Cambridge University Press.

Bolles, R. C. (1962). The difference between statistical hypotheses and scientific hypotheses. *Psychological Reports, 11,* 639–645.

Bonett, D. G., & Wright, T. A. (2007). Comments and recommendations regarding the hypothesis testing controversy. *Journal of Organizational Behavior, 28,* 647–659.

Boos, D. D., & Stefanski, L. A. (2011). *P*-value precision and reproducibility. *The American Statistician, 65,* 213–221.

Borenstein, M., Rothstein, H., & Cohen, J. (1997). *Power and precision: A computer program for statistical power analysis and confidence intervals.* Teaneck, NJ: Biostat.

Borkowski, S. C., Welsh, M. J., & Zhang, Q. (2001). An analysis of statistical power in behavioral accounting research. *Behavioral Research in Accounting, 13*(1), 63–84.

Bottomley, P. A., & Holden, S. J. S. (2001). Do we really know how consumers evaluate brand extensions? Empirical generalizations based on secondary analysis of eight studies. *Journal of Marketing Research, 38,* 494–500.

Bound, J. A., & Ehrenberg, A. S. C. (1989). Significant sameness. *Journal of the Royal Statistical Society, Ser. A, 152,* 241–247.

Bowling, A. (2002). *Research methods in health: Investigating health and health services* (2nd ed.). Maidenhead, UK: Open University Press.

Box, G. E. P. (1994). Statistics and quality improvement. *Journal of the Royal Statistical Society, Ser. A, 157,* 209–229.

Boyd, B. K., Gove, S., & Hitt, M. A. (2005). Construct measurement in strategic management research: Illusion or reality? *Strategic Management Journal, 26,* 239–257.

Bozarth, J. D., & Roberts, R. R., Jr. (1972). Signifying significant significance. *American Psychologist, 27,* 774–775.

Brace, N., Kemp, R., & Snelgar, R. (2006). *SPSS for psychologists.* New York, NY: Palgrave Macmillan.

Braithwaite, R. B. (1953). *Scientific explanation.* Cambridge, UK: Cambridge University Press.

Brewer, J. K. (1972). On the power of statistical tests in the *American Educational Research Journal. American Educational Research Journal, 9,* 391–401.

Brinberg, D., & McGrath, J. E. (1985). *Validity and the research process.* Newbury Park, CA: Sage.

Broad, W. (1981). The publishing game: Getting more for less. *Science, 211,* 1137–1139.

Broad, W., & Wade, N. (1982). *Betrayers of the truth: Fraud and deceit in the halls of science.* New York, NY: Simon and Shuster.

Brooks, C. (2002). *Introductory econometrics for finance*. Cambridge, UK: Cambridge University Press.

Broverman, I. K., Broverman, D. M., Clarkson, F. E., Rosenkrantz, P. S., & Vogel, S. R. (1970). Sex role stereotypes and clinical judgments of mental health. *Journal of Consulting and Clinical Psychology, 34*, 1–7.

Brown, R. (1963). *Explanation in social science*. Chicago, IL: Aldine.

Brown, S. W., & Coney, K. A. (1976). Building a replication tradition in marketing. In K. L. Bernhardt (Ed.), *Marketing 1776–1976 and beyond* (pp. 622–625). Chicago, IL: American Marketing Association.

Bruner, G. C., II, & Hensel, P. J. (1994). *Marketing scales handbook: A compilation of multi-item measures*. Chicago, IL: American Marketing Association.

Bryman, A. (2004). *Social research methods*. Oxford, UK: Oxford University Press.

Bullock, R. J., & Svyantek, D. J. (1985). Analyzing meta-analysis: Potential problems, an unsuccessful replication, and evaluation criteria. *Journal of Applied Psychology, 70*(1), 108–115.

Bunge, M. (1997). Mechanism and explanation. *Philosophy of the Social Sciences, 24*, 410–465.

Bunge, M. (2004). How does it work? The search for explanatory mechanisms. *Philosophy of the Social Sciences, 34*(2), 182–210.

Burns, A. C., & Bush, R. F. (2010). *Marketing research* (6th ed.). New York, NY: Prentice Hall.

Busche, K., & Kennedy, P. (1984). On economists' belief in the law of small numbers. *Economic Inquiry, 22*, 602–603.

Cafri, G., Kromrey, J. D., & Brannick, M. T. (2010). A meta-meta-analysis: Empirical review of statistical power, type I error rates, effect sizes, and model selection of meta-analyses published in psychology. *Multivariate Behavioral Research, 45*, 239–270.

Cahill, D. J. (1994, September). A reply to David Aaker: "Yes you are talking to yourselves, I'm afraid." *ACR Newsletter*, p. 11.

Campanario, J. M. (1995). On influential books and journal articles initially rejected because of negative referees' evaluations. *Science Communication, 16*, 304–325.

Campanario, J. M. (1996). Have referees rejected some of the most-cited articles of all times? *Journal of the American Society for Information Science, 47*, 302–310.

Campbell, D. T., & Fiske, D. W. (1959). Convergent and discriminant validation by the multitrait-multimethod matrix. *Psychological Bulletin, 56*, 81–105.

Campbell, D. T., & Stanley, J. C. (1966). *Experimental and quasi-experimental designs for research*. Chicago, IL: Rand McNally.

Campbell, K. E., & Jackson, T. T. (1979). The role of and need for replication research in social psychology. *Replications in Social Psychology, 1*(1), 3–14.

Card, D., & Krueger, A. B. (1994). Minimum wages and employment: A case study of the fast-food industry in New Jersey and Pennsylvania. *American Economic Review, 84*, 772–793.

Card, D., & Krueger, A. B. (1998). *A reanalysis of the effect of the New Jersey minimum wage increase on the fast-food industry with representative payroll data* (Working Paper 6386). Washington, DC: National Bureau of Economic Research.

Card, D., & Krueger, A. B. (2000). Minimum wages and employment: A case study of the fast-food industry in New Jersey and Pennsylvania: Reply. *American Economic Review, 90,* 1397–1420.

Carlson, R. (1976). The logic of tests of significance. *Philosophy of Science, 43,* 116–128.

Carpenter, S. (2012). Psychology's bold initiative. *Science, 335,* 558, 1559–1561.

Carson, D. (1995). A comment on: The commodification of marketing knowledge. *Journal of Marketing Management, 11,* 661–664.

Carver, R. P. (1978). The case against statistical significance testing. *Harvard Educational Review, 48,* 378–399.

Casella, G., & Berger, R. L. (1987). Reconciling Bayesian and frequentist evidence in the one-sided testing problem (with discussion). *Journal of the American Statistical Association, 82,* 106–139.

Cashen, L. H., & Geiger, S. W. (2004). Statistical power and the testing of null hypotheses: A review of contemporary management research and recommendations for future studies. *Organizational Research Methods, 7*(2), 151–167.

Cassidy, J. (2009). *How markets fail: The logic of economic calamity.* New York, NY: Farrar, Straus and Giroux.

Ceci, S. J., & Walker, E. (1983). Private archives and public needs. *American Psychologist, 38,* 414–423.

Chalmers, A. F. (1999). *What is this thing called science?* (3rd ed.). Maidenhead, UK: Open University Press.

Chase, L. J., & Baran, S. J. (1976). An assessment of quantitative research in mass communication. *Journalism Quarterly, 53,* 308–311.

Chase, L. J., & Chase, R. B. (1976). A statistical power analysis of applied psychological research. *Journal of Applied Psychology, 61,* 234–237.

Chase, L. J., & Tucker, R. K. (1975). A power-analytic examination of contemporary communication research. *Speech Monographs, 42*(1), 29–41.

Chatfield, C. (1985). The initial examination of data (with discussion). *Journal of the Royal Statistical Society, Ser. A, 148,* 214–253.

Chatfield, C. (1995). Model uncertainty, data mining and statistical inference (with discussion). *Journal of the Royal Statistical Society, Ser. A, 158,* 419–466.

Chatfield, C. (2002). Confessions of a pragmatic statistician. *The Statistician, 51,* 1–20.

Christensen, C. M., & Carlile, P. R. (2009). Course research: Using the case method to build and teach management theory. *Academy of Management Learning and Education, 8,* 240–251.

Christensen, C. M., & Raynor, M. E. (2003a). *The innovator's solution: Creating and sustaining successful growth.* Cambridge, MA: Harvard Business School Press.

Christensen, C. M., & Raynor, M. E. (2003b, September). Why hard-nosed executives should care about management theory. *Harvard Business Review,* 66–74.

Christensen, R. (2005). Testing Fisher, Neyman, Pearson, and Bayes. *The American Statistician, 59,* 1–6.

Churchill, G. A., Jr. (1988). Comment on the AMA Task Force study. *Journal of Marketing, 52*(4), 26–31.

Cicchetti, D. V. (1998). Role of null hypothesis significance testing (NHST) in the design of neuropsychological research. *Journal of Clinical and Experimental Neuropsychology, 20*, 293–295.

Clark, D. A., & Beck, A. T. (1999). *Scientific foundations of cognitive theory and therapy of depression*. New York, NY: Wiley.

Clark, T., Floyd, S. W., & Wright, M. (2006). On the review process and journal development. *Journal of Management Studies, 43*, 655–664.

Clarke, Y., & Soutar, G. N. (1982). Consumer acquisition patterns for durable goods: Australian evidence. *Journal of Consumer Research, 8*, 456–460.

Clegg, S. (2005). Evidence-based practice in educational research: A critical realist critique of systematic review. *British Journal of Sociology of Education, 26*, 415–428.

Clegg, S. R., & Ross-Smith, A. (2003). Revising the boundaries: Management education and learning in a postpositivist world. *Academy of Management Learning and Education, 2*(1), 85–99.

Cochran, W. G. (1974). Footnote to an appreciation of R. A. Fisher. *Science, 156*, 1460–1462.

Cohen, D. J. (2007). The very separate worlds of academic and practitioner publications in human resource management: Reasons for the divide and concrete solutions for bridging the gap. *Academy of Management Journal, 50*, 1013–1019.

Cohen, I. B. (1985). *Revolution in science*. Cambridge, MA: Harvard University Press.

Cohen, J. (1962). The statistical power of abnormal-social psychological research: A review. *Journal of Abnormal and Social Psychology, 65*, 145–153.

Cohen, J. (1969). *Statistical power analysis for the behavioral sciences*. New York, NY: Academic Press.

Cohen, J. (1988). *Statistical power analysis for the behavioral sciences* (2nd ed.). Hillsdale, NJ: Lawrence Erlbaum.

Cohen, J. (1990). Things I have learned (so far). *American Psychologist, 45*, 1304–1312.

Cohen, J. (1992). A power primer. *Psychological Bulletin, 112*, 155–159.

Cohen, J. (1994). The earth is round ($p < .05$). *American Psychologist, 49*, 997–1003.

Cohen, J., & Cohen, P. (1983). *Applied multiple regression/correlation analysis for the behavioral sciences* (2nd ed.). Hillsdale, NJ: Lawrence Erlbaum.

Cole, P. (1979). The evolving case-control study. *Journal of Chronic Diseases, 32*(1–2), 15–27.

Collier, J. E., & Bienstock, C. C. (2007). An analysis of how nonresponse error is assessed. *Marketing Theory, 7*, 163–183.

Collins, H. M. (1985). *Changing order: Replication and induction in scientific practice*. Beverly Hills, CA: Sage.

Collins, R. (1989). Sociology: Proscience or antiscience? *American Sociological Review, 54*(1), 124–139.

Combs, J. G. (2010). Big samples and small effects: Let's not trade relevance and rigor for power. *Academy of Management Journal, 53*, 9–13.

Cook, T. D. (1983). Quasi-experimentation: Its ontology, epistemology, and methodology. In G. Morgan (Ed.), *Beyond method* (pp. 74–94). Beverly Hills, CA: Sage.

Cook, T. D. (1985). Postpositivist critical multiplism. In R. L. Shotland & M. M. Mark (Eds.), *Social science and social policy* (pp. 21–62). Beverly Hills, CA: Sage.

Cook, T. D., & Campbell, D. T. (1979). *Quasi-experimentation: Design and analysis issues for field settings*. Boston, MA: Houghton Mifflin.

Cook, T. D., Gruder, C. L., Hennigan, K. M., & Flay, B. R. (1979). History of the sleeper effect: Some logical pitfalls in accepting the null hypothesis. *Psychological Bulletin, 86*, 662–679.

Cooper, D. R., & Schindler, P. (2006). *Marketing research*. New York, NY: McGraw-Hill.

Cooper, H. (2003). Editorial. *Psychological Bulletin, 129*, 3–9.

Cornelissen, J. (2002). Academic and practitioner theories of marketing. *Marketing Theory, 2*, 133–143.

Cornfield, J. (1966). Sequential trials, sequential analysis, and the likelihood principle. *The American Statistician, 20*(6), 18–23.

Cortina, J. M., & Dunlap, W. P. (1997). On the logic and purpose of significance testing. *Psychological Methods, 2*, 161–172.

Cortina, J. M., & Folger, R. G. (1998). When is it acceptable to accept a null hypothesis: No way, Jose? *Organizational Research Methods, 1*, 334–350.

Cote, J. A., & Buckley, M. R. (1987). Estimating trait, method, and error variance: Generalizing across 70 construct validation studies. *Journal of Marketing Research, 24*, 315–318.

Coursol, A., & Wagner, E. E. (1986). Effect of positive findings on submission and acceptance rates: A note on meta-analysis bias. *Professional Psychology: Research and Practice, 17*, 136–137.

Cowles, M. (2001). *Statistics in psychology: An historical perspective*. Mahwah, NJ: Lawrence Erlbaum.

Cox, D. R. (1977). The role of significance tests (with discussion). *Scandinavian Journal of Statistics, 4*, 49–70.

Cox, D. R. (1982). Statistical significance tests. *British Journal of Clinical Pharmacology, 14*, 325–331.

Cox, D. R. (1986). Some general aspects of the theory of statistics. *International Statistical Review, 54*, 117–126.

Craig, J. R., & Reese, S. C. (1973). Retention of raw data: A problem revisited. *American Psychologist, 28*, 723.

Cronbach, L. J., & Meehl, P. E. (1955). Construct validity in psychological tests. *Psychological Bulletin, 52*, 281–302.

Cronin, B. (2013, April 17). Seminal economic paper on debt draws criticism. *Wall Street Journal*, p. A2.

Crosier, K. (2004). How effectively do marketing journals transfer useful learning from scholars to practitioners? *Marketing Intelligence and Planning, 22*, 540–556.

Cumming, G. (2008). Replication and *p* intervals: *p* values predict the future only vaguely, but confidence intervals do much better. *Perspectives on Psychological Science, 3*, 286–300.

Cumming, G. (2012). *Understanding the new statistics: Effect sizes, confidence intervals, and meta-analysis*. New York, NY: Routledge.

Cumming, G., Fidler, F., Leonard, M., Kalinowski, P., Christiansen, A., Kleinig, A., . . . Wilson, S. (2007). Statistical reform in psychology: Is anything changing? *Psychological Science, 18,* 230–232.

Cumming, G., & Finch, S. (2001). A primer on the understanding, use and calculation of confidence intervals that are based on central and noncentral distributions. *Educational and Psychological Measurement, 61,* 532–574.

Cumming, G., & Finch, S. (2005). Inference by eye: Confidence intervals and how to read pictures of data. *American Psychologist, 60,* 170–180.

Cumming, G., & Maillardet, R. (2006). Confidence intervals and replication: Where will the next mean fall? *Psychological Methods, 11,* 217–227.

Cummings, T. G. (2007). Quest for an engaged academy. *Academy of Management Review, 32,* 355–360.

Daft, R. L., & Lewin, A. Y. (1990). Can organizational studies begin to break out of the normal science straightjacket? An editorial essay. *Organization Science, 1*(1), 1–9.

Dancey, C. P., & Reidy, J. (2004). *Statistics without maths for psychology: Using SPSS for Windows* (3rd ed.). Harlow, UK: Pearson Prentice Hall.

Danermark, B., Ekström, M., Jakobsen, L., & Karlsson, J. C. (2002). *Explaining society: Critical realism in the social sciences.* London, UK: Routledge.

Darley, W. K. (2000). Status of replication studies in marketing: A validation and extension. *Marketing Management Journal, 10*(2), 121–132.

Darnell, A. C., & Evans, J. L. (1990). *The limits of econometrics.* Aldershot, UK: Edward Elgar.

Davis, S. F., & Smith, R. A. (2005) *An introduction to statistics and research methods: Becoming a psychological detective.* Upper Saddle River, NJ: Pearson Prentice Hall.

Davis, S. W., & Ketz, J. E. (1991). Fraud and accounting research. *Accounting Horizons, 5*(September), 106–109.

Dawes, R. M. (1994). *House of cards: Psychology and psychotherapy built on myth.* New York, NY: Free Press.

DeLone, W. H., & McClean, E. R. (1992). Information systems success: The quest for the dependent variable. *Information Systems Research, 3,* 60–95.

De Long, J. B., & Lang, K. (1992). Are all economic hypotheses false? *Journal of Political Economy, 100,* 1257–1272.

Deming, W. E. (1975). On probability as a basis for action. *The American Statistician, 29,* 146–152.

Denning, S. (2012, November 20). What killed Michael Porter's Monitor Group? The one force that really matters. *Forbes.* Retrieved from http://www.forbes.com

Denrell, J. (2003). Vicarious learning, undersampling of failure, and the myths of management. *Organization Science, 14,* 227–243.

Denrell, J. (2005, April). Selection bias and the perils of benchmarking. *Harvard Business Review,* 114–119.

Denton, F. T. (1985). Data mining as an industry. *Review of Economics and Statistics, 67*(1), 124–127.

Denton, F. T. (1988). The significance of significance: Rhetorical aspects of statistical hypothesis testing in economics. In A. Klamer, D. N. McCloskey, & R. M. Solow (Eds.), *The consequences of economic rhetoric* (pp. 163–183). Cambridge, UK: Cambridge University Press.

Dewald, W. G., Thursby, J. G., & Anderson, R. G. (1986). Replication in empirical economics: *The Journal of Money, Credit and Banking* project. *American Economic Review, 76*, 587–603.

Dickersin, K., Chan, S., Chalmers, T. C., Sacks, H. S., & Smith, H., Jr. (1987). Publication bias and clinical trials. *Controlled Clinical Trials, 8*, 343–353.

Dijksterhuis, A., & van Knippenberg, A. (1998). The relation between perception and behavior, or how to win a game of Trivial Pursuit. *Journal of Personality and Social Psychology, 74*, 865–877.

Dirnagl, U., & Lauritzen, M. (2010). Fighting publication bias: Introducing the negative results section. *Journal of Cerebral Blood Flow and Metabolism, 30*, 1263–1264.

Doucouliagos, C., & Stanley, T. D. (2013). Are all economic facts greatly exaggerated? Theory competition and selectivity. *Journal of Economic Surveys, 27*, 316–339.

Dougherty, C. (2007). *Introduction to econometrics* (3rd ed.). Oxford, UK: Oxford University Press.

Dunn, D. S. (2009). *Research methods for social psychology.* Chichester, UK: Wiley–Blackwell.

Dyer, I. (1997). The significance of statistical significance. *Intensive Critical Care Nursing, 13*, 259–265.

Easley, R. W., Madden, C. S., & Dunn, M. G. (2000). Conducting marketing science: The role of replication in the research process. *Journal of Business Research, 48*, 83–92.

Easley, R. W., Madden, C. S., & Gray, B. (2013). A tale of two cultures: Revisiting journal editors' views of replication research. *Journal of Business Research, 66*, 1457–1459.

Easton, G. (2002). Marketing: A critical realist approach. *Journal of Business Research, 55*, 103–109.

Eaton, W. O. (1984). On obtaining data for research investigations. *American Psychologist, 39*, 1325–1326.

Eden, D. (2002). Replication, meta-analysis, scientific progress, and *AMJ*'s publication policy. *Academy of Management Journal, 45*, 841–846.

Edwards, A. W. F. (1992). *Likelihood* (Exp. ed.). Baltimore, MD: Johns Hopkins University Press.

Edwards, J. R. (2008). To prosper, organizational psychology should . . . overcome methodological barriers to progress. *Journal of Organizational Behavior, 29*, 469–491.

Edwards, J. R., & Berry, J. W. (2010). The presence of something or the absence of nothing: Increasing theoretical precision in management research. *Organizational Research Methods, 13*, 668–689.

Efron, B. (1998). R. A. Fisher in the 21st century (with discussion). *Statistical Science, 13,* 95–122.

Ehrenberg, A. S. C. (1968). The elements of lawlike relationships. *Journal of the Royal Statistical Society, Ser. A, 131,* 280–302.

Ehrenberg, A. S. C. (1975). *Data reduction* (Rev. reprint). Chichester, UK: Wiley.

Ehrenberg, A. S. C. (1983). We must preach what is practiced. *The American Statistician, 37,* 248–250.

Ehrenberg, A. S. C. (1988). *Repeat buying: Facts, theory, and applications* (2nd ed.). London, UK: Charles Griffin.

Ehrenberg, A. S. C. (1990). A hope for the future of statistics: MSOD. *The American Statistician, 44,* 195–196.

Ehrenberg, A. S. C. (1993a). Even the social sciences have laws. *Nature, 365,* 385.

Ehrenberg, A. S. C. (1993b). Theory or well-based results: Which comes first? In G. Laurent, G. Lilien, & B. Pras (Eds.), *Research traditions in marketing* (pp. 79–108). Boston, MA: Kluwer.

Ehrenberg, A. S. C. (1995). Empirical generalisations, theory, and method. *Marketing Science, 14*(3), G20–G28.

Ehrenberg, A. S. C. (2004). My research in marketing: How it happened. *Marketing Research, 16*(Winter), 34–41.

Ehrenberg, A. S. C., & Bound, J. A. (1993). Predictability and prediction (with discussion). *Journal of the Royal Statistical Society, Ser. A, 156,* 167–206.

Ehrenberg, A. S. C., & England, L. R. (1990). Generalising a pricing effect. *Journal of Industrial Economics, 39*(1), 47–68.

Ehrenberg, A. S. C., Goodhardt, G. J., & Barwise, T. P. (1990). Double jeopardy revisited. *Journal of Marketing, 54*(3), 82–91.

Eichenbaum, M. (1995). Some comments on the role of econometrics in economic theory. *Economic Journal, 105,* 1609–1621.

Eisenhardt, K. M. (1989). Building theories from case study research. *Academy of Management Review, 14,* 532–550.

Eisenhardt, K. M., & Graebner, M. E. (2007). Theory building from cases: Opportunities and challenges. *Academy of Management Journal, 50,* 25–32.

Elms, A. C. (1975). The crisis of confidence in social psychology. *American Psychologist, 30,* 967–976.

Enders, W., & Hoover, G. A. (2004). Whose line is it? Plagiarism in economics. *Journal of Economic Literature, 42,* 487–493.

Enders, W., & Hoover, G. A. (2006). Plagiarism in the economics profession: A survey. *Challenge, 49*(September/October), 92–107.

Epstein, W. M. (2004). Confirmational response bias and the quality of the editorial processes among American social work journals," *Research on Social Work Practice, 14,* 450–458.

Erdfelder, E., Faul, F., & Buchner, A. (1996). GPOWER: A general power analysis program. *Behavior Research Methods, Instruments, & Computers, 28*(1), 1–11.

Estes, W. K. (1997). On the communication of information by displays of standard errors and confidence intervals. *Psychonomic Bulletin & Review, 4,* 330–341.

Evans, J. T., Nadjari, H. I., & Burchell, S. A. (1990). Quotational and reference accuracy in surgical journals. *Journal of the American Medical Association, 263,* 1353–1354.

Evanschitzky, H., & Armstrong, J. S. (2010). Replications of forecasting research. *International Journal of Forecasting, 26,* 4–8.

Evanschitzky, H., & Armstrong, J. S. (2013). Research with in-built replications: Comment and further suggestions for replication research. *Journal of Business Research, 66,* 1406–1408.

Evanschitzky, H., Baumgarth, C., Hubbard, R., & Armstrong, J. S. (2007). Replication research's disturbing trend. *Journal of Business Research, 46,* 411–415.

Fagley, N. S. (1985). Applied statistical power analysis and the interpretation of nonsignificant results by research consumers. *Journal of Counseling Psychology, 32,* 391–396.

Falk, R., & Greenbaum, C. W. (1995). Significance tests die hard: The amazing persistence of a probabilistic misconception. *Theory & Psychology, 5,* 75–98.

Fanelli, D. (2009). How many scientists fabricate and falsify research? A systematic review and meta-analysis of survey data. *PLoS ONE, 4*(May), 1–11.

Fanelli, D. (2010). Do pressures to publish increase scientists' bias? An empirical support from US states data. *PLoS ONE, 5*(April), 1–7.

Fanelli, D. (2012). Negative results are disappearing from most disciplines and countries. *Scientometrics, 90,* 891–904.

Farley, J. U., & Lehmann, D. R. (1986). *Meta-analysis in marketing: Generalization of response models.* Lexington, MA: Lexington Books.

Faust, D. (1984). *The limits of scientific reasoning.* Minneapolis, MN: University of Minnesota Press.

Fay, B. (1996). *Contemporary philosophy of social science.* Oxford, UK: Blackwell.

Feige, E. L. (1975). The consequences of journal editorial policies and a suggestion for revision. *Journal of Political Economy, 83,* 1291–1295.

Feigenbaum, S., & Levy, D. M. (1993). The market for (ir)reproducible econometrics. *Social Epistemology, 7,* 215–232.

Felson, D. T. (1992). Bias in meta-analytic research. *Journal of Clinical Epidemiology, 45,* 885–892.

Ferguson, C. J., & Brannick, M. T. (2012). Publication bias in psychological science: Prevalence, methods for identifying and controlling, and implications for the use of meta-analyses. *Psychological Methods, 17,* 120–128.

Ferguson, C. J., & Heene, M. (2012). A vast graveyard of undead theories: Publication bias and psychological science's aversion to the null. *Perspectives on Psychological Science, 7,* 555–561.

Ferguson, G. A. (1959). *Statistical analysis in psychology and education.* New York, NY: McGraw-Hill.

Ferguson, T. D., & Ketchen, D. J., Jr. (1999). Organizational configurations and performance: The role of power in extant research. *Strategic Management Journal, 20,* 385–395.

Fidler, F. (2002). The fifth edition of the APA *Publication Manual*: Why its statistics recommendations are so controversial. *Educational and Psychological Measurement, 62,* 749–770.

Fidler, F., Thomason, N., Cumming, G., Finch, S., & Leeman, J. (2004). Editors can lead researchers to confidence intervals, but can't make them think: Statistical reform lessons from medicine. *Psychological Science, 15,* 119–126.

Fidler, F., Thomason, N., Cumming, G., Finch, S., & Leeman, J. (2005). Still much to learn about confidence intervals: Reply to Rouder and Morey. *Psychological Science, 16,* 494–495.

Fife-Shaw, C. (2006). Principles of statistical inference tests. In G. M. Breakwell, S. Hammond, C. Fife-Shaw, & J. A. Smith (Eds.), *Research methods in psychology* (3rd ed., pp. 388–413). Thousand Oaks, CA: Sage.

Finch, S., Cumming, G., & Thomason, N. (2001). Reporting of statistical inference in the *Journal of Applied Psychology:* Little evidence of reform. *Educational and Psychological Measurement, 61,* 181–210.

Fishe, R. P. H., & Wohar, M. (1990). The adjustment of expectations to a change in regime: Comment. *American Economic Review, 80,* 968–976.

Fisher, R. A. (1922). On the mathematical foundations of theoretical statistics. *Philosophical Transactions of the Royal Society of London, Ser. A, 222,* 309–368.

Fisher, R. A. (1925). *Statistical methods for research workers.* Edinburgh, Scotland: Oliver and Boyd.

Fisher, R. A. (1926). The arrangement of field experiments. *Journal of the Ministry of Agriculture of Great Britain, 33,* 503–513.

Fisher, R. A. (1935). *The design of experiments.* Edinburgh, Scotland: Oliver and Boyd.

Fisher, R. A. (1947). Development of the theory of experimental design. *Proceedings of the International Statistical Conferences, 3,* 434–439.

Fisher, R. A. (1955). Statistical methods and scientific induction. *Journal of the Royal Statistical Society, Ser. B, 17,* 69–78.

Fisher, R. A. (1959). *Statistical methods and scientific inference* (2nd rev. ed.). Edinburgh, Scotland: Oliver and Boyd.

Fisher, R. A. (1966). *The design of experiments* (8th ed.). Edinburgh, Scotland: Oliver and Boyd.

Fisher, R. A. (1970). *Statistical methods for research workers* (14th ed.). New York, NY: Hafner.

Fisher, R. A. (1971). *The design of experiments* (9th ed.). New York, NY: Hafner.

Fisher, R. A. (1973). *Statistical methods and scientific inference* (3rd rev. ed.). New York, NY: Hafner.

Fisher Box, J. (1978). *R. A. Fisher: The life of a scientist.* New York, NY: Wiley.

Fiske, D. W. (1986). Specificity of method and knowledge in social science. In D. W. Fiske & R. A. Shweder (Eds.), *Metatheory in social science: Pluralisms and subjectivities* (pp. 61–82). Chicago, IL: University of Chicago Press.

Flyvbjerg, B. (2001). *Making social science matter: Why social inquiry fails and how it can succeed again.* Cambridge, UK: Cambridge University Press.

Forster, N. (2007). CEO's readership of business and management journals in Australia: Implications for research and teaching. *Journal of Management and Organization, 13*(1), 24–40.

Fowler, J., Cohen, L., & Jarvis, P. (1998). *Practical statistics for field biology* (2nd ed.). Chichester, UK: Wiley.

Fox, J. (2009). *The myth of the rational market: A history of risk, reward, and delusion on Wall Street.* New York, NY: Harper Business.

Francis, G. (2012a). The psychology of replication and replication in psychology. *Perspectives on Psychological Science, 7,* 585–594.

Francis, G. (2012b). Too good to be true: Publication bias in two prominent studies from experimental psychology. *Psychonomic Bulletin & Review, 19,* 151–156.

Frank, M. C., & Saxe, R. (2012). Teaching replication. *Perspectives on Psychological Science, 7,* 600–604.

Franke, R. H. (1980). Worker productivity at Hawthorne. *American Sociological Review, 45,* 1006–1027.

Franke, R. H., & Kaul, J. D. (1978). The Hawthorne experiments: First statistical interpretation. *American Sociological Review, 43,* 623–643.

Freedman, D. (1999). From association to causation: Some remarks on the history of statistics. *Statistical Science, 14,* 243–258.

Freedman, D., Pisani, R., & Purves, R. (1978). *Statistics.* New York, NY: W. W. Norton.

Freeman, J. (1986). Data quality and the development of organizational social science: An editorial essay. *Administrative Science Quarterly, 31,* 298–303.

Freeman, P. R. (1993). The role of *P*-values in analysing trial results. *Statistics in Medicine, 12,* 1443–1452.

Freese, J. (2007). Replication standards for quantitative social science: Why not sociology? *Sociological Methods & Research, 36,* 153–172.

Freund, J. E., & Perles, B. M. (1993). Observations on the definition of p-values. *Teaching Statistics, 15,* 8–9.

Frick, R. W. (1996). The appropriate use of null hypothesis testing. *Psychological Methods, 1,* 379–390.

Friedrich, J., Buday, E., & Kerr, D. (2000). Statistical training in psychology: A national survey and commentary on undergraduate programs. *Teaching of Psychology, 27,* 248–257.

Fuchs, H. M., Jenny, M., & Fiedler, S. (2012). Psychologists are open to change, yet wary of rules. *Perspectives on Psychological Science, 7,* 639–642.

Fuess, S. M. Jr. (1996). On replications in business and economics research: The *QJBE* case. *Quarterly Journal of Business and Economics, 35*(2), 3–13.

Furse, D. H., & Stewart, D. W. (1982). Monetary incentives versus promised contribution to charity: New evidence on mail survey response. *Journal of Marketing Research, 19,* 375–380.

Gage, N. L. (1996). Confronting counsels of despair for the behavioral sciences. *Educational Researcher, 25*(3), 5–15, 22.

Galak, J., LeBoeuf, R. A., Nelson, L. D., & Simmons, J. P. (2012). Correcting the past: Failures to replicate psi. *Journal of Personality and Social Psychology, 103,* 933–948.

Gans, J. S., & Shepherd, G. B. (1994). How are the mighty fallen: Rejected classic articles by leading economists. *Journal of Economic Perspectives, 8*(1), 165–179.

Garda, R. A. (1988). Comment on the AMA Task Force study. *Journal of Marketing, 52*(4), 32–41.

Gasparikova-Krasnec, M., & Ging, N. (1987). Experimental replication and professional cooperation. *American Psychologist, 42,* 266–267.

Gauch, H. G., Jr. (2003). *Scientific method in practice.* Cambridge, UK: Cambridge University Press.

Gelman, A., & Stern, H. (2006). The difference between "significant" and "not significant" is not itself statistically significant. *The American Statistician, 60,* 328–331.

Gendall, P., Hoek, J., & Brennan, M. (1998). The tea bag experiment: More evidence on incentives in mail surveys. *Journal of the Market Research Society, 40,* 347–351.

George, M. L., Rowlands, D., Price, M., & Maxey, J. (2005). *The lean Six Sigma pocket toolbook: A quick reference guide to 100 tools for improving quality and speed.* New York, NY: McGraw-Hill.

Gerber, A. S., & Malhotra, N. (2008). Publication bias in empirical sociological research—Do arbitrary significance levels distort published results? *Sociological Methods & Research, 37,* 3–30.

Gescheider, G. A. (1976). *Psychophysics: Method and theory.* Hillsdale, NJ: Lawrence Erlbaum.

Geuens, M. (2011). Where does business research go from here? Food-for-thought on academic papers in business research. *Journal of Business Research, 64,* 1104–1107.

Ghoshal, S. (2005). Bad management theories are destroying good management practices. *Academy of Management Learning and Education,* 4(1), 75–91.

Gibbons, J. D. (1986). *P*-values. In S. Kotz & N. L. Johnson (Eds.), *Encyclopedia of statistical sciences* (pp. 366–368). New York, NY: Wiley.

Gibbons, J. D., & Pratt, J. W. (1975). *P*-values: Interpretation and methodology. *The American Statistician, 29,* 20–25.

Giere, R. N. (1999). *Science without laws.* Chicago, IL: University of Chicago Press.

Gigerenzer, G. (1993). The superego, the ego, and the id in statistical reasoning. In G. Keren & C. A. Lewis (Eds.), *A handbook for data analysis in the behavioral sciences: Methodological issues* (pp. 311–339). Hillsdale, NJ: Lawrence Erlbaum.

Gigerenzer, G. (2004). Mindless statistics. *Journal of Socio-Economics, 33,* 587–606.

Gigerenzer, G., Krauss, S., & Vitouch, O. (2004). The null ritual: What you always wanted to know about significance testing but were afraid to ask. In David W. Kaplan (Ed.), *The Sage handbook of quantitative methodology for the social sciences* (pp. 391–408). Thousand Oaks, CA: Sage.

Gigernezer, G., & Marewski, J. N. (2014). Surrogate science: The idol of a universal method for scientific inference. *Journal of Management, 41* (2), 421-440.

Gigerenzer, G., & Murray, D. J. (1987). *Cognition as intuitive statistics.* Hillsdale, NJ: Lawrence Erlbaum.

Gigerenzer, G., Swijtink, Z., Porter, T., Daston, L., Beatty, J., & Krüger, L. (1989). *The empire of chance: How probability changed science and everyday life.* Cambridge, UK: Cambridge University Press.

Gilbert, W. (2012). The journal editor's view. *The Psychologist, 25,* 354–355.

Gill, J. (1999). The insignificance of null hypothesis significance testing. *Political Research Quarterly, 52,* 647–674.

Gill, J., & Whittle, S. (1993). Management by panacea: Accounting for transience. *Journal of Management Studies, 30*(2), 281–295.

Gill, T. G. (2010). *Informing business: Research and education on a rugged landscape.* Santa Rosa, CA: Informing Science Press.

Giner-Sorolla, R. (2012). Science or art? How aesthetic standards grease the way through the publication bottleneck but undermine science. *Perspectives on Psychological Science, 7,* 562–571.

Glymour, C. (1980a). Hypothetico-deductivism is hopeless. *Philosophy of Science, 47,* 322–325.

Glymour, C. (1980b). *Theory and evidence.* Princeton, NJ: Princeton University Press.

Golden, M. A. (1995). Replication and non-quantitative research. *PS: Political Science and Politics, 28,* 481–483.

Goldstein, H., & Healy, M. J. R. (1995). The graphical presentation of a collection of means. *Journal of the Royal Statistical Society, Ser. A, 158,* 175–177.

Gollwitzer, P. M. (1999). Implementation intentions: Strong effects of simple plans. *American Psychologist, 54,* 493–503.

Goodman, S. N. (1992). A comment on replication, *p*-values and evidence. *Statistics in Medicine, 11,* 875–879.

Goodman, S. N. (1993). *p* values, hypothesis tests, and likelihood: Implications for epidemiology of a neglected historical debate. *American Journal of Epidemiology, 137,* 485–496.

Goodman, S. N. (1999). Toward evidence-based medical statistics. 1: The *P* value fallacy. *Annals of Internal Medicine, 130,* 995–1004.

Goodman, S. N. (2001). Of *P*-values and Bayes: A modest proposal. *Epidemiology, 12,* 295–297.

Goodman, S. N., & Royall, R. M. (1988). Evidence and scientific research. *American Journal of Public Health, 78,* 1568–1574.

Gordon, H. R. D. (2001). American Vocational Education Research Association members' perceptions of statistical significance tests and other statistical controversies. *Journal of Vocational Education Research, 26*(2), 1–18.

Gordon, R. A., & Howell, J. E. (1959). *Higher education for business.* New York, NY: Columbia University Press.

Gordon, S. (1991). *The history and philosophy of social science.* London, UK: Routledge.

Gorn, G. J. (1982). The effects of music in advertising on choice behavior: A classical conditioning approach. *Journal of Marketing, 46*(1), 94–101.

Gorski, P. S. (2004). The poverty of deductivism: A constructive realist model of sociological explanation. *Sociological Methodology, 34*(1), 1–33.
Gower, B. (1997). *Scientific method: An historical and philosophical introduction.* London, UK: Routledge.
Grant, R. M. (1998). *Contemporary strategy analysis* (3rd ed.). Oxford, UK: Blackwell.
Gratton, C., & Jones, I. (2004). *Research methods for sports studies.* Abingdon, UK: Routledge.
Gray, D. E. (2009). *Doing research in the real world.* Thousand Oaks, CA: Sage.
Green, H. A. J. (1976). *Consumer theory.* London, UK: Macmillan.
Greenwald, A. G. (1975). Consequences of prejudice against the null hypothesis. *Psychological Bulletin, 82,* 1–20.
Greenwald, A. G. (1976). An editorial. *Journal of Personality and Social Psychology, 83,* 1–7.
Greenwald, A. G., Pratkanis, A. R., Leippe, M. R., & Baumgardner, M. H. (1986). Under what conditions does theory obstruct research progress? *Psychological Review, 93,* 216–229.
Gribbin, J. (2004). *The scientists: A history of science told through the lives of its greatest inventors.* New York, NY: Random House.
Guest, D. E. (2007). Don't shoot the messenger: A wake-up call for academics. *Academy of Management Journal, 50,* 1020–1026.
Guildford, J. P. (1942). *Fundamental statistics in psychology and education.* New York, NY: McGraw-Hill.
Gujarati, D. (2003). *Basic econometrics* (4th ed.). New York, NY: McGraw-Hill.
Guttman, L. (1977). What is not what in statistics. *The Statistician, 26,* 81–107.
Guttman, L. (1985). The illogic of statistical inference for cumulative science. *Applied Stochastic Models and Data Analysis, 1,* 3–10.
Haavelmo, T. (1944). The probability approach in econometrics. *Econometrica, 12*(Suppl), iii–iv, 1–115.
Habel, C., & Lockshin, L. (2013). Realizing the value of extensive replication: A theoretically robust portrayal of double jeopardy. *Journal of Business Research, 66,* 1448–1456.
Hackett, E. J. (2005). Essential tensions: Identity, control, and risk in research. *Social Studies of Science, 35,* 787–826.
Hacking, I. (1965). *Logic of statistical inference.* Cambridge, UK: Cambridge University Press.
Hacking, I. (1983). *Representing and intervening.* Cambridge, UK: Cambridge University Press.
Hacking, I. (1990). *The taming of chance.* Cambridge, UK: Cambridge University Press.
Hahn, G. J., & Meeker, W. Q. (1993). Assumptions for statistical inference. *The American Statistician, 47,* 1–11.
Haig, B. D. (2005). An abductive theory of scientific method. *Psychological Methods, 10,* 371–388.
Haig, B. D. (2013a). Detecting psychological phenomena: Taking bottom-up research seriously. *American Journal of Psychology, 126,* 135–153.

Haig, B. D. (2013b). The philosophy of quantitative methods. In T. D. Little (Ed.), *The Oxford handbook of quantitative methods* (Vol. 1, pp. 6–30). Oxford, UK: Oxford Library of Psychology.

Haig, B. D. (2014). *Investigating the psychological world: Scientific method in the behavioral sciences.* Cambridge, MA: MIT Press.

Haire, M. (1950). Projective techniques in marketing research. *Journal of Marketing, 14,* 649–652.

Hall, D. (with Stamp, J.). (2004). *Meaningful marketing: 100 data-proven truths and 402 practical ideas for selling more with less effort.* Covington, KY: Clerisy Press.

Halpin, P. F., & Stam, H. J. (2006). Inductive inference or inductive behavior: Fisher and Neyman–Pearson approaches to statistical testing in psychological research (1940–1960). *American Journal of Psychology, 119,* 625–653.

Hambrick, D. C. (1994). What if the academy actually mattered? *Academy of Management Review, 19,* 11–16.

Hambrick, D. C. (2007). The field of management's devotion to theory: Too much of a good thing? *Academy of Management Journal, 50,* 1346–1352.

Hamilton, D. P. (1990). Publishing by—and for?—the numbers. *Science, 250,* 1331–1332.

Hanssens, D. H. (Ed.). (2009). *Empirical generalizations about marketing impact: What we have learned from academic research.* Cambridge, MA: Marketing Science Institute.

Hardin, G. (1968). The tragedy of the commons. *Science, 162,* 1243–1248.

Harlow, L. L., Mulaik, S. A., Steiger, J. H. (Eds.). (1997). *What if there were no significance tests?* Mahwah, NJ: Lawrence Erlbaum.

Harman, G. H. (1965). Inference to the best explanation. *Philosophical Review, 74,* 88–95.

Harris, C. R., Coburn, N., Rohrer, D., & Pashler, H. (2013). Two failures to replicate high-performance-goal priming effects. *PLoS One, 8*(August), 1–9.

Hartshorne, C., & Weiss, P. (Eds.). (1934). *Collected papers of Charles Sanders Peirce* (Volumes 1–5). Cambridge, MA: Harvard University Press.

Hartshorne, J. K., & Schachner, A. (2012). Tracking replicability as a method of post-publication open evaluation. *Frontiers in Computational Neuroscience, 6,* 1–14.

Harvey, D. (1969). *Explanation in geography.* London, UK: Edward Arnold.

Hauck, W. W., & Anderson, S. (1986). A proposal for interpreting and reporting negative results. *Statistics in Medicine, 5,* 203–209.

Hauser, R. M. (1987). Sharing data: It's time for ASA journals to follow the folkways of a scientific sociology. *American Sociological Review, 52*(6), vi–viii.

Hausman, D. M. (1992). *The inexact and separate science of economics.* Cambridge, UK: Cambridge University Press.

Hayek, F. A. von. (1989, December). The pretence of knowledge: Nobel Memorial Lecture, December 11, 1974. *American Economic Review, 79,* 3–7.

Healey, J. F. (2007). *The essentials of statistics: A tool for social research.* Belmont, CA: Thomson Wadsworth.

Hedges, L. V. (1987). How hard is hard science, how soft is soft science: The empirical cumulativeness of research. *American Psychologist, 42,* 443–455.

Heilprin, J. (2012, July 5). "God particle" gives scientists reasons to celebrate, cheer. *Des Moines Register,* p. 5A.

Helfat, C. E. (2007). Stylized facts, empirical research and theory development in management. *Strategic Organization, 5*(2), 185–192.

Helmig, B., Spraul, K., & Tremp, K. (2012). Replication studies in nonprofit research: A generalization and extension of findings regarding the media publicity of nonprofit organizations. *Nonprofit and Voluntary Sector Quarterly, 41,* 360–385.

Hempel, C. G. (1942). The function of general laws in history. *Journal of Philosophy, 39,* 35–48.

Hempel, C. G. (1965). *Aspects of scientific explanation.* New York, NY: Free Press.

Hendry, D. F. (1980). Econometrics—Alchemy or science? *Economica, 47,* 387–406.

Henrich, J., Norenzayan, A., & Heine, S. J. (2010). The weirdest people in the world. *Behavioral and Brain Sciences, 33,* 61–83.

Henry, G. T. (1990). *Practical sampling.* Newbury Park, CA: Sage.

Herndon, T., Ash, M., & Pollin, R. (2013). *Does high public debt consistently stifle economic growth? A critique of Reinhart and Rogoff* (Working Paper No. 322). Amherst: University of Massachusetts, Amherst, Political Economy Research Institute.

Heyes, S., Hardy, M., Humphreys, P., & Rookes, P. (1993). *Starting statistics in psychology and education: A student handbook.* Oxford, UK: Oxford University Press.

Hicks, J. (1979). *Causality in economics.* Oxford, UK: Blackwell.

Hill, I. D. (1984). Statistical Society of London–Royal Statistical Society. The first 100 years: 1834–1934. *Journal of the Royal Statistical Society, Ser. A, 147,* 130–139.

Hinings, C. R., & Lee, G. L. (1971). Dimensions of organization structure and their context: A replication. *Sociology, 5,* 83–93.

Hinkley, D. V. (1987). Comment. *Journal of the American Statistical Association, 82,* 128–129.

Hoekstra, R., Finch, S., Kiers, H. A. L., & Johnson, A. (2006). Probability as certainty: Dichotomous thinking and the misuse of *p* values. *Psychonomic Bulletin & Review, 13,* 1033–1037.

Hogben, L. (1957). *Statistical theory.* New York, NY: W. W. Norton.

Holcombe, A. O., & Pashler, H. (2012). Making it quick and easy to report replications. *The Psychologist, 25,* 355–356.

Holtzman, P. S. (1986). Similarity and collaboration within the sciences. In D. W. Fiske & R. A. Shweder (Eds.), *Metatheory in social science: Pluralisms and subjectivities* (pp. 347–352). Chicago, IL: University of Chicago Press.

Honig, B., & Bedi, A. (2012). The fox in the hen house: A critical examination of plagiarism among members of the Academy of Management. *Academy of Management Learning & Education, 11,* 101–123.

Honig, B., Lampel, J., Siegel, D., & Drnevich, P. (2013). Ethics in the production and dissemination of management research: Institutional failure or individual fallibility? *Journal of Management Studies, 51,* 118–142.

Hossein-zadeh, I. (2014). *Beyond mainstream explanations of the financial crisis.* London, UK: Routledge.

Hotelling, H. (1927). Review of R. A. Fisher's *Statistical Methods for Research Workers. Journal of the American Statistical Association, 22,* 411–412.

Hotz, R. L. (2007, September 14). Most science studies appear to be tainted by sloppy analysis. *Wall Street Journal,* p. B1.

How science goes wrong. (2013, October 19). *The Economist,* pp. 13, 26–30.

Howard, G. S., Hill, T. L., Maxwell, S. E., Baptista, T. M., Farias, M. H., Coelho, C., . . . Coulter-Kern, R. (2009). What's wrong with research literatures? And how to make them right. *Review of General Psychology, 13*(2), 146–166.

Howell, D. C. (2002). *Statistical methods for psychology* (5th ed.). Belmont, CA: Wadsworth.

Howell, D. C. (2008). *Fundamental statistics for the behavioral sciences.* Belmont, CA: Thomson Wadsworth.

Howie, D. (2002). *Interpreting probability: Controversies and developments in the early twentieth century.* Cambridge, UK: Cambridge University Press.

Hubbard, R. (1990). *An empirical analysis of the publication frequency of replications in the management and organizational behavior literature.* Unpublished manuscript.

Hubbard, R. (1992). Comment on "The factors influencing academic research productivity: A survey of management scientists." *Interfaces, 22*(5), 33–35.

Hubbard, R. (1994). The dangers of generalizing from published marketing studies. *Journal of the Market Research Society, 36,* 257–260.

Hubbard, R. (1995a). Comment on "Model Uncertainty, Data Mining and Statistical Inference." *Journal of the Royal Statistical Society, Ser. A, 158,* 458.

Hubbard, R. (1995b). The commodification of marketing knowledge: It's not enough to count the numbers. *Journal of Marketing Management, 11,* 671–673.

Hubbard, R. (1995c). The earth is highly significantly round (p < .0001). *American Psychologist, 50,* 1098.

Hubbard, R. (2004). Alphabet soup: Blurring the distinctions between *p*'s and α's in psychological research. *Theory & Psychology, 14,* 295–327.

Hubbard, R. (2011a). P-values. In M. Lovric (Ed.), *The international encyclopedia of statistical science* (pp. 1144–1145). Berlin, Germany: Springer-Verlag.

Hubbard, R. (2011b). The widespread misinterpretation of *p*-values as error probabilities. *Journal of Applied Statistics, 38,* 2617–2626.

Hubbard, R., & Armstrong, J. S. (1991). *Replication in marketing* (Working Paper No. 90-008R). Philadelphia: University of Pennsylvania, Wharton School.

Hubbard, R., & Armstrong, J. S. (1992). Are null results becoming an endangered species in marketing? *Marketing Letters, 3,* 127–136.

Hubbard, R., & Armstrong, J. S. (1994). Replications and extensions in marketing: Rarely published but quite contrary. *International Journal of Research in Marketing, 11,* 233–248.

Hubbard, R., & Armstrong, J. S. (1997). Publication bias against null results. *Psychological Reports, 80,* 337–338.

Hubbard, R., & Armstrong, J. S. (2006). Why we don't really know what *statistical significance* means: Implications for educators. *Journal of Marketing Education,*

28(2), 114–120. (Reprinted in *Fundamentals of regression modeling,* Vol. 1, pp. 75–86, by S. Babones, Ed., 2013, London, UK: Sage)

Hubbard, R., & Bayarri, M. J. (2003). Confusion over measures of evidence (*p*'s) versus errors (α's) in classical statistical testing (with discussion). *The American Statistician, 57,* 171–182. (Reprinted in *Fundamentals of Regression Modeling,* Vol. 1, pp. 57–74, by S. Babones, Ed., 2013, London, UK: Sage)

Hubbard, R., & Bayarri, M. J. (2005). Comment on "Testing Fisher, Neyman, Pearson, and Bayes." *The American Statistician, 59,* 353.

Hubbard, R., Brodie, R. J., & Armstrong, J. S. (1992). Knowledge development in marketing: The role of replication research. *New Zealand Journal of Business, 14,* 1–12.

Hubbard, R., & Lindsay, R. M. (1995). *Caveat emptor* applies to the consumption of published empirical research results, too. *Management Research News, 18*(10/11), 49–55.

Hubbard, R., & Lindsay, R. M. (2002). How the emphasis on "original" empirical marketing research impedes knowledge development. *Marketing Theory, 2,* 381–402.

Hubbard, R., & Lindsay, R. M. (2008). Why *P* values are not a useful measure of evidence in statistical significance testing. *Theory & Psychology, 18,* 69–88.

Hubbard, R., & Lindsay, R. M. (2013a). From significant difference to significant sameness: Proposing a paradigm shift in business research. *Journal of Business Research, 66,* 1377–1388.

Hubbard, R., & Lindsay, R. M. (2013b). The significant difference paradigm promotes bad science. *Journal of Business Research, 66,* 1393–1397.

Hubbard, R., & Little, E. L. (1988). Promised contributions to charity and mail survey responses: Replication with extension. *Public Opinion Quarterly, 52,* 223–230.

Hubbard, R., & Little, E. L. (1997). Share and share alike? A review of empirical evidence concerning information sharing among researchers. *Management Research News, 20,* 41–49.

Hubbard, R., & Meyer, C. K. (2013). The rise of statistical significance testing in public administration research and why this is a mistake. *Journal of Business and Behavioral Sciences, 25*(1), 4–20.

Hubbard, R., & Norman, A. T. (2007). What impact has practitioner research had in the marketing academy? *Management Research News, 30*(1), 25–33.

Hubbard, R., Norman, A. T., & Parsa, R. A. (2010). Marketing's "Oscars": A citation analysis of award-winning articles. *Marketing Intelligence and Planning, 28,* 669–684.

Hubbard, R., Parsa, R. A., & Luthy, M. R. (1997). The spread of statistical significance testing in psychology: The case of the *Journal of Applied Psychology,* 1917–1994. *Theory & Psychology, 7,* 545–554.

Hubbard, R., & Ryan, P. A. (2000). The historical growth of statistical significance testing in psychology—and its future prospects (with discussion). *Educational and Psychological Measurement, 60,* 661–696.

Hubbard, R., & Vetter, D. E. (1991). Replications in the finance literature: An empirical study. *Quarterly Journal of Business and Economics, 30*(4), 70–81.

Hubbard, R., & Vetter, D. E. (1992). The publication incidence of replications and critical commentary in economics. *The American Economist, 36*(1), 29–34.

Hubbard, R., & Vetter, D. E. (1996). An empirical comparison of published replication research in accounting, economics, finance, management, and marketing. *Journal of Business Research, 35,* 153–164.

Hubbard, R., & Vetter, D. E. (1997). Journal prestige and the publication frequency of replication research in the finance literature. *Quarterly Journal of Business and Economics, 36*(4), 3–14.

Hubbard, R., Vetter, D. E., & Little, E. L. (1998). Replication in strategic management: Scientific testing for validity, generalizability, and usefulness," *Strategic Management Journal, 19,* 243–254.

Huber, J. (2007). *Journal of Marketing Research* in the new competitive journalistic environment. *Journal of Marketing Research, 44,* 1–3.

Huberty, C. J. (1993). Historical origins of statistical testing practices: The treatment of Fisher versus Neyman–Pearson views in textbooks. *Journal of Experimental Education, 61,* 317–333.

Huberty, C. J., & Lowman, L. L. (2000). Group overlap as a basis for effect size. *Educational and Psychological Measurement, 60,* 543–563.

Huberty, C. J., & Pike, C. J. (1999). On some history regarding statistical testing. In B. Thompson (Ed.), *Advances in social science methodology* (pp. 1–22). Stamford, CT: JAI Press.

Hull, D. L. (1988). *Science as a process: An evolutionary account of the social and conceptual development of science.* Chicago, IL: University of Chicago Press.

Humphreys, L. G. (1980). The statistics of failure to replicate: A comment on Buriel's (1978) conclusions. *Journal of Educational Psychology, 72,* 71–75.

Hung, H. M. J., O'Neill, R. T., Bauer, P., & Köhne, K. (1997). The behavior of the *p*-value when the alternative hypothesis is true. *Biometrics, 53*(1), 11–22.

Hunt, E. (2013). Calls for replicability must go beyond motherhood and apple pie. *European Journal of Personality, 27,* 126–127.

Hunt, S. D. (1991). *Modern marketing theory: Critical issues in the philosophy of marketing science.* Cincinnati, OH: South-Western.

Hunt, S. D. (2003). *Controversy in marketing theory: For reason, realism, truth, and objectivity.* Armonk, NY: M. E. Sharpe.

Hunter, J. E. (1997). Needed: A ban on the significance test. *Psychological Science, 8,* 3–7.

Hunter, J. E. (2001). The desperate need for replications. *Journal of Consumer Research, 28,* 149–158.

Iacobucci, D. (2005). From the editor. *Journal of Consumer Research, 32,* 6–11.

Iacobucci, D., & Churchill, G. A., Jr. (2010). *Marketing research: Methodological foundations* (10th ed.). Mason, OH: South-Western.

Ijzerman, H., Brandt, M. J., & van Wolferen, J. (2013). Rejoice! In replication. *European Journal of Personality, 27,* 128–129.

Imrey, P. B. (1994). Statistical values, quality, and certification. *The American Statistician, 48*, 65–70.

Inman, H. F. (1994). Karl Pearson and R. A. Fisher on statistical tests: A 1935 exchange in *Nature. The American Statistician, 48*, 2–11.

Ioannidis, J. P. A. (2005a). Contradicted and initially stronger effects in highly cited clinical research. *Journal of the American Medical Association, 294*, 218–228.

Ioannidis, J. P. A. (2005b). Why most published research findings are false. *PLoS Medicine, 2*, 0101–0106.

Ioannidis, J. P. A. (2011). An epidemic of false claims. *Scientific American, 304*(6), 16.

Ioannidis, J. P. A. (2012). Why science is not necessarily self-correcting. *Perspectives on Psychological Science, 7*, 645–654.

Ioannidis, J. P. A., & Doucouliagos, C. (2013). What's to know about the credibility of empirical economics? *Journal of Economic Surveys, 27*, 997–1004.

Ioannidis, J. P. A., & Trikalinos, T. A. (2007). An exploratory test for an excess of significant findings. *Clinical Trials, 4*, 245–253.

Iyengar, S. (1991). Much ado about meta-analysis. *Chance: New Directions for Statistics and Computing, 4*(1), 33–40.

Jacobs, R. N., & Glass, D. J. (2002). Media publicity and the voluntary sector: The case of nonprofit organizations in New York City. *Voluntas: International Journal of Voluntary and Nonprofit Organizations, 13*, 235–252.

Jasny, B. R., Chin, G., Chong, L., & Vignieri, S. (2011). Again, and again, and again . . . *Science, 334*, 1225.

Jaworski, B. J. (2011). On managerial relevance. *Journal of Marketing, 75*, 211–224.

Jaworski, B. J., & Kohli, A. K. (1993). Market orientation: Antecedents and consequences. *Journal of Marketing, 57*(3), 53–70.

Jeffreys, H. (1939). *Theory of probability.* Oxford, UK: Clarendon.

Jiambalvo, J. (1982). Measures of accuracy and congruence in the performance evaluation of CPA personnel: Replication and extensions. *Journal of Accounting Research, 20*, 152–161.

Johansson, T. (2011). Hail the impossible: *P*-values, evidence, and likelihood. *Scandinavian Journal of Psychology, 52*, 113–125.

John, L. K., Loewenstein, G., & Prelec, D. (2012). Measuring the prevalence of questionable research practices with incentives for truth-telling. *Psychological Science, 23*, 524–532.

Johnstone, D. J. (1986). Tests of significance in theory and practice (with discussion). *The Statistician, 35*, 491–504.

Jones, L. V., & Tukey, J. W. (2000). A sensible formulation of the significance test. *Psychological Methods, 5*, 411–414.

Jones, S. R. G. (1992). Was there a Hawthorne effect? *American Journal of Sociology, 98*, 451–468.

Jones, T. C., & Dugdale, D. (2002). The ABC bandwagon and the juggernaut of modernity. *Accounting, Organizations and Society, 27*, 121–163.

Judson, H. F. (2004). *The great betrayal: Fraud in science.* New York, NY: Harcourt.

Kahneman, D. (2011). *Thinking, fast and slow.* New York, NY: Farrar, Straus and Giroux.

Kalbfleisch, J. G., & Sprott, D. A. (1976). On tests of significance. In W. L. Harper & C. A. Hooker (Eds.), *Foundations of probability theory, statistical inference, and statistical theories of science* (Vol. II, pp. 259–272). Dordrecht, Netherlands: Reidel.

Kane, E. J. (1984). Why journal editors should encourage the replication of applied econometric research. *Quarterly Journal of Business and Economics, 23*(1), 3–8.

Kaplan, A. (1964). *The conduct of inquiry: Methodology for behavioral science.* San Francisco, CA: Chandler.

Kasanen, E., Lukka, K., & Siitonen, A. (1993). The constructive approach in management accounting research. *Journal of Management Accounting Research, 5,* 243–264.

Kass, R. E. (2011). Statistical inference: The big picture. *Statistical Science, 26,* 1–9.

Kasulis, J. J., Lusch, R. F., & Stafford, E. F., Jr. (1979). Consumer acquisition patterns for durable goods. *Journal of Consumer Research, 6,* 47–57.

Keen, S. (2001). *Debunking economics: The naked emperor of the social sciences.* London, UK: Zed Books.

Kellaris, J. J., & Cox, A. D. (1989). The effect of background music in advertising: A reassessment. *Journal of Consumer Research, 16,* 113–118.

Kelley, L. P., & Blashfield, R. K. (2009). An example of psychological science's failure to self-correct. *Review of General Psychology, 13*(2), 122–129.

Kendall, M. G. (1943). *The advanced theory of statistics* (Vol. 1). New York, NY: Lippincott.

Kendall, M. G. (1961). Natural law in the social sciences. *Journal of the Royal Statistical Society, Ser. A, 124*(1), 1–16.

Kennedy, P. (2008). *A guide to econometrics* (6th ed.). Oxford, UK: Blackwell.

Kent, D., & Hayward, R. (2007). When averages hide individual differences in clinical trials. *American Scientist, 95,* 60–68.

Kent, R. (2007). *Marketing research: Approaches, methods and applications in Europe.* London, UK: Thomson.

Kerin, R. A. (1996). In pursuit of an ideal: The editorial and literary history of the *Journal of Marketing. Journal of Marketing, 60*(1), 1–13.

Kerin, R. A., Hartley, S. W., & Rudelius, W. (2013). *Marketing* (11th ed.). New York, NY: McGraw-Hill.

Kerin, R. A., & Sethuraman, R. (1999). "Revisiting marketing's lawlike generalizations": A comment. *Journal of the Academy of Marketing Science, 27,* 101–104.

Kerlinger, F. N. (1970). A social attitude scale: Evidence on reliability and validity. *Psychological Reports, 26,* 379–383.

Kerlinger, F. N. (1973). *Foundations of behavioral research* (2nd ed.). New York, NY: Holt, Rinehart and Winston.

Kerr, N. L. (1998). HARKing: Hypothesizing after the results are known. *Personality and Social Psychology Review, 2*(3), 196–217.

Kerr, S., Tolliver, J., & Petree, D. (1977). Manuscript characteristics which influence acceptance for management and social science journals. *Academy of Management Journal, 20,* 132–141.

Ketokivi, M., & Mantere, S. (2010). Two strategies for inductive reasoning in organizational research. *Academy of Management Review, 35,* 315–333.

Keuzenkamp, H. A. (2000). *Probability, econometrics and truth: The methodology of econometrics.* Cambridge, UK: Cambridge University Press.

Keuzenkamp, H. A., & Magnus, J. R. (1995). On tests and significance in econometrics. *Journal of Econometrics, 67,* 5–24.

Keuzenkamp, H. A., & McAleer, M. (1995). Simplicity, scientific inference and econometric modelling. *Economic Journal, 105*(1), 1–21.

Khurana, R. (2007). *From higher aims to hired hands: The social transformation of American business schools and the unfulfilled promise of management as a profession.* Princeton, NJ: Princeton University Press.

Kieffer, K. M., Reese, R. J., & Thompson, B. (2001). Statistical techniques employed in *AERJ* and *JCP* articles from 1988 to 1997: A methodological review. *Journal of Experimental Education, 69,* 280–310.

Kilpatrick, S. J. (1992). Environmental tobacco smoke and lung cancer: A critique of a meta-analysis. *The Statistician, 41,* 331.

Kincaid, H. (1996). *Philosophical foundations of the social sciences: Analyzing controversies in social research.* Cambridge, UK: Cambridge University Press.

King, G. (1995). Replication, replication. *PS: Political Science and Politics, 28,* 444–452.

Kingman, J. F. C. (1989). Statistical responsibility. *Journal of the Royal Statistical Society, Ser. A, 152,* 277–285.

Kinnear, T. C. (1992). From the editor. *Journal of Marketing, 56,* 1–3.

Kinney, M. R., & Wempe, W. F. (2002). Further evidence on the extent and origin of JIT's profitability effects. *Accounting Review, 77,* 203–225.

Kirchner, W. K., & Dunnette, M. D. (1954). Attitudes toward older workers. *Personnel Psychology, 7,* 257–265.

Kirk, R. E. (1996). Practical significance: A concept whose time has come. *Educational and Psychological Measurement, 56,* 746–759.

Kirk, R. E. (2001). Promoting good statistical practices: Some suggestions. *Educational and Psychological Measurement, 61,* 213–218.

Kirk, R. E. (2003). The importance of effect magnitude. In S. F. Davis (Ed.), *Handbook of research methods in experimental psychology* (pp. 83–105). Oxford, UK: Blackwell.

Kline, R. B. (2004). *Beyond significance testing: Reforming data analysis methods in behavioral research.* Washington, DC: American Psychological Association.

Kmenta, J. (1971). *Elements of econometrics.* New York, NY: Macmillan.

Kmetz, J. L. (1998). *The information processing theory of organization.* Aldershot, UK: Ashgate.

Kmetz, J. L. (2011). Fifty lost years: Why international business scholars must not emulate the US social-science research model. *World Journal of Management, 3*(2), 172–200.

Koenker, R., & Zeileis, A. (2009). On reproducible econometric research. *Journal of Applied Econometrics, 24,* 833–847.

Kohli, A. K., & Jaworski, B. J. (1990). Market orientation: The construct, research propositions, and managerial implications. *Journal of Marketing, 54*(2), 1–18.

Koole, S. L., & Lakens, D. (2012). Rewarding replications: A sure and simple way to improve psychological science. *Perspectives on Psychological Science, 7,* 608–614.

Krämer, W., & Gigerenzer, G. (2005). How to confuse with statistics or: The use and misuse of conditional probabilities. *Statistical Science, 20,* 223–230.

Krantz, D. H. (1999). The null hypothesis testing controversy in psychology. *Journal of the American Statistical Association, 94,* 1372–1381.

Krebs, R. E. (2001). *Scientific laws, principles, and theories: A reference guide.* Westport, CT: Greenwood Press.

Kroll, R. M., & Chase, L. J. (1975). Communication disorders: A power analytic assessment of recent research. *Journal of Communication Disorders, 8,* 237–247.

Krueger, J. (2001). Null hypothesis significance testing: On the survival of a flawed method. *American Psychologist, 56,* 16–26.

Kruskal, W. H. (1978). Tests of significance. In W. H. Kruskal & J. M. Tanur (Eds.), *International encyclopedia of statistics* (pp. 944–958). New York, NY: Free Press.

Kruskal, W., & Majors, R. (1989). Concepts of relative importance in recent scientific literature. *The American Statistician, 43,* 2–6.

Kuhn, T. S. (1970). *The structure of scientific revolutions* (2nd ed.). Chicago, IL: University of Chicago Press.

Kupfersmid, J., & Fiala, M. (1991). A survey of attitudes and behaviors of authors who publish in psychology and education journals. *American Psychologist, 46,* 249–250.

Laband, D. N. (1986). Article popularity. *Economic Inquiry, 24,* 173–180.

Laband, D. N. (1990). Is there value-added from the review process in economics? Preliminary evidence from authors. *Quarterly Journal of Economics, 105,* 341–352.

Ladyman, J. (2002). *Understanding philosophy of science.* London, UK: Routledge.

Lakatos, I. (1970). Falsification and the methodology of research programmes. In I. Lakatos & A. Musgrave (Eds.), *Criticism and the growth of knowledge* (pp. 91–196). Cambridge, UK: Cambridge University Press.

Lamal, P. A. (1990). On the importance of replication. *Journal of Social Behavior and Personality, 5*(4), 31–35.

Lambdin, C. (2012). Significance tests as sorcery: Science is empirical—Significance tests are not. *Theory & Psychology,* 22, 67–90.

Lane, D. A. (1980). Fisher, Jeffreys, and the nature of probability. In S. E. Feinberg & D. V. Hinkley (Eds.), *R. A. Fisher: An appreciation* (pp. 148–160). New York, NY: Springer–Verlag.

Lane, D. M., & Dunlap, W. P. (1978). Estimating effect size: Bias resulting from the significance criterion in editorial decisions. *British Journal of Mathematical and Statistical Psychology, 31*(2), 107–112.

Lane, G. S., & Watson, G. L. (1975). A Canadian replication of Mason Haire's "shopping list" study. *Journal of the Academy of Marketing Science, 3,* 48–59.

Lang, L. H. P., Stultz, R. M., & Walking, R. A. (1989). Managerial performance, Tobin's *q* and the gains from successful tender offers. *Journal of Financial Economics, 24*, 137–154.

Latham, G. P., Erez, M., & Locke, E. A. (1988). Resolving scientific disputes by the joint design of crucial experiments by the antagonists: Application to the Erez–Latham dispute regarding participation in goal setting. *Journal of Applied Psychology, 73*, 753–772.

Laurent, G. (2000). Improving the external validity of marketing models: A plea for more qualitative input. *International Journal of Research in Marketing, 17*, 177–182.

Lawson, T. (1997). *Economics and reality.* London, UK: Routledge.

Lawson, T. (2003). *Reorienting economics.* London, UK: Routledge.

Lawson, T. (2009). The current economic crisis: Its nature and the course of academic economics. *Cambridge Journal of Economics, 33*, 759–777.

Leahey, E. (2005). Alphas and asterisks: The development of statistical significance testing standards in sociology. *Social Forces, 84*(1), 1–24.

Leamer, E. E. (1978). *Specification searches: Ad hoc inference with nonexperimental data.* New York, NY: Wiley.

Leamer, E. E. (1983). Let's take the con out of econometrics. *American Economic Review, 73*(1), 31–43.

LeBel, E. P., & Peters, K. R. (2011). Fearing the future of empirical psychology: Bem's (2011) evidence of psi as a case study of deficiencies in modal research practice. *Review of General Psychology, 15*, 371–379.

Lecoutre, M.-P., Poitevineau, J., & Lecoutre, B. (2003). Even statisticians are not immune to misrepresentations of null hypothesis significance tests. *International Journal of Psychology, 38*(1), 37–45.

Lee, A. S., & Baskerville, R. L. (2003). Generalizing generalizability in information systems research. *Information Systems Research, 14*, 221–243.

Lehmann, D. R., McAlister, L., & Staelin, R. (2011). Sophistication in research in marketing. *Journal of Marketing, 75*(4), 155–165.

Lehmann, E. L. (1993). The Fisher, Neyman–Pearson theories of testing hypotheses: One theory or two? *Journal of the American Statistical Association, 88*, 1242–1249.

Lenhard, J. (2006). Models and statistical inference: The controversy between Fisher and Neyman–Pearson. *British Journal for the Philosophy of Science, 57*, 69–91.

Leone, R. P., & Schultz, R. L. (1980). A study of marketing generalizations. *Journal of Marketing, 44*(1), 10–18.

Leung, K. (2011). Presenting post hoc hypotheses as a priori: Ethical and theoretical issues. *Management and Organization Review, 7*, 471–479.

Levine, D. M., & Stephan, D. F. (2010). *Even you can learn statistics: A guide for everyone who has ever been afraid of statistics* (2nd ed.). Upper Saddle River, NJ: FT Press.

Leviton, L. C. (2001). External validity. In N. J. Smelser & P. B. Baltes (Eds.), *International encyclopedia of the social and behaviour sciences* (Vol. 8, pp. 5195–5200). Oxford, UK: Elsevier.

Levitt, T. (1960, July-August). Marketing myopia. *Harvard Business Review,* 45–56.

Levy, J. (2010). *Scientific feuds.* London, UK: New Holland.

Lindley, D. V. (1957). A statistical paradox. *Biometrika, 44,* 187–192.

Lindley, D. V. (1999). Comment on Bayarri and Berger. In J. M. Bernardo, J. O. Berger, A. P. Dawid, & A. F. M. Smith (Eds.), *Bayesian statistics* (Vol. 6, p. 75). Oxford, UK: Clarendon.

Lindquist, E. F. (1940). *Statistical analysis in educational research.* Boston, MA: Houghton Mifflin.

Lindsay, R. M. (1993a). Achieving scientific knowledge: The rationality of scientific method. In M. Mumford & K. V. Peasnell (Eds.), *Philosophical perspectives on accounting: Essays in honour of Edward Stamp* (pp. 221–254). London, UK: Routledge.

Lindsay, R. M. (1993b). Incorporating statistical power into the test of significance procedure: A methodological and empirical inquiry. *Behavioral Research in Accounting, 5,* 211–236.

Lindsay, R. M. (1994). Publication system biases associated with the statistical testing paradigm. *Contemporary Accounting Research, 11*(1), 33–57.

Lindsay, R. M. (1995). Reconsidering the status of tests of significance: An alternative criterion of adequacy. *Accounting, Organizations and Society, 20,* 35–53.

Lindsay, R. M. (1997). Lies, damned lies, and *more* statistics: The neglected issue of multiplicity in accounting research. *Accounting and Business Research, 27,* 243–258.

Lindsay, R., M., & Ehrenberg, A. S. C. (1993). The design of replicated studies. *The American Statistician, 47,* 217–228.

Lindsay, R. M., & Hubbard, R. (2011). *Redefining the game of research in management accounting: Thinking of case studies in a largely unexplained way following from a critical realist perspective.* Working Paper, University of Lethbridge, Canada.

Lipsey, M. W., & Wilson, D. B. (1993). The efficacy of psychological, educational, and behavioral treatment: Confirmation from meta-analysis. *American Psychologist, 48,* 1181–1209.

Lipton, P. (2004). *Inference to the best explanation* (2nd ed.). London, UK: Routledge.

List, J. A., Bailey, C. D., Euzent, P. J., & Martin, T. L. (2001). Academic economists behaving badly? A survey on three areas of unethical behavior. *Economic Inquiry, 39,* 162–170.

Little, D. (1993). On the scope and limits of generalizations in the social sciences. *Synthese, 97,* 183–207.

Livio, M. (2013). *Brilliant blunders from Darwin to Einstein: Colossal mistakes by great scientists that changed our understanding of life and the universe.* New York, NY: Simon & Schuster.

Locke, E. A. (1986). *Generalizing from laboratory to field settings.* Lexington, MA: Lexington Books.

Locke, E. A. (2007). The case for inductive theory building. *Journal of Management, 33,* 867–890.

Locke, E. A., & Latham, G. P. (1990). *A theory of goal setting and task performance.* Englewood Cliffs, NJ: Prentice Hall.

Locke, E. A., & Latham, G. P. (2002). Building a practically useful theory of goal setting and task motivation. *American Psychologist, 57,* 705–717.

Loftus, G. R. (1993). Editorial comment. *Memory & Cognition, 21*(1), 1–3.

Loftus, G. R. (1996). Psychology will be a much better science when we change the way we analyze data. *Psychological Science, 5*(6), 161–171.

Losee, J. (2001). *A historical introduction to the philosophy of science* (4th ed.). Oxford, UK: Oxford University Press.

Lovell, M. C. (1983). Data mining. *Review of Economics and Statistics, 65*(1), 1–12.

Lovie, A. D. (1979). The analysis of variance in experimental psychology: 1934–1945. *British Journal of Mathematical and Statistical Psychology, 32*(2), 151–178.

Lucas, R. E. (1972). Expectations and the neutrality of money. *Journal of Economic Theory, 4,* 103–124.

Lykken, D. T. (1968). Statistical significance in psychological research. *Psychological Bulletin, 70,* 151–159.

Lynch, J. G., Jr. (1999). Theory and external validity. *Journal of the Academy of Marketing Science, 27,* 367–376.

Lynch, J. G., Jr., Alba, J. W., Krishna, A., Morwitz, V. G., & Gürhan-Canli, Z. (2012). Knowledge creation in consumer research: Multiple routes, multiple criteria. *Journal of Consumer Psychology, 22,* 473–485.

Macdonald, S., & Kam, J. (2007). Ring a ring o' roses: Quality journals and gamesmanship in management studies. *Journal of Management Studies, 44,* 640–655.

Macdonald, S., & Simpson, M. (2001). Learning from management consultants: The lesson for management researchers. *Prometheus, 19,* 117–133.

MacInnis, D. J. (2011). A framework for conceptual contributions in marketing. *Journal of Marketing, 75*(4), 136–154.

Maddala, G. S. (1983). *Limited dependent and qualitative variables in econometrics.* Cambridge, UK: Cambridge University Press.

Madden, C. S., Easley, R. W., & Dunn, M. G. (1995). How journal editors view replication research. *Journal of Advertising, 24*(4), 77–87.

Madden, C. S., Franz, L. S., & Mittelstaedt, R. A. (1979). The replicability of research in marketing: Reported content and author cooperation. In O. C. Ferrell, S. W. Brown, & C. W. Lamb (Eds.), *Conceptual and theoretical developments in marketing* (pp. 76–85). Chicago, IL: American Marketing Association.

Madrick, J. (2014). *Seven bad ideas: How mainstream economists have damaged America and the world.* New York, NY: Alfred A. Knopf.

Magnani, L. (2001). *Abduction, reason, and science: Processes of discovery and explanation.* Dordrecht, Netherlands: Kluwer.

Maher, M. W., Ramanathan, K. V., & Peterson, R. B. (1979). Preference congruence, information accuracy, and employee performance: A field study. *Journal of Accounting Research, 17,* 476–503.

Maher, P. (1988). Prediction, accommodation, and the logic of discovery. In M. Martin & L. C. McIntyre (Eds.), *Readings in the philosophy of social science* (pp. 5–19). Cambridge, MA: MIT Press.

Mahoney, M. J. (1985). Open exchange and epistemic progress. *American Psychologist, 40,* 29–39.

Makel, M. C., & Plucker, J. A. (2014). Facts are more important than novelty: Replication in the education sciences. *Educational Researcher, 43*, 304–316.

Makel, M. C., Plucker, J. A., & Hegarty, B. (2012). Replications in psychology research: How often do they really occur? *Perspectives on Psychological Science, 7*, 537–542.

Manicas, P. T., & Secord, P. F. (1983). Implications for psychology of the new philosophy of science. *American Psychologist, 38*, 399–413.

Mankiw, N. G., Miron, J. A., & Weil, D. N. (1987). The adjustment of expectations to a change in regime: A study of the founding of the federal reserve. *American Economic Review, 77*, 358–374.

Marden, J. I. (2000). Hypothesis testing: From *p* values to Bayes factors. *Journal of the American Statistical Association, 95*, 1316–1320.

Margenau, H. (1961). *Open vistas.* New Haven, CT: Yale University Press.

Martin, R. (2012). The price of actionability. *Academy of Management Learning & Education, 11*, 293–299.

Martin, S. A. (2011). Not as the crow flies: "Styles" of educational measurement in the reception of inferential statistics at Iowa and Minnesota. *History of Science, 49*, 187–215.

Martinson, B. C., Anderson, M. A., & de Vries, R. (2005). Scientists behaving badly. *Nature, 435*, 737–738.

Masicampo, E. J., & Lalande, D. R. (2012). A peculiar prevalence of *p* values just below .05," *Quarterly Journal of Experimental Psychology, 65*, 2271–2279.

Maxwell, J. A. (1992). Understanding and validity in qualitative research. *Harvard Educational Review, 62*, 279–300.

Maxwell, S. E. (2004). The persistence of underpowered studies in psychological research: Causes, consequences, and remedies. *Psychological Methods, 9*, 147–163.

Mayer, T. (1980). Economics as a hard science: Realistic goal or wishful thinking? *Economic Inquiry, 18*, 165–178.

Mayer, T. (1993). *Truth versus precision in economics.* Aldershot, UK: Edward Elgar.

Mayo, D. G. (1996). *Error and the growth of experimental knowledge.* Chicago, IL: University of Chicago Press.

Mayo, D. G. (2013). Discussion: Bayesian methods: Applied? Yes. Philosophical defense? In flux. *The American Statistician, 67*, 11–15.

Mazen, A. M., Hemmasi, M., & Lewis, M. F. (1987). Assessment of statistical power in contemporary strategy research. *Strategic Management Journal, 8*, 403–410.

Mazen, A. M., Kellog, C. E., & Hemmasi, M. (1987). Statistical power in contemporary management research. *Academy of Management Journal, 30*, 369–380.

McCabe, G. (1984). Editor's note. *Quarterly Journal of Business and Economics, 23*(1), 79.

McCloskey, D. (2002). *The secret sins of economics.* Chicago, IL: Prickly Paradigm Press.

McCloskey, D. N., & Ziliak, S. T. (1996). The standard error of regressions. *Journal of Economic Literature, 34*, 97–114.

McCullough, B. D. (2007). Got replicability? The *Journal of Money, Credit and Banking* archive. *Econ Journal Watch, 4*, 326–337.

McCullough, B. D., McGeary, K. A., & Harrison, T. D. (2006). Lessons from the *JMCB* archive. *Journal of Money, Credit and Banking, 38,* 1093–1107.

McGrath, R. G. (2007). No longer a stepchild: How the management field can come into its own. *Academy of Management Journal, 50,* 1365–1378.

McGrayne, S. B. (2011). *The theory that would not die: How Bayes' rule cracked the enigma code, hunted down Russian submarines, and emerged triumphant from two centuries of controversy.* New Haven, CT: Yale University Press.

McGuire, W. J. (1983). A contextualist theory of knowledge: Its implications for innovation and reform in psychological research. *Advances in Experimental Social Psychology, 16,* 2–47.

McHugh, R. K., & Barlow, D. H. (2010). The dissemination and implementation of evidence-based psychological treatments: A review of current efforts. *American Psychologist, 65,* 73–84.

McKelvey, B. (1978). Organizational systematics: Taxonomic lessons from biology. *Management Science, 24,* 1428–1440.

McKenzie, C. J., Wright, S., Ball, D. F., & Baron, P. J. (2002). The publications of marketing faculty: Who are we really talking to? *European Journal of Marketing, 36,* 1196–1208.

McLeod, B. D., & Weisz, J. R. (2004). Using dissertations to examine potential bias in child and adolescent clinical trials. *Journal of Consulting and Clinical Psychology, 72,* 235–251.

McNatt, D. B., & Judge, T. A. (2004). Boundary conditions of the Galatea effect: A field experiment and constructive replication. *Academy of Management Journal, 47,* 550–565.

McPhee, W. N. (1963). *Formal theories of mass behavior.* New York, NY: Free Press.

McQuarrie, E. F. (2004). Integration of construct and external validity by means of proximal similarity: Implications for laboratory experiments in marketing. *Journal of Business Research, 57,* 142–153.

Meehl, P. E. (1967). Theory-testing in psychology and physics: A methodological paradox. *Philosophy of Science, 34,* 103–115.

Meehl, P. E. (1978). Theoretical risks and tabular asterisks: Sir Karl, Sir Ronald, and the slow progress of soft psychology. *Journal of Consulting and Clinical Psychology, 46,* 806–834.

Meehl, P. E. (1990). Why summaries of research on psychological theories are often uninterpretable. *Psychological Reports, 66,* 195–244.

Meehl, P. E. (1997). The problem is epistemology, not statistics: Replace significance tests by confidence intervals and quantify accuracy of risky numerical predictions. In L. L. Harlow, S. A. Mulaik, & J. H. Steiger (Eds.), *What if there were no significance tests?* (pp. 393–425). Mahwah, NJ: Lawrence Erlbaum.

Mellers, B., Hertwig, R., & Kahneman, D. (2001). Do frequency representations eliminate conjunction effects? An exercise in adversarial collaboration. *Psychological Science, 12,* 269–275.

Metaphysicians: Combating bad science. (2014, March 15). *The Economist,* p. 74.

Mezias, S. J., & Regnier, M. O. (2007). Walking the walk as well as talking the talk: Replication and the normal science paradigm in strategic management research. *Strategic Organization, 5,* 283–296.

Michaels, R. E., & Day, R. L. (1985). Measuring customer orientation of salespeople: A replication with industrial buyers. *Journal of Marketing Research, 22,* 43–446.

Mick, D. G. (2001). From the editor. *Journal of Consumer Research, 28*(1), iii–vii.

Micklethwait, J., & Wooldridge, A. (1996), *The witch doctors: Making sense of management gurus.* New York, NY: Time Books.

Miles, J., & Banyard, P. (2007). *Understanding and using statistics in psychology: An introduction.* Thousand Oaks, CA: Sage.

Miller, D. (1993). The architecture of simplicity. *Academy of Management Review, 18,* 116–138.

Miller, J. (2009). What is the probability of replicating a statistically significant effect? *Psychonomic Bulletin & Review, 16,* 617–640.

Miner, J. B. (2003). The rated importance, scientific validity and practical usefulness of organizational behavior theories: A quantitative review. *Academy of Management Learning and Education, 2,* 250–267.

Mirowski, P., & Sklivas, S. (1991). Why econometricians don't replicate (although they do reproduce). *Review of Political Economy, 3,* 146–163.

Mitchell, G. (2012). Revisiting truth or triviality: The external validity of research in the psychological laboratory. *Perspectives on Psychological Science, 7,* 109–117.

Mittag, K. C., & Thompson, B. (2000). A national survey of AERA members' perceptions of statistical significance tests and other statistical issues. *Educational Researcher, 29*(4), 14–20.

Mittelstaedt, R. A., & Zorn, T. S. (1984). Econometric replication: Lessons from the experimental sciences. *Quarterly Journal of Business and Economics, 23*(1), 9–15.

Mol, M., & Birkinshaw, J. (2008). *Giant steps in management: Innovations that change the way we work.* Harlow, UK: Prentice Hall/Financial Times.

Mone, M. A., Mueller, G. C., & Mauland, W. (1996). The perceptions and usage of statistical power in applied psychology and management research. *Personnel Psychology, 49*(1), 103–120.

Monroe, K. B. (1992a). On replications in consumer research: Part 1. *Journal of Consumer Research, 19*(1), i–ii.

Monroe, K. B. (1992b). On replications in consumer research: Part 2. *Journal of Consumer Research, 19*(2), i–ii.

Monterde-i-Bort, H., Frias-Navarro, D., & Pascual-Llobell, J. (2010). Uses and abuses of statistical significance tests and other statistical resources: A comparative study. *European Journal of Psychology of Education, 25,* 429–447.

Montgomery, C. A. (1982). The measurement of firm diversification: Some more empirical evidence. *Academy of Management Journal, 25,* 299–307.

Mook, D. G. (1983). In defense of external invalidity. *American Psychologist, 38,* 379–387.

Moonesinghe, R., Khoury, M. J., & Janssens, C. J. W. (2007). Most published research findings are false—But a little replication goes a long way. *PLoS Medicine, 4*(February), 0218–0221.

Moore, T. E. (1982). Subliminal advertising: What you see is what you get. *Journal of Marketing, 46*(2), 38–47.

Morgan, M. S. (1990). *The history of econometric ideas.* Cambridge, UK: Cambridge University Press.

Morris, D. J., Jr. (1990). The railroad and movie industries: Were they myopic? *Journal of the Academy of Marketing Science, 18,* 279–283.

Morrison, D. E., & Henkel, R. E. (Eds.). (1970). *The significance test controversy: A reader.* Chicago, IL: Aldine.

Mouncey, P. (2011). Editorial. *International Journal of Market Research, 53,* 4.

Mulkay, M., & Gilbert, G. N. (1986). Replication and mere replication. *Philosophy of the Social Sciences, 16,* 21–37.

Müller, H., Kroll, E. B., & Vogt, B. (2012). Do real payments really matter? A re-examination of the compromise effect in hypothetical and binding choice settings. *Marketing Letters, 23,* 73–92.

Murdoch, D. J., Tsai, Y.-L., & Adcock, J. (2008). *P*-values are random variables. *The American Statistician, 62,* 242–245.

Nagel, E. (1961). *The structure of science: Problems in the logic of scientific explanation.* New York, NY: Harcourt Brace.

Narver, J. C., & Slater, S. F. (1990). The effect of a market orientation on business profitability. *Journal of Marketing, 54*(4) 20–35.

Nash, L. K. (1963). *The nature of the natural sciences.* Boston, MA: Little, Brown.

Natrella, M. G. (1960). The relation between confidence intervals and tests of significance. *The American Statistician, 14*(1), 20–23.

Neale, J. M., & Liebert, R. M. (1986). *Science and behavior: An introduction to methods of research* (3rd ed.). Englewood Cliffs, NJ: Prentice Hall.

Nelder, J. A. (1985). Comment on Chatfield. *Journal of the Royal Statistical Society, Ser. A, 148,* 238.

Nelder, J. A. (1986). Statistics, science and technology (with discussion). *Journal of the Royal Statistical Society, Ser. A, 149,* 109–121.

Nelder, J. A. (1999). Statistics for the millennium: From statistics to statistical science. *The Statistician, 48,* 257–269.

Nelson, C. R., & Plosser, C. R. (1982). Trends and random walks in macroeconomic time series: Some evidence and implications. *Journal of Monetary Economics, 10,* 139–162.

Nester, M. R. (1996). An applied statistician's creed. *Applied Statistics, 45,* 401–410.

Netemeyer, R. G., Durvasula, S., & Lichtenstein, D. R. (1991). A cross-national assessment of the reliability and validity of the CETSCALE. *Journal of Marketing Research, 28,* 320–327.

Neuliep, J. W., & Crandall, R. (1990). Editorial bias against replication research. *Journal of Social Behavior and Personality, 5*(4), 85–90.

Neuliep, J. W., & Crandall, R. (1993). Reviewer bias against replication research. *Journal of Social Behavior and Personality, 8*(6), 21–29.

Neyman, J. (1934). On the two different aspects of the representative method: The method of stratified sampling and the method of purposive selection. *Journal of the Royal Statistical Society, 97,* 558–625.

Neyman, J. (1937). Outline of a theory of estimation based on the classical theory of probability. *Philosophical Transactions of the Royal Society, A, 236*, 333–380.

Neyman, J. (1971). Foundations of behavioristic statistics (with discussion). In V. P. Godambe & D. A. Sprott (Eds.), *Foundations of statistical inference* (pp. 1–19). Toronto, Canada: Holt, Rinehart and Winston.

Neyman, J., & Pearson. E. S. (1928a). On the use and interpretation of certain test criteria for purposes of statistical inference. Part I. *Biometrika, 20A*, 175–240.

Neyman, J., & Pearson, E. S. (1928b). On the use and interpretation of certain test criteria for purposes of statistical inference. Part II. *Biometrika, 20A*, 263–294.

Neyman, J., & Pearson, E. S. (1933). On the problem of the most efficient tests of statistical hypotheses. *Philosophical Transactions of the Royal Society of London, Ser. A, 231*, 289–337.

Nicholson, J. M., & Ioannidis, J. P. A. (2012). Research grants: Conform and be funded. *Nature, 492*, 34–36.

Nickerson, R. S. (2000). Null hypothesis significance testing: A review of an old and continuing controversy. *Psychological Methods, 5*, 241–301.

Nola, R., & Sankey, H. (2007). *Theories of scientific method.* Montreal, Canada: McGill–Queen's University Press.

Nolan, S. A., & Heinzen, T. E. (2008). *Statistics for the behavioral sciences.* New York, NY: Worth.

Nosek, B. A., Spies, J. R., & Motyl, M. (2012). Scientific utopia: II. Restructuring incentives and practices to promote truth over publishability. *Perspectives on Psychological Science, 7*, 615–631.

Note to contributors: Statistical significance. (2008). *Journal of Advertising Research, 48*, 165.

November, P. (2004). Seven reasons why marketing practitioners should ignore marketing academic research. *Australasian Marketing Journal, 12*(2), 39–50.

Oakes, M. (1986). *Statistical inference: A commentary for the social and behavioural sciences.* Chichester, UK: Wiley.

Okasha, S. (2002). *Philosophy of science.* Oxford, UK: Oxford University Press.

Open Science Collaboration. (2012). An open, large-scale, collaborative effort to estimate the reproducibility of psychological science. *Perspectives on Psychological Science, 7*, 657–660.

Orlitzky, M. (2012). How can significance tests be deinstitutionalized? *Organizational Research Methods, 15*, 199–228.

Ormerod, P. (1997). *The death of economics.* Chichester, UK: Wiley.

Ortinau, D. J. (2011). Writing and publishing important scientific articles: A reviewer's perspective. *Journal of Business Research, 64*, 150–156.

Otley, D. T. (2003). Management control and performance management: Whence and whither? *British Accounting Review, 35*, 309–326.

Ottenbacher, K. J. (1996). The power of replications and replications of power. *The American Statistician, 50*, 271–275.

Pagano, R. R. (2001). *Understanding statistics in the behavioral sciences* (6th ed.). Belmont, CA: Wadsworth.

Parasuraman, A., Grewal, D., & Krishnan, R. (2004). *Marketing research.* Boston, MA: Houghton Mifflin.

Parker, S. (1990). A note on the growth of the use of statistical tests in *Perception & Psychophysics. Bulletin of the Psychonomic Society, 28,* 565–566.

Pashler, H., & Harris, C. R. (2012). Is the replicability crisis overblown? Three arguments examined. *Perspectives on Psychological Science, 7,* 531–536.

Pashler, H., & Wagenmakers, E.-J. (2012). Editors' introduction to the special section on replicability in psychological science: A crisis of confidence. *Perspectives on Psychological Science, 7,* 528–530.

Pearce, J. L., & Huang, L. (2012). The decreasing value of our research to management education. *Academy of Management Learning & Education, 11,* 247–262.

Pearson, K. (1900). On the criterion that a given system of deviations from the probable in the case of a correlated system of variables is such that it can be reasonably supposed to have arisen from random sampling. *London, Edinburgh and Dublin Philosophical Magazine and Journal of Science, 50,* 157–175.

Perlman, M. D., & Wu, L. (1999). The emperor's new tests (with discussion). *Statistical Science, 4,* 355–381.

Peter, J. P. (1981). Construct validity: A review of basic issues and marketing practices. *Journal of Marketing Research, 18,* 133–145.

Peterson, R. A. (2001). On the use of college students in social science research: Insights from a second-order meta-analysis. *Journal of Consumer Research, 28,* 450–461.

Peterson, R. A., & Merunka, D. R. (2014). Convenience samples of college students and research reproducibility. *Journal of Business Research, 67,* 1035–1041.

Peterson, R. R. (1996a). A re-evaluation of the economic consequences of divorce. *American Sociological Review, 61,* 528–536.

Peterson, R. R. (1996b). Statistical errors, faulty conclusions, misguided policy: Reply to Weitzman. *American Sociological Review, 61,* 539–540.

Pettigrew, A. M. (2005). The character and significance of management research on the public services. *Academy of Management Journal, 48,* 973–977.

Pfeffer, J. (1993). Barriers to the advance of organizational science: Paradigm development as a dependent variable. *Academy of Management Review, 18,* 599–620.

Pfeffer, J. (2007). A modest proposal: How we might change the process and product of managerial research. *Academy of Management Journal, 50,* 1334–1345.

Pfeffer, J., & Sutton, R. I. (2006a, January). Evidence-based management. *Harvard Business Review,* 63–74.

Pfeffer, J., & Sutton, R. I. (2006b). *Hard facts, dangerous half-truths, and total nonsense: Profiting from evidence-based management.* Boston, MA: Harvard Business School Press.

Phelan, T. J. (1999). A compendium of issues for citation analysis. *Scientometrics, 45*(1), 117–136.

Phillips, R., D., & Gilroy, F. D. (1985). Sex-role stereotypes and clinical judgments of mental health: The Brovermans' findings reexamined. *Sex Roles, 12,* 179–193.

Phlips, L. (1974). *Applied consumption analysis.* New York, NY: Elsevier.

Pierson, F. C. (1959). *The education of American businessmen.* New York, NY: McGraw-Hill.

Piketty, T. (2014). *Capital in the twenty-first century.* Cambridge, MA: Harvard University Press.

Pindyck, R. S., & Rubinfeld, D. L. (1998). *Econometric models and economic forecasts* (4th ed.). New York, NY: McGraw-Hill.

Poincaré, H. (2004). *Science and method.* New York, NY: Barnes and Noble Books. (Original work published 1908)

Pollard, P., & Richardson, J. T. (1987). On the probability of making Type I errors. *Psychological Bulletin, 102,* 159–163.

Poole, C. (2001). Low P-values or narrow confidence intervals: Which are more durable? *Epidemiology, 12,* 291–294.

Poole, M. S., & Van de Ven, A. H. (1989). Using paradox to build management and organizational theories. *Academy of Management Review, 14,* 562–578.

Popper, K. R. (1959). *The logic of scientific discovery.* London, UK: Hutchinson.

Popper, K. R. (1963). *Conjectures and refutations.* London, UK: Routledge.

Porter, M. (1980). *Competitive strategy: Techniques for analyzing industries and competitors.* New York, NY: Free Press.

Porter, T. M. (1986). *The rise of statistical thinking 1820–1900.* Princeton, NJ: Princeton University Press.

Preece, D. A. (1987). Good statistical practice. *The Statistician, 36,* 397–408.

Prinz, F., Schlange, T., & Asadullah, K. (2011). Believe it or not: How much can we rely on potential drug targets? *Nature Reviews Drug Discovery, 10,* 712–713.

Psillos, S. (1999). *Scientific realism: How science tracks truth.* London, UK: Routledge.

Quinlan, C. (2011). *Business research methods.* Andover, UK: South-Western.

Raison, T. (Ed.). (1963). *The founding fathers of social science.* Middlesex, UK: Penguin Books.

Ravetz, J. R. (1971). *Scientific knowledge and its social problems.* Oxford, UK: Oxford University Press.

Ray, J. L., & Valeriano, B. (2003). Barriers to replication in systematic empirical research on world politics. *International Studies Perspectives, 4,* 79–85.

Reibstein, D. J., Day, G., & Wind, J. (2009, July). Guest editorial: Is marketing academia losing its way? *Journal of Marketing, 73* (3), 1–3.

Reid, L. N., Rotfeld, H. J., & Wimmer, R. D. (1982). How researchers respond to replication requests. *Journal of Consumer Research, 9,* 216–218.

Reid, L. N., Soley, L. C., & Wimmer, R. D. (1981). Replication in advertising research: 1977, 1978, 1979. *Journal of Advertising, 10*(1), 3–13.

Reif, F. (1961). The competitive world of the pure scientist. *Science, 134,* 1957–1962.

Reinhart, C. M., & Rogoff, K. S. (2010). Growth in a time of debt. *American Economic Review, 100,* 573–578.

Ries, A., & Trout, J. (1993). *The 22 immutable laws of marketing: Violate them at your own risk.* New York, NY: Harper Business.

Ritchie, S. J., Wiseman, R., & French, C. C. (2012a). Failing the future: Three unsuccessful attempts to replicate Bem's "retroactive facilitation of recall" effect. *PLoS One, 7*(3), e33423.

Ritchie, S. J., Wiseman, R., & French, C. C. (2012b). Replication, replication, replication. *The Psychologist, 25,* 346–348.

Roberts, R. M. (1989). *Serendipity: Accidental discoveries in science.* New York, NY: Wiley.

Robertson, T. S. (1971). *Innovative behavior and communication.* New York, NY: Holt, Rinehart, and Winston.

Rogers, G., & Soopramanien, D. (2009). The truth is out there! How external validity can lead to better marketing decisions. *International Journal of Market Research, 51*(2), 163–180.

Rosen, D. L., & Singh, S. N. (1992). An investigation of subliminal embedded effect on multiple measures of advertising effectiveness. *Psychology & Marketing, 9*(2), 157–173.

Rosenbaum, P. R. (1999). Choice as an alternative to control in observational studies (with discussion). *Statistical Science, 14,* 259–304.

Rosenbaum, P. R. (2001). Replicating effects and biases. *The American Statistician, 55,* 223–227.

Rosenbaum, P. R. (2002). *Observational studies* (2nd ed.). New York, NY: Springer–Verlag.

Rosenberg, A. (1986). Philosophy of science and the potentials for knowledge in the social sciences. In D. W. Fiske & R. A. Shweder (Eds.), *Metatheory in social science: Pluralisms and subjectivities* (pp. 339–346). Chicago, IL: University of Chicago Press.

Rosenthal, R. (1966). *Experimenter effects in behavioral research.* New York, NY: Appleton-Century-Crofts.

Rosenthal, R. (1979). The "file drawer problem" and tolerance for null results. *Psychological Bulletin, 86,* 638–641.

Rosenthal, R. (1990). Replication in behavioral research. *Journal of Social Behavior and Personality, 5*(4), 1–30.

Rosenthal, R., & Rosnow, R. L. (1984). *Essentials of behavioral research: Methods and data analysis.* New York, NY: McGraw–Hill.

Rosner, B. (2006). *Fundamentals of biostatistics* (6th ed.). Belmont, CA: Thomson.

Rosnow, R. L., & Rosenthal, R. (1989). Statistical procedures and the justification of knowledge in psychological science. *American Psychologist, 44,* 1276–1284.

Ross, D. (1991). *The origins of American social science.* Cambridge, UK: Cambridge University Press.

Rossi, J. S. (1990). Statistical power of psychological research: What have we gained in 20 years? *Journal of Consulting and Clinical Psychology, 58,* 646–656.

Rossi, J. S. (1997). A case study in the failure of psychology as a cumulative science: The spontaneous recovery of verbal learning. In L. L. Harlow, S. A. Mulaik, & J. H. Steiger (Eds.), *What if there were no significance tests?* (pp. 175–197). Mahwah, NJ: Lawrence Erlbaum.

Rourke, B. P., & Costa, L. (1979). Editorial policy II. *Journal of Clinical Neuropsychology, 1,* 93–95.

Rousseau, D. M. (2006). Is there such a thing as "evidence-based management"? *Academy of Management Review, 31,* 256–269.

Rousseau, D. M., Manning, J., & Denyer, D. (2008). Evidence in management and organizational science: Assembling the field's full weight of scientific knowledge through syntheses. *Academy of Management Annals, 2,* 475–515.

Rousseeuw, P. J. (1991). Why the wrong papers get published. *Chance: New Directions for Statistics and Computing, 4*(1), 41–43.

Rowney, J. A., & Zenisek, T. J. (1980). Manuscript characteristics influencing reviewers' decisions. *Canadian Psychologist, 21*, 17–21.

Royall, R. M. (1986). The effect of sample size on the meaning of significance tests. *The American Statistician, 40*, 313–315.

Royall, R. M. (1997). *Statistical evidence: A likelihood paradigm*. New York, NY: Chapman and Hall.

Rozeboom, W. W. (1960). The fallacy of the null-hypothesis significance test. *Psychological Bulletin, 57*, 416–428.

Rozeboom, W. W. (1997). Good science is abductive, not hypothetico-deductive. In L. L. Harlow, S. A. Mulaik, & J. H. Steiger (Eds.), *What if there were no significance tests?* (pp. 335–391). Mahwah, NJ: Lawrence Erlbaum.

Rozin, P. (2009). What kind of empirical research should we publish, fund, and reward? *Perspectives on Psychological Science, 4*, 435–439.

Rubin, D. R. (2007). The design *versus* the analysis of observational studies for causal effects: Parallels with the design of randomized trials. *Statistics in Medicine, 26*, 20–36.

Rucci, A. J., & Tweney, R. D. (1980). Analysis of variance and the "second discipline" of scientific psychology: A historical account. *Psychological Bulletin, 87*, 166–184.

Rumelt, R. P. (1974). *Strategy, structure and economic performance*. Boston, MA: Harvard Graduate School of Business Administration.

Rynes, S. L. (2007). Afterward: To the next 50 years. *Academy of Management Journal, 50*, 1379–1383.

Rynes, S. L., Bartunek, J. M., & Daft, R. L. (2001). Across the great divide: Knowledge creation and transfer between practitioners and academics. *Academy of Management Journal, 44*, 340–355.

Rynes, S. L., Colbert, A. E., & Brown, K. G. (2002). HR professionals' beliefs about effective human resource practices: Correspondence between research and practice. *Human Resource Management, 41*, 149–174.

Rynes, S. L., Giluk, T. L., & Brown, K. G. (2007). The very separate worlds of academic and practitioner publications in human resource management: Implications for evidence-based management. *Academy of Management Journal, 50*, 987–1008.

Salkind, N. J. (2008). *Statistics for people who (think they) hate statistics* (3rd ed.). Thousand Oaks, CA: Sage.

Salsburg, D. S. (1985). The religion of statistics as practiced in medical journals. *The American Statistician, 39*, 220–223.

Samelson, F. (1980). J. B. Watson's Little Albert, Cyril Burt's twins and the need for a critical science. *American Psychologist, 35*, 619–625.

Sandberg, J. (2005). How do we justify knowledge produced within interpretive approaches? *Organizational Research Methods, 8*(1), 41–68.

Sarasohn, J. (1993). *Science on trial: The whistle-blower, the accused, and the Nobel laureate*. New York, NY: St. Martin's Press.

Sauley, K. S., & Bedeian, A. G. (1989). .05: A case of the tail wagging the distribution. *Journal of Management, 15*, 335–344.

Saunders, M., Lewis, P., & Thornhill, A. (2012). *Research methods for business students* (6th ed.). Harlow, UK: Financial Times Press.

Savage, L. J. (1954). *The foundations of statistics.* New York, NY: Wiley.

Savage, L. J. (1961). The foundations of statistics reconsidered. In J. Neyman (Ed.), *Proceedings of the Fourth Berkeley Symposium on Mathematics and Probability* (Vol. 1, pp. 575–586). Berkeley: University of California Press.

Savage, L. J. (1976). On rereading R. A. Fisher (with discussion). *Annals of Statistics, 4,* 441–500.

Sawyer, A. G., & Ball, A. D. (1981). Statistical power and effect size in marketing research. *Journal of Marketing Research, 18,* 275–290.

Saxe, R., & Weitz, B. A. (1982). The SOCO scale: A measure of the customer orientation of salespeople. *Journal of Marketing Research, 19,* 343–351.

Sayer, A. (1992). *Method in social science: A realist approach* (2nd ed.). London, UK: Routledge.

Sayer, A. (2000). *Realism and social science.* London, UK: Sage.

Scandura, T. A., & Williams, E. A. (2000). Research methodology in management: Current practices, trends, and implications for future research. *Academy of Management Journal, 43,* 1248–1264.

Schenker, N., & Gentleman, J. F. (2001). On judging the significance of differences by examining the overlap between confidence intervals. *The American Statistician, 55,* 182–186.

Schervish, M. J. (1996). *P* values: What they are and what they are not. *The American Statistician, 50,* 203–206.

Schimmack, U. (2012). The ironic effect of significant results on the credibility of multiple-study articles. *Psychological Methods, 17,* 551–566.

Schmidt, F. L. (1992). What do data really mean? Research findings, meta-analysis, and cumulative knowledge in psychology. *American Psychologist, 47,* 1173–1181.

Schmidt, F. L. (1996). Statistical significance testing and cumulative knowledge in psychology: Implications for the training of researchers. *Psychological Methods, 1,* 115–129.

Schmidt, F. L., & Hunter, J. E. (1997). Eight common but false objections to the discontinuation of significance testing in the analysis of research data. In L. L. Harlow, S. A. Mulaik, & J. H. Steiger (Eds.), *What if there were no significance tests?* (pp. 37–64). Mahwah, NJ: Lawrence Erlbaum.

Schmidt, F. L., & Hunter, J. E. (2002). Are there benefits from NHST? *American Psychologist, 57,* 65–66.

Schmidt, S. (2009). Shall we really do it again? The powerful concept of replication is neglected in the social sciences. *Review of General Psychology, 13*(2), 90–100.

Searle, S. R. (1989). Statistical computing packages: Some words of caution. *The American Statistician, 43,* 189–190.

Sechrest, L., & Sidani, S. (1995). Qualitative and quantitative methods: Is there an alternative? *Evaluation and Program Planning, 18,* 77–87.

Sedlmeier, P., & Gigerenzer, G. (1989). Do studies of statistical power have an effect on the power of studies?" *Psychological Bulletin, 105,* 309–316.

Seidenfeld, T. (1979). *Philosophical problems of statistical inference: Learning from R. A. Fisher.* Dordrecht, Netherlands: Reidel.

Sellke, T., Bayarri, M. J., & Berger, J. O. (2001). Calibration of p values for testing precise null hypotheses. *The American Statistician, 55,* 62–71.

Selltiz, C., Wrightsman, L. S., & Cook, S. W. (1976). *Research methods in social relations* (3rd ed.) New York, NY: Random House.

Serlin, R. C. (2002). Constructive criticism. *Journal of Modern Applied Statistical Methods, 1,* 202–227.

Servaes, H. (1991). Tobin's Q and the gains from takeovers. *Journal of Finance, 46,* 409–419.

Shadish, W. R., & Cook, T. D. (1999). Comment—Design rules: More steps toward a complete theory of quasi-experimentation. *Statistical Science, 14,* 294–300.

Shadish, W. R., Cook, T. D., & Campbell, D. T. (2002). *Experimental and quasi-experimental designs for generalized causal inference.* Boston, MA: Houghton Mifflin.

Shadish, W. R., Doherty, M., & Montgomery, L. M. (1989). How many studies are in the file drawer? An estimate from the family/marital psychotherapy literature. *Clinical Psychology Review, 9,* 589–603.

Shanks, D. R., Newell, B. R., Lee, E. H., Balakrishnan, D., Ekelund, L., Cenac, Z., . . . Moore, C. (2013). Priming intelligent behavior: An elusive phenomenon. *PLoS One, 8*(April), 1–10.

Shapin, S. (1995). Here and everywhere: Sociology of scientific knowledge. *Annual Review of Sociology, 21,* 289–321.

Sharp, B. (2004). Book review of Douglas Hall and Jeffrey Stamp (2003), *Meaningful marketing: 100 data proven truths and 402 practical ideas for selling more with less.* Ohio, Brain Brew Books. *Australasian Marketing Journal, 12*(3), 104–105.

Sharp, B., & Wind, Y. (2009). Today's advertising laws: Will they survive the digital revolution? *Journal of Advertising Research, 49,* 120–126.

Sharpe, W. (1964). Capital asset prices: A theory of equilibrium under conditions of risk. *Journal of Finance, 19,* 425–442.

Shaver, J. P., & Norton, R. S. (1980a). Populations, samples, randomness, and replication in two social studies journals. *Theory and Research in Social Education, 8*(2), 1–10.

Shaver, J. P., & Norton, R. S. (1980b). Randomness and replication in ten years of the *American Educational Research Journal. Educational Researcher, 9*(1), 9–15.

Sheth, J. N., & Sisodia, R. S. (1999). Revisiting marketing's lawlike generalizations. *Journal of the Academy of Marketing Science, 27,* 71–87.

Shimp, T. A., & Sharma, S. (1987). Consumer ethnocentrism: Construction and validation of the CETSCALE. *Journal of Marketing Research, 24,* 280–289.

Shrout, P. E. (1997). Should significance tests be banned? *Psychological Science, 8,* 1–2.

Shweder, R. A., & Fiske, D. W. (1986). Introduction: Uneasy social science. In D. W. Fiske & R. A. Schweder (Eds.), *Metatheory in social science: Pluralisms and subjectivities* (pp. 1–18). Chicago, IL: University of Chicago Press.

Sieber, J. E. (1991). Introduction: Sharing social science data. In J. Sieber (Ed.), *Sharing social science data: Advantages and challenges* (pp. 1–18). Newbury Park, CA: Sage.

Siegfried, T. (2010). Odds are, it's wrong: Science fails to face the shortcomings of statistics. *Science News, 177*(March 27), 26.

Silva-Aycaguer, L. C., Suarez-Gil, P., & Fernandez-Somoano, A. (2010). The null hypothesis significance test in health sciences research (1995–2006): Statistical analysis and interpretation. *BMC Medical Research Methodology, 10*(44).

Simes, R. J. (1986). Publication bias: The case for an international registry of clinical trials. *Journal of Clinical Oncology, 4,* 1529–1541.

Simmons, J. P., Nelson, L. D., & Simonsohn, U. (2011). False-positive psychology: Undisclosed flexibility in data collection and analysis allows presenting anything as significant. *Psychological Science, 22,* 1359–1366.

Simon, H. A. (1990). Invariants of human behavior. *Annual Review of Psychology, 41,* 1–19.

Simon, J. L. (1979). What do Zielske's real data really show about pulsing? *Journal of Marketing Research, 16,* 415–420.

Simon, J. L., & Burstein, P. (1985). *Basic research methods in social science* (3rd ed.). New York, NY: Random House.

Singh, K., Ang, S. H., & Leong, S. M. (2003). Increasing replication for knowledge accumulation in strategy research. *Journal of Management, 29,* 533–549.

Skidelsky, R. (2009). *Keynes: The return of the master.* New York, NY: Public Affairs.

Smart, R. G. (1964). The importance of negative results in psychological research. *Canadian Psychologist, 5,* 225–232.

Smith, L. D., Best, L. A., Cylke, V. A., & Stubbs, D. A. (2000). Psychology without *p* values: Data analysis at the turn of the 19th century. *American Psychologist, 55,* 260–263.

Smith, M. L. (1980). Publication bias and meta-analysis. *Evaluation in Education, 4,* 22–24.

Smith, M. L. (2006). Overcoming theory-practice inconsistencies: Critical realism and information systems research. *Information Organization, 16,* 191–211.

Smithson, M. (2003). *Confidence intervals.* Thousand Oaks, CA: Sage.

Snedecor, G. W. (1934). *Calculation and interpretation of the analysis of variance and covariance.* Ames, IA: Collegiate Press.

Snedecor, G. W. (1937). *Statistical methods.* Ames, IA: Collegiate Press.

Sobh, R., & Perry, C. (2006). Research design and data analysis in realism research. *European Journal of Marketing, 40,* 1194–1209.

Sohn, D. (1996). Meta-analysis and science. *Theory & Psychology, 6,* 229–246.

Sohn, D. (1998). Statistical significance and replicability: Why the former does not presage the latter. *Theory & Psychology, 8,* 291–311.

Sovacool, B. K. (2008). Exploring scientific misconduct: Isolated individuals, impure institutions, or an inevitable idiom of modern science? *Bioethical Inquiry, 5,* 271–282.

Spanos, A. (1986). *Statistical foundations of econometric modelling.* Cambridge, UK: Cambridge University Press.

Spellman, B A. (2013a, March 5). But I don't want people to try to replicate my research. *MyPoPS.* Retrieved from https://morepops.wordpress.com/2013/03/05/but-i-dont-want-people-to-try-to-replicate-my-research

Spellman, B. A. (2013b). There is no such thing as replication, but we should do it anyway. *European Journal of Personality, 27*, 136–137.

Stang, A., Poole, C., & Kuss, O. (2010). The ongoing tyranny of statistical significance testing in biomedical research. *European Journal of Epidemiology, 25*, 225–230.

Starbuck, W. H. (1993). Keeping a butterfly and an elephant in a house of cards: The elements of exceptional success. *Journal of Management Studies, 30*, 885–921.

Starbuck, W. H. (2006). *The production of knowledge: The challenge of social science research.* Oxford, UK: Oxford University Press.

Steckler, A., & McLeroy, K. R. (2008). The importance of external validity. *American Journal of Public Health, 98*(1), 9–10.

Steiger, J. H. (1990). Structural model evaluation and modification: An interval estimation approach. *Multivariate Behavioral Research, 25*, 173–180.

Stephan, P. (2012). *How economics shapes science.* Cambridge, MA: Harvard University Press.

Sterling, T. D. (1959). Publication decisions and their possible effects on inferences drawn from tests of significance—or vice versa. *Journal of the American Statistical Association, 54*, 30–34.

Sterling, T. D., Rosenbaum, W. L., & Weinkam, J. J. (1995). Publication decisions revisited: The effect of the outcome of statistical tests on the decision to publish and vice versa. *The American Statistician, 49*, 108–112.

Stewart, D. W. (2000). Testing statistical significance testing: Some observations of an agnostic. *Educational & Psychological Measurement, 60*, 685–690.

Stewart, D. W. (2002). Getting published: Reflections of an old editor. *Journal of Marketing, 66*(4), 1–6.

Stewart, W. W., & Feder, N. (1987). The integrity of the scientific literature. *Nature, 325*, 207–214.

Stigler, S. M. (1986). *The history of statistics: The measurement of uncertainty before 1900.* Cambridge, MA: Harvard University Press.

Stigler, S. M. (1999). *Statistics on the table: The history of statistical concepts and methods.* Cambridge, MA: Harvard University Press.

Stiglitz, J. E. (2010). *Freefall: America, free markets, and the sinking of the world economy.* New York, NY: W. W. Norton.

Storey, J. D. (2011). False discovery rate. In M. Lovric (Ed.), *International encyclopedia of statistical science* (pp. 504–508). Berlin, Germany: Springer-Verlag.

Strasak, A. M. Zaman, Q., Marinell, G., Pfeiffer, K. P., & Ulmer, H. (2007). The use of statistics in medical research: A comparison of *The New England Journal of Medicine* and *Nature Medicine. The American Statistician, 61*, 47–55.

Strathern, P. (2001). *A brief history of economic genius.* New York, NY: Texere.

Straub, D. W. (2008). Editor's comments: Why do top journals reject good papers? *MIS Quarterly, 32*(3), iii–vii.

Stremersch, S., & Lehmann, D. R. (2007), Editorial. *International Journal of Research in Marketing, 24*, 1–2.

Stremersch, S., Verniers, I., & Verhoef, P. C. (2007). The quest for citations: Drivers of article impact. *Journal of Marketing, 71*(3), 171–193.

Stricker, G. (1977). Implications of research for psychotherapeutic treatment of women. *American Psychologist, 32,* 14–22.

Stroebe, W., Postmes, T., & Spears, R. (2012). Scientific misconduct and the myth of self-correction in science. *Perspectives on Psychological Science, 7,* 670–688.

Student. (1938). Comparison between balanced and random arrangements of field plots. *Biometrika, 29,* 363–378.

Summers, L. H. (1991). The scientific illusion in empirical macroeconomics. *Scandinavian Journal of Economics, 93*(2), 129–148.

Suppe, F. (1977). Critical introduction and afterword. In F. Suppe (Ed.), *The structure of scientific theories* (2nd ed., pp. 3–241, 617–730). Urbana: University of Illinois Press.

Taagepera, R. (2008). *Making social sciences more scientific: The need for predictive models.* Oxford, UK: Oxford University Press.

Tadajewski, M. (2008). Incommensurable paradigms, cognitive bias and the politics of marketing theory. *Marketing Theory, 8,* 273–297.

Tankard, J. W., Jr. (1984). *The statistical pioneers.* Cambridge, MA: Schenkman.

Taylor, S. (2007). *Business statistics for non-mathemeticians* (2nd ed.). New York, NY: Palgrave Macmillan.

Teigen, K. H. (2002). One hundred years of laws in psychology. *American Journal of Psychology, 115*(1), 103–118.

Terpstra, D. E. (1981). Relationship between methodological rigor and reported outcomes in organizational development evaluation research. *Journal of Applied Psychology, 66,* 541–543.

Theil, H. (1971). *Principles of econometrics.* New York, NY: Wiley.

Theus, K. T. (1994). Subliminal advertising and the psychology of processing unconscious stimuli: A review of the research. *Psychology & Marketing, 11,* 271–290.

Thompson, B. (1994). The pivotal role of replication in psychological research: Empirically evaluating the replicability of sample results. *Journal of Personality, 62,* 157–176.

Thompson, B. (2002). What future quantitative social science research could look like: Confidence intervals for effect sizes. *Educational Researcher, 31*(April), 25–32.

Thompson, B., & Snyder, P. A. (1997). Statistical significance testing practices in *The Journal of Experimental Education. Journal of Experimental Education, 66,* 75–83.

Thompson, H. A. (1981). *The great writings in marketing* (2nd ed.). Tulsa, OK: PennWell Books.

Those who can't, teach. (2014, February 8). *The Economist,* p. 66.

Titus, S. L., Wells, J. A., & Rhoades, L. J. (2008). Repairing research integrity. *Nature, 453,* 980–982.

Toulmin, S. (2001). *Return to reason.* Cambridge, MA: Harvard University Press.

Tranfield, D., Denyer, D., & Smart, P. (2003). Towards a methodology for developing evidence-informed management knowledge by means of systematic review. *British Journal of Management, 14,* 207–222.

Trusted, J. (1979). *The logic of scientific inference.* London, UK: Macmillan.

Tryon, W. W. (1998). The inscrutable null hypothesis. *American Psychologist, 53,* 796.

Tryon, W. W. (2001). Evaluating statistical difference, equivalence, and indeterminacy using inferential confidence intervals: An integrated alternative method of conducting null hypothesis statistical tests. *Psychological Methods, 6,* 371–386.

Tsang, E. W. K., & Kwan, K.-M. (1999). Replication and theory development in organizational science: A critical realist perspective. *Academy of Management Review, 24,* 759–780.

Tucker, W. H. (1994). Fact and fiction in the discovery of Sir Cyril Burt's flaws. *Journal of the History of the Behavioral Sciences, 30*(October), 335–347.

Tukey, J. W. (1960). Conclusions vs. decisions. *Technometrics, 2,* 423–433.

Tukey, J. W. (1969). Analyzing data: Sanctification or detective work? *American Psychologist, 24,* 83–91.

Tukey, J. W. (1980). We need both exploratory and confirmatory. *The American Statistician, 34,* 23–25.

Tukey, J. W. (1991). The philosophy of multiple comparisons. *Statistical Science, 6,* 100–116.

Turner, M. B. (1967). *Philosophy and the science of behavior.* New York, NY: Appleton-Century-Crofts.

Tversky, A., & Kahneman, D. (1971). Belief in the law of small numbers. *Psychological Bulletin, 76,* 105–110.

Tversky, A., & Kahneman, D. (1974). Judgment under uncertainty. *Science,* 185, 1124–1131.

Uncles, M. (2011). Publishing replications in marketing. *International Journal of Market Research, 53,* 579–582.

Uncles, M. D., & Kwok, S. (2013). Designing research with in-built differentiated replication. *Journal of Business Research, 66,* 1398–1405.

Uncles, M. D., & Wright, M. (2004). Empirical generalisation in marketing. *Australasian Marketing Journal, 12*(3), 5–18.

Utts, J. (1991). Replication and meta-analysis in parapsychology. *Statistical Science, 6,* 363–403.

Vacha-Haase, T., Nilsson, J. E., Reetz, D. R., Lance, T. S., & Thompson, B. (2000). Reporting practices and APA editorial policies regarding statistical significance and effect size. *Theory & Psychology, 10,* 413–425.

van Dalen, H. P., & Henkens, K. (2001). What makes a scientific article influential? The case of demographers. *Scientometrics, 50,* 455–482.

Van de Ven, A. H. (2007). *Engaged scholarship: A guide for organizational and social research.* New York, NY: Oxford University Press.

Van de Ven, A. H., & Johnson, P. E. (2006). Knowledge for theory and practice. *Academy of Management Review, 31,* 802–821.

van Poppel, F., & Day, L. H. (1996). A test of Durkheim's theory of suicide—Without committing the "ecological fallacy." *American Sociological Review, 61,* 500–507.

Van Rooy, D. L., Rotton, J., & Burns, T. L. (2006). Convergent, discriminant, and predictive validity of aggressive driving inventories: They drive as they live. *Aggressive Behavior 32*(April), 89–98.

Vickers, A. (2010). *What is a P-value anyway? 34 stories to help you actually understand statistics.* Boston, MA: Addison-Wesley.

Wagenmakers, E.-J., Wetzels, R., Borsboom, D., & van der Maas, H. L. J. (2011). Why psychologists must change the way they analyze their data: The case of psi: Comment on Bem (2011). *Journal of Personality and Social Psychology, 100,* 426–432.

Wainer, H. (1999). One cheer for null hypothesis significance testing. *Psychological Methods, 4,* 212–213.

Walsh, J. P. (2011). Embracing the sacred in our secular scholarly world. *Academy of Management Review, 36,* 215–234.

Walster, G. W., & Cleary, T. A. (1970). A proposal for a new editorial policy in the social sciences. *The American Statistician, 24*(2), 16–19.

Waters, L. (2004). *Enemies of promise: Publishing, perishing, and the eclipse of scholarship.* Chicago, IL: Prickly Paradigm Press.

Webb, E. J., Campbell, D. T., Schwartz, R. D., & Sechrest, L. (2000). *Unobtrusive measures* (Rev. ed.). Thousand Oaks, CA: Sage.

Webster, F. E., Jr., & von Pechmann, F. (1970). A replication of the "shopping list" study. *Journal of Marketing, 34*(2), 61–63.

Weick, K. E. (1995). What theory is *not,* theorizing *is. Administrative Science Quarterly, 40,* 385–390.

Weitzman, L. J. (1985). *The divorce revolution: The unexpected social and economic consequences for women and children in America.* New York, NY: Free Press.

Weitzman, L. J. (1996). The economic consequences of divorce are still unequal: Comment on Peterson. *American Sociological Review, 61,* 537–538.

Wells, W. D. (1993). Discovery-oriented consumer research. *Journal of Consumer Research, 19,* 489–504.

Wells, W. D. (2001). The perils of $N = 1$. *Journal of Consumer Research, 28,* 494–498.

What's wrong with science. (2013, December 14). *The Economist,* pp. 86–87.

Wicherts, J. M., Bakker, M., & Molenaar, D. (2011). Willingness to share research data is related to the strength of the evidence and the quality of reporting of statistical results. *PLoS One, 6*(November), 1–7.

Wicherts, J. M., Borsboom, D., Kats, J., & Molenaar, D. (2006). The poor availability of psychological research data for reanalysis. *American Psychologist, 61,* 726–728.

Widiger, T. A., & Settle, S. A. (1987). Broverman et al. revisited: An artifactual sex bias. *Journal of Personality and Social Psychology, 53,* 463–469.

Wigner, E. P. (1964). Events, laws of nature, and invariance principles. *Science, 145,* 995–999.

Wikipedia. (2015). *Flynn effect.* Retrieved from http://en.wikipedia.org/wiki/Flynn_effect

Wilk, R. R. (2001). The impossibility and necessity of re-inquiry: Finding middle ground in social science. *Journal of Consumer Research, 28,* 308–312.

Wilkie, W. L. (1986). *Consumer behavior.* New York, NY: Wiley.

Wilkinson, L., & American Psychological Association Task Force on Statistical Inference. (1999). Statistical methods in psychology journals: Guidelines and explanations. *American Psychologist, 54,* 594–604.

Williamson, O. E. (1975). *Markets and hierarchies: Analysis and antitrust implications.* New York, NY: Free Press.

Wilson, F. D., Smoke, G. L., & Martin, J. D. (1973). The replication problem in sociology: A report and a suggestion. *Sociological Inquiry, 43*(2), 141–149.

Wind, Y., & Sharp, B. (Eds.). (2009). What we know about advertising: Lessons from empirical generalizations [Special issue]. *Journal of Advertising Research, 49*(2).

Winer, R. S. (1998). From the editor. *Journal of Marketing Research, 35*(1), iii–v.

Winer, R. S. (1999). Experimentation in the 21st century: The importance of external validity. *Journal of the Academy of Marketing Science, 27*, 349–358.

Witmer, J. A., & Clayton, M. K. (1986). On objectivity and subjectivity in statistical inference: A response to Mayo. *Synthese, 67*, 369–379.

Wolins, L. (1962). Responsibility for raw data. *American Psychologist, 17*, 657–658.

Woodman, R. W., & Wayne, S. J. (1985). An investigation of positive-findings bias in evaluation of organization development interventions. *Academy of Management Journal, 28*, 889–913.

Wrege, C. D., & Perroni, A. G. (1974). Taylor's pig-tale: A historical analysis of Frederick W. Taylor's pig-iron experiments. *Academy of Management Journal, 17*, 6–27.

Yadav, M. (2010). The decline of conceptual articles and implications for knowledge development. *Journal of Marketing, 74*(1), 1–19.

Yates, F. (1951). The influence of *Statistical Methods for Research Workers* on the development of the science of statistics. *Journal of the American Statistical Association, 46*, 19–34.

Yeung, H. W. C. (1997). Critical realism and realist research in human geography. *Progress in Human Geography, 21*, 51–74.

Young, S. S. (2008). Everything is dangerous: A controversy. Retrieved from http://nisla05.niss.org/talks/Young_Safety_June_2008.pdf

Yu, C. H. (2006). *Philosophical foundations of quantitative research methodology.* Lanham, MD: University Press of America.

Zabell, S. L. (1992). R. A. Fisher and the fiducial argument. *Statistical Science, 7*, 369–387.

Zielske, H. A. (1959). The remembering and forgetting of advertising. *Journal of Marketing, 23* (1), 239–243.

Ziliak, S. T., & McCloskey, D. N. (2004). Size matters: The standard error of regressions in the *American Economic Review. Journal of Socio-Economics, 33*, 527–546.

Ziliak, S. T., & McCloskey, D. N. (2008). *The cult of statistical significance: How the standard error costs us jobs, justice, and lives.* Ann Arbor: University of Michigan Press.

Ziman, J. (1978). *Reliable knowledge: An exploration of the grounds for belief in science.* Cambridge, UK: Cambridge University Press.

Zimmer, C. (2012. April 16). A sharp rise in retractions prompts calls for reform. *New York Times*, pp. 1–5.

Zinkhan, G. M., Jones, M. Y., Gardial, S., & Cox, K. K. (1990). Methods of knowledge development in marketing and macromarketing. *Journal of Macromarketing, 10*(2), 3–17.

Zuckerman, H., & Merton, R. K. (1971). Patterns of evaluation in science: Institutionalization, structure, and functions of the referee system. *Minerva, 9*(1), 66–100.

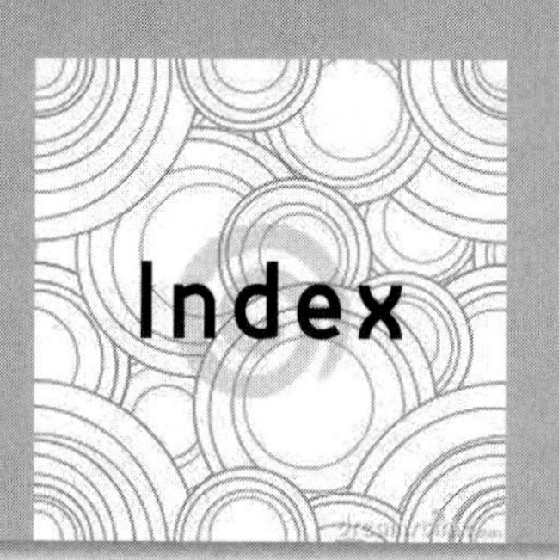

Index